The Whale Problem

SCHEVILL, William E., ed. The whale problem; a status report. Harvard, 1974. 419p il map tab 73-88056. 12.50. SBN 674-95075-5
Proceedings of an international conference of cetologists called together by the U.S. Secretary of the Interior to consider the status, biology, conservation, and methods of study of the great whales. The proceedings contain 19 papers, 12 of which consider the worldwide status of whales. These reports, with few exceptions, indicate that commercially important whales have been seriously overharvested. The section on biology indicates how little is known of the biology of cetaceans. Immediate management recommendations are simple: a moratorium on harvest of the most seriously depleted populations and minimal harvest of less seriously depleted populations. Less immediately obvious, but in the long-run even more insidious than harvesting because it is not as apparently rectifiable, is the destruction and pollution of estuaries — irreplaceable calving and nursery habitats. *The whale problem* is an important contribution by leading international cetologists. The papers are objective, unemotional, and technical. Recommended for upper-division and graduate students.

"Cétacé," dit la baleine, "je me cachalot."

(Blue whale. Photograph © by Kenneth C. Balcomb)

THE WHALE PROBLEM

A Status Report

William E. Schevill, editor

G. Carleton Ray and Kenneth S. Norris
consulting editors

Harvard University Press

Cambridge, Massachusetts

1974

PREFACE

On 24 November 1970 the United States Secretary of the Interior, Walter Hickel, announced that an international meeting of leading cetologists would be called to consider the plight of the great whales and their biology as related to their conservation. He did this as he placed eight commercially sought-after whales on the Endangered Species List, which made it illegal to import any parts or products of these animals into the United States. The eight whales listed were the bowhead, right, blue, sperm, finback, sei, humpback, and gray.

The resulting meeting, the International Conference on the Biology of Whales, was held on 10-12 June 1971 in Shenandoah National Park on the Blue Ridge of Virginia, primarily sponsored by the Department of the Interior, the Smithsonian Institution, the New York Zoological Society, and the International Biological Program (Marine Productivity (PM) Section). Cosponsors were the Department of Commerce, the National Science Foundation, the National Audubon Society, the Environmental Defense Fund, the World Wildlife Fund, the Conservation Foundation, and the National Wildlife Federation. Thus funded, the conference was organized and managed by a committee of the Marine Mammal Council of the United States International Biological Program: G. C. Ray, W. E. Schevill, and K. S. Norris, specialists in marine mammals then unconnected with either whaling or sealing or their regulation. Ten countries were represented at the conference (the Russians were invited but did not attend) by thirty-four participants and twenty observers, who are listed at page 21. The cetologists invited to the conference had been concerned with the conservation of whales, and many had been active in calling the attention of the International Whaling Commission to the plight of the whales. Although their objectivity was not and is not in question (and I emphasize this), it was felt appropriate to have protectionists represented at the conference, and so also in this book (Chapter 17).

The meeting was planned, not primarily to consider matters of

international law or problems of resource management as such, but
rather to define what we know and do not know about whale biology and
numbers, especially in relevance to their conservation as healthy
members of the ecosystem. We were concerned with the basic biology
needed for proper management rather than with management itself. We
hoped that the conference would show where our cetology is based on
shaky premises, and point to new departures to help us learn as
quickly as may be what we need to place management on a sound footing.

At the end of the Shenandoah conference, the participants were
somewhat informally divided into three working groups, (1) on biology
and natural history, (2) on regional populations, and (3) on biometrics and management. Their conclusions are presented at page 5
following a digest prepared immediately after the conference. Although some of the points we then raised have since been acted upon
(for example, the blue whale unit), we give the reports as they were
submitted on 12 June 1971. The acceleration and expansion of cetacean
research that we then anticipated has scarcely occurred, although it
was unanimously approved by the International Whaling Commission at
its 1972 meeting; we have merely a name: the International Decade
of Cetacean Research.

The chapters of this book are revisions of presentations made
during the three-day conference. It had been hoped to publish it
within the year, but a series of delays has prevented this. First,
the tardiness of a few contributors slowed the assembly of the
material until July 1972, and during the next several months there
were further delays following submission of the manuscript to the
Press. It must be understood that these chapters were completed in
1971 (a few in early 1972), and that the authors, with one exception,
have had no opportunity to include new data that may have become
available in the meantime. The exception, which was sought by the
editor, is Dr. McHugh's historical account of the International
Whaling Commission (Chapter 13). The two IWC meetings following the
Shenandoah conference saw important changes and marked progress
toward whale conservation, and it seemed useful to have Dr. McHugh,
who was Chairman of the Commission in 1972, report on this.

Although in many areas whaling was prosecuted at excessive rates
during the first years of the International Whaling Commission, it
has not been entirely reasonable to scold this body for not saving
the world's whales from depletion. Many of its critics seem unaware
of the actual provisions of the Convention under which the Commission
was set up a quarter century ago. The Convention states, *inter alia*,
"that the whale stocks are susceptible of natural increases if whaling
is properly regulated and that increases in the size of whale stocks
will permit increases in the numbers of whales which may be captured
without endangering these natural resources; . . . [and] that it is
in the common interest to achieve the optimum level of whale stocks
as rapidly as possible without causing widespread economic and nutritional distress." The Convention, in Article V, also protects the
national and industrial independence of its members, thus providing
for the weakening or evasion of amendments to the Schedule. The results have been that overfishing of some species has continued and
that the recommendations of concerned scientists have been repeatedly
ignored.

Nevertheless, the International Whaling Commission has taken some
positive actions toward conservation, and it should receive credit for
them. Long before the 1972 United Nations Conference on the Human
Environment at Stockholm, it imposed moratoria on certain species of
whales. At the time of its establishment, the Commission forbade
the taking or killing of gray and right whales. Subsequently, blue
and humpback whales were added to the list—though admittedly after
the species had been extremely depleted. The "right" whale, in the
language of the IWC, includes the bowhead, northern right, southern
right, and pygmy right whales. Thus seven zoological species are protected by IWC moratoria.

The excellent book by Dr. N. A. Mackintosh, "The Stocks of
Whales," published in London in 1965, may be considered a foundation
and point of departure for this book of ours. Ours is an attempt to
extend somewhat the basic work of Dr. Mackintosh, but we by no means
supplant it. The interested reader should have a copy at hand for
general information; besides, there are numerous references to it

throughout these pages.

In preparing these chapters for publication, the editor has been
at some pains to avoid meddling with the authors' mathematics and
English style, his concern being only that the meaning be apparent.
Of course, each chapter is the statement and opinion of its author.
Consistency (sometimes useful but often tiresome) has been attempted
in the technical nomenclature of cetaceans and in the references to
the literature (where, to save space, the reports of the International
Whaling Commission are cited as Rpt. IWC). We follow the example of
the International Commission on Whaling in speaking of it as the
International Whaling Commission or IWC. In the use of vernacular
and trade names for whales we have let each author make his own choice,
and append a glossary of most of these, and other technical jargon.

The editor thanks his colleagues G. Carleton Ray and Kenneth S.
Norris for substantial help with certain chapters, and most particularly Suzanne M. Contos, executive secretary of the Marine Mammal
Council, for active assistance throughout the project. Especial
thanks are due the photographers who supplied the photographs of
whales at sea, following page 302; these photographs are not covered
by the general copyright of the book, but remain the property of
the photographers.

W. E. S.

CONTENTS

Plates following page 302

PART ONE

Conference Summary

A DIGEST OF THE MAJOR CONCLUSIONS OF THE INTERNATIONAL CONFERENCE

ON THE BIOLOGY OF WHALES, 10-12 JUNE 1971,

SHENANDOAH NATIONAL PARK, VIRGINIA

by the Conference Management Committee and Messrs. K. R. Allen,

L. K. Boerema, D. G. Chapman, R. Gambell, and J. A. Gulland

(1) Many whale populations have been drastically reduced by commercial whaling. Some have been reduced to the point at which continued harvesting as a separate venture is no longer economic; others are below the level at which they could produce the maximum yield sustainable for long periods, although they can still economically support commercial whaling; others again are at or above the level which can sustain the maximum yield. Stocks in the first category are generally already protected by the International Whaling Commission, and protection should be continued until they have recovered to about the level giving maximum sustainable yield. It should also be management policy to limit catches so as to allow stocks in the second category to return to this level.

(2) While it is unlikely that any whale species is on the verge of extinction, there are a number of highly localized stocks of some of these species which could possibly be wiped out by whaling operations undertaken either by nations outside the International Whaling Commission or in breach of its regulations. Regulations of the Commission prohibit capture of those species that are in any possible danger. Institution of the International Observer Scheme would help to insure that these regulations are carried out.

(3) The effective biological unit in whales is not the species but the breeding population of local stock. While these are rarely entirely discrete, every effort should be made to manage the harvesting of each stock separately in accordance with the concepts of Paragraph (1) above. It is still more important that the species should be managed individually in each area and not as groups. Therefore the

blue whale unit, which is used to establish a single quota for all species of baleen whales in the Antarctic, should be abandoned and each species managed separately.

(4) More knowledge is needed for fully effective management of individual whale stocks. Therefore, certain specific lines of research are required and the development of some new research approaches is needed.

(5) New research approaches that appear to have special merit include (not necessarily in order of priority): (1) tracking and biological studies by use of radio-telemetry, recoverable instrument packages, transponders, aircraft and satellite monitoring; (2) the development of age-specific marking techniques and of improved techniques for the direct aging of whales; (3) the allocation of ships for whale study outside the normal catching grounds; and (4) biochemical and other methods for population determination.

(6) Studies of whales should be carried out throughout their ranges and at all seasons. The distinction and relationships of stocks should be defined. The relationships of whales to their ecosystems should be clarified.

(7) The present scientific manpower and resources devoted to whale research are inadequate and should be expanded. Through the Bureau of International Whaling Statistics and the United Nations Food and Agriculture Organization, there is already valuable centralization of data on whaling operations and on whales caught, but further expansion, particularly as regards biological data, would be beneficial. Further expansion of international cooperation in whale research is also required, and this should be done under the auspices of an appropriate international body. Either an existing body, such as the International Whaling Commission, or one established for the purpose should be considered.

(8) Small toothed whales (including porpoises) represent an almost unknown but potentially important worldwide resource. Biological studies relating to their future management need to be instituted as soon as possible.

REPORT OF THE WORKING GROUP ON BIOLOGY AND NATURAL HISTORY

The Relation of Natural History to Whale Management

Many populations of large whales have been drastically reduced by
commercial whaling. It is a truism that effective management of wild
species requires that harvesting procedures be based on accurate knowledge of their natural history and their population structure and dynamics. Substantial information on these topics has been developed
from data derived from analysis of commercially caught whales. However, to produce a picture sufficiently complete to be the basis of
biologically sound harvesting, it will be necessary to fill in substantial gaps in our knowledge. The purpose of this report is to
identify some of the more conspicuous of the gaps and to suggest
methods that can be used to fill them. To achieve the goals that are
outlined will require the establishment of two innovations in research
methodology:

(1) A program of pelagic research captures from specially equipped
whaling vessels such as the Norwegian *Peder Huse*, which is a combined
catcher and factory ship.

(2) The development of methods for the study of individual whales
through radio-telemetry.

Identification of Whale Stocks

Effective management requires that the distribution, size, and
dynamics of the population in question be known. It seems clear that
the effective biological unit in most whales is not the species, but
the individual stocks or breeding population. Therefore, the unequivo-
cal identification of stocks of all commercial species (and others as
well) must be given high priority.

Methods

(1) In addition to a continuation of the sampling methods used
on the commercial catch, it is important to obtain detailed information

on the movements of individually marked animals over long periods of
time. To augment the traditional methods of marking, such as Discovery tags, tracking by radio-telemetry must be undertaken to give
continuous records of individuals. (2) The methods of population
genetics must be brought to bear—serological studies employing electrophoretic and immunological techniques, enzymatic studies, amino
acid sequencing, and karyotype analyses. (3) Identification of parasites, diatoms, and other biological indicators. (4) Effective prose-
cution of these studies requires the use of small specialized whaling
vessels on which a biologist must be in charge.

Population Analyses

It is essential that population analysis be uncoupled from the
past and present dependence on the capture of animals on the traditional whaling grounds. An adequate picture of population structure
can be obtained only if sampling is carried on throughout the year in
all parts of the population's range and all classes in the population.
To do this will require that the breeding area be found by exploration
and that some selective whaling (under license) be done on a basis of
scientific rather than economic need. It will be helpful if such
breeding areas could be designated as *research control areas*, where
whaling would be by license only. Population analyses of the sort
proposed will involve the use of ships such as the *Peder Huse* which
can serve as catchers, biological laboratories, and as factory ships,
and are not confined to coastal areas. This will make it possible
for whales taken for biological purposes to be commercially utilized.
Animals captured for study and subsequent processing should be subjected to at least the following tests: (1) detailed analysis of
reproductive state and reproductive history; (2) age determination
using teeth, ear plugs, ovarian histology, jawbone structure, baleen,
and biological samples prepared for later biophysical and biochemical
study (analysis of seasonal variations of bone tissues by electron
microscopy, X-ray diffraction, ultra-sound, etc.); and (3) determina-
of hormone levels where practical.

Trophic Analysis for Whales

Although knowledge of the biology of whales when they are away
from the main feeding grounds must be sought (see above), detailed
quantitative information related to energetics of whale populations
must be obtained. A number of topics invite attention. A few of the
more obvious are listed below:

(1) Trophic effects of the removal of a major consumer, such as
the blue or fin whale on (a) other baleen whales, and (b) other consumers of the same trophic level (seabirds, pelagic seals).

(2) Seasonal and geographic localization of feeding and the role
of prolonged fasting.

(3) Effects of age on food habits.

(4) Role of patchiness of plankton on food searching and the
local distribution of whales. Does patchy distribution of whale food
such as krill allow uncoupling of the mean productivity of a region
from whale energetics? To answer this would require simultaneous
sampling of water and plankton at areas where whales are taken and
comparison of the observed productivity with average values in the
region.

(5) Development of models for estimating metabolic rates of
large whales and methods of direct measurement of metabolism on small
ones.

(6) Use of aerial (and possibly satellite) photography or remote
sensing to determine patchiness and distribution of plankton.

The patterns of distribution of the feeding of whales are widespread geographically, while at the same time stocks are sufficiently
circumscribed geographically that whales should be useful monitors
of pollution of the oceans by heavy metals, pesticide residues, radio-
nuclides, and other fallout products.

The local distribution of whales and porpoises has already been
affected by the environmental deterioration associated with human
activities; for example, their numbers in the southern parts of the
North Sea have decreased markedly in recent years. Since whales are
among the few marine animals that can be seen from above the surface

of the sea and since they are at the top of the trophic pyramid, it
should be possible to use them as indicators of changes in the environmental quality of the seas.

Natural History of Poorly Known Species

Work on the taxonomy and natural history of poorly known species
should be continued and expanded. The smaller cetaceans, including
porpoises, invite attention. As previously indicated, they represent
a biological resource concerning which our ignorance is almost complete.

Study of Unexploited Stocks of Large Whales

A few relatively unexploited stocks of large whales remain. They
are too small to be of major economic importance, but knowledge of
them can serve as baselines with which to compare stocks which have
long been exploited or overexploited. The sperm whale population of
the western South Atlantic and the gray whale population of the eastern
North Pacific offer attractive possibilities.

Whale Movements in Relation to
Hydrographic Factors and Submarine Geology

To marine mammals, the sea is surely an environment with many
local variations. It is possible that bottom topography—submarine
canyons, ridges, flat-topped seamounts, etc.—plays an important role
in their distribution and ecology. It is clear that oceanographic
factors are important to them. The role of these elements of the submarine environment could be explored by the conventional means of
correlating them with the occurrence of whales, or by detailed tracking by means of telemetry of individuals carrying radio packages that
have sensors equipped for depth and temperature measurements (see below).

Telemetric Studies of Free-Ranging Individuals

Because we cannot easily observe whales directly in their natural environment, we have had to develop our knowledge of cetacean natural history by inferences drawn from dead animals and brief observations of live animals when they were near the surface. The development of radio telemetry offers the possibility of directly following known individuals for long periods of time, while also obtaining data on the environmental conditions they meet and measurements of physiological parameters from which behavioral responses can be deduced. It is technically possible to equip a whale with a telemetry and recording package which would allow the determination of location, abdominal temperature, water pressure, light intensity, velocity, time, magnetic heading, water temperature, heart rate, and breathing. From these data, it would be possible to follow an individual (either alone or in a group) and obtain information about location, direction, and speed of travel; frequency, duration, and depth of dives; times and depths of feeding; and the nature of the water through which it passes.

The Smaller Whales

It is important that we consider not only the "commercial" species—the baleen whales and the sperm whale—but also the smaller whales, including the porpoises. These smaller cetaceans, although for the most part not commercially exploited, represent a resource of major potential economic importance. This has been clearly demonstrated by recent developments in the American tuna fishery, which has produced a problem of great economic and biological importance affecting delphinids of the genera *Stenella* and *Delphinus*. Tuna fishermen locate schools of tuna by watching for porpoise and seabird activity. Having sighted a school, they enclose it in a long purse seine which captures not only the tuna but also the cetaceans. There is at present no practical method for effectively separating these delphinids from the tuna. As a result, the setting of one tuna net may result in the death of scores or even hundreds of porpoises. Over most of the world, there is no good market for porpoise meat, so the dead porpoises are discarded. It has been estimated that in the eastern Pacific alone as many as 250,000 porpoises are wasted annually

as a result of this situation. It is obviously only a matter of time
until this extremely efficient method of tuna fishing becomes widespread. The resultant wastage of porpoises will represent a major
problem in all oceans. Aside from this specific instance, it is clear
that in view of the worldwide shortage of animal protein, in the near
future porpoise hunting as an end in itself will become an important
enterprise. At present, we do not have the biological information
necessary to apply effective management to a porpoise fishery—even
the alpha-taxonomy of the smaller cetaceans is fragmentary. Knowledge
of their natural history and distribution—not to mention their behavior and reproductive biology—is equally incomplete.

Chairman: George A. Bartholomew

Members: A. Aguayo L.
           W. Dawbin
           R. Gambell
           T. Ichihara
           Å. Jonsgård
           K. Kenyon
           C. Ray
           V. Scheffer
           W. Schevill
12 June 1971   R. Vaz-Ferreira

REPORT OF THE WORKING GROUP ON REGIONAL POPULATIONS

Stock identification is an important prerequisite for rational exploitation of a fishery. The working group considers that there are five main approaches to this problem for cetaceans: studies of density distribution, marking, biochemical techniques, morphometrics and morphology, and biological indicators.

The working group considered these five main approaches in relation to three categories of exploited species: large cetaceans (the larger mysticetes and *Physeter*), medium-sized cetaceans (minke, pilot, killer, belugas, narwhals, and all species of beaked whales), and small cetaceans (porpoises in general).

Before reasonable estimates of density distribution can be obtained, comprehensive catch and effort data have to be available. At present, certain statistical data are generally unavailable for large cetaceans caught off Spain, Chile, Peru, Korea, China, Brazil, and for the operations of a fleet of unlicensed whalers in the Atlantic. There are also inadequate data for animals caught in native and aboriginal fisheries. Better biological figures (length, sex, fetus, and month of capture) and effort information are needed for all minke, bottlenose (*Hyperoodon* and *Berardius* separately), killer, pilot, and beluga fisheries. There are incomplete data for beluga and narwhal fisheries in the Arctic, and no data for pilot whale catches at Okinawa, for example. Some of these figures are already handled by the Bureau of International Whaling Statistics, and it is recommended that this bureau or some other central agency should handle this extra work. In areas where a fishery involving small cetaceans is in operation, provisions should be made for the collection by regional bodies of catch data such as "species," numbers, and, if possible, tonnages killed.

To obtain density estimates, more sighting work should be carried out by scouting vessels and those engaged in marking, especially in regions where no catching is done. Data for minke, killer, pilot, and all bottlenose species should be collected. In addition, annual trends in the abundance of protected species should be monitored in

as many areas as possible. Plots of regions of high density give indications of areas within which a stock may be found. These preliminary
hypotheses can then be tested using one of the other four techniques.
The correlation of high densities with existing oceanographic information may also assist in provisional estimates of stock boundaries.

The working group gives high priority to the use of biochemical
methods, such as electrophoresis, immunology, and enzyme typing; once
the techniques have been established, relatively rapid results can be
obtained. The group assigns the following priorities for this work:

(1) Sperm whales worldwide, especially in the southern hemisphere

(2) Sei whales, Antarctic and North Pacific

(3) Fin whales, Antarctic and North Pacific

(4) Fin whales, North Atlantic

(5) Medium-sized cetaceans, North Atlantic

(6) Minke whales, Antarctic

(7) Bowheads, Pacific and Atlantic Arctic.

This technique is highly desirable for other medium-sized and for small
cetaceans where fresh material may be readily available.

Independently of any biochemical method, it is recommended that
marking of all cetaceans should be developed and intensified. The use
of standard Discovery-type marks for large mysticetes should be continued. Further trials of more efficient streamer marks using larger
and more conspicuously colored streamers are required. There is a
need for test firings of increased-charge marks into sperm whale carcasses. The .410 Discovery-type marks also should be brought into
wider use for calves and minke whales. This type of marking could also
be carried out by nonwhaling scientists from recognized institutions,
and restricted to bulk marking of medium-sized and small cetaceans,
especially in areas where there is a fishery. Experimental studies
using radio-telemetry packages, freeze-branding, and laser beams should
be encouraged. Whenever possible, double marking, particularly using
the Discovery-type .410 or other suitable tags, should be carried out
on these animals. It is essential to consider the question of a central agency to coordinate the marking information. Priorities for
species and areas are the same as those assigned for biochemical methods.

In areas where other methods are not available, the use of morphometrics is recommended. Certain simple morphological characters such as right whale callosities, flipper and baleen colorations in minke whales, the shape of Bryde's whale baleen plates, and the detailed documentation of color patterns in Delphinidae may all prove useful. Biological indicators such as external parasites and scarring, viral and bacterial flora, and possible dialects in vocalization may have a limited value.

In conclusion, the working group on regional populations recommends that research efforts be concentrated on the following problems:

(1) The development of biochemical techniques for identification of populations.

(2) The intensification of marking of large cetaceans and the development of new marks for all cetaceans.

(3) The collection of more comprehensive catch and effort statistics for all fisheries where unavailable at present.

Chairman: Peter B. Best

Members: J. Bannister
S. Brown
W. Evans
E. Mitchell
S. Ohsumi
H. Omura
D. Rice
W. van Utrecht

12 June 1971

REPORT OF THE WORKING GROUP ON BIOMETRICS AND MANAGEMENT

The group concludes as follows:

(1) The northern and southern seas are among the most biologically
productive regions on earth. In these regions, macroplanktonic animals
in most striking abundance are euphausids and copepods—the diet of the
baleen whales. Although direct harvesting by man of these zooplankton
resources, especially of euphausids, has been proposed, and some research is in progress to this end, practical and economic techniques
have not yet been devised. Harvesting of whales is, at present, by far
the most effective way of tapping these resources. Even if and when
practical techniques are devised, they could supplement rather than
replace the whale harvest since the optimum whale stock sizes (that is,
those giving maximum sustainable yields) are at about half the unexploited levels, and we would expect there to be a large harvestable
zooplankton surplus in these circumstances.

(2) The herds of the great whales could at optimum stock levels
contribute significantly to human needs for protein. We therefore
concur in a policy for whale management which will insure that no stock
is reduced below its optimum level or, when this has already happened,
that exploitation is restricted to permit the stock to recover to its
optimum.

(3) The history of the past centuries has been one of successive
overexploitation of many of the major whale species. Right, bowhead,
gray, blue, humpback, and some stocks of fin whales have been depleted
to well below their optimum levels.

The present scientific evidence is sufficiently good to provide
guidance in general terms for the management of most stocks of large
whales. Action has been taken by the International Whaling Commission
to rebuild stocks of the most depleted species. Reduction of fin and
sei whale stocks that had been occurring at a very rapid rate has been
slowed or stopped. But this is not good enough, since no deliberate
action is yet being taken to rebuild the very important Antarctic fin
whale stock. The more conservative estimates suggest that this stock,

and some other stocks of fin, sei, and sperm whales, are still being depleted, although rather slowly. No action has yet been taken to prevent the overexploitation of the remaining unexploited stocks of cetaceans, particularly the Antarctic minke whale.

The group affirms that it is not enough to seek protection for a species only after its numbers have been so reduced as to threaten its existence. This minimum action is not resource management. Restraint should be exercised early enough that the species remains sufficiently abundant to fulfill its role in the ecosystem. It seems reasonable to assume that a species held at the level of maximum sustainable yield is still a major element in the ecosystem and that the policy of harvesting at maximum sustainable yield is therefore not at variance with considerations of ecology.

Existence of means of harvesting and of markets for the products must, in the absence of regulation, be seen as an implicit danger to the well-being of a species or stock. Thus, we recognize as a general principle of resource management that controls must evolve as new areas and species are opened to exploitation and as major technological innovations appear. In the past when industries have been allowed to expand without restraint, economic considerations have made it difficult or impossible for the industry to abide by the necessary reductions. Application of regulations after a period of overexploitation has usually meant a period of painful readjustment to a more rational pattern. Increasingly rapid rates of exploitation may otherwise be presumed to lead to depletion before conclusive scientific evidence on the optimum stock size could be obtained and regulation secured. A possible remedy to protect newly exploited species might be to set an arbitrarily low quota which can be raised as data provided by capture indicate how high sustainable yields can be set. Specifically, it is recommended that conservative catch quotas be set forthwith for the Antarctic minke whale stocks. Better scientific information would permit a more responsible attitude on these matters to be adopted.

As scientists, we feel that the burden of proof that any given rate of harvesting will not endanger a stock must rest with those who stand to benefit from such harvesting. The industry should accept the judgment of a competent group of scientists in this matter. Further, it should be the responsibility of those exploiting the resource to provide the statistical and other information regarding their operations which the scientists require to make their assessments.

Although scientific efforts in the past have been invaluable, they have been limited both in species and in geographical areas. For example, information on the abundance and distribution of all whale stocks in the Indian, South Atlantic, and South Pacific Oceans, and their relationships with other stocks, particularly the Antarctic stocks, is almost totally lacking. In the case of small whales in all oceans, it is almost absent.

Thus, we emphasize the need for a quantum jump in research personnel and facilities for research on a far broader front to elucidate the unknown potential of this important marine mammal research.

(4) Because of the importance of studying whales that are not exploited (either due to protection or to lack of an industry), it is of prime importance that techniques be developed which would produce useful information from live, free-ranging whales. The objectives of research, in relation to management, should be to produce the following information:

(a) The identity of individual stocks, their location, movement, and possible mixing.

(b) The number of animals being caught.

(c) The number of animals in the stock, both catch and stock being broken down by age, sex, and size.

(d) The relation between the number of mature animals and the number of young produced per unit time which survive to recruit to the fishery, and the density-dependent changes in this relationship.

For (b) as regards the large whales good information is available, mainly through the Bureau of International Whaling Statistics. However, the meeting believes that (1) the introduction of an international observer scheme would add greatly to the reliance that can be placed on their data (without the scheme, fears of incorrect reporting undermine the confidence that can be placed on the scientific assessments), and that (2) similar data on numbers and sizes caught should be collected under a suitable international arrangement for all small cetaceans. Stock identity has, in the past, been studied mainly by traditional marking techniques, which are slow and often inconclusive. A number of new techniques offer good promise and should be investigated and used where possible. These include attached telemetering devices for following by ships, planes, satellites, etc., and biochemical methods. One of the principal needs is for a tag that could be read without killing the animal.

Age data are available for most stocks of the large species, and collection should be started for other stocks. Some doubt still surrounds the precise interpretation of ear plugs. New techniques should be tried, such as tags that will mark the animal's bones, teeth, or eye-lens. Direct catch-per-unit-effort data need careful checking. Possible changes in the ship's equipment or tactics used by whalers should be examined to see whether the efficiency is likely to have changed. Sighting techniques should be improved and they should be used systematically. Observations on the changes, if any, in the abundance of the presently protected whales (blue, right, gray, bowhead, and humpback) by sightings or other means are needed. Other techniques which could yield estimates of stock abundance, either of the whole stock or in a particular area, and which deserve further examination, include:

(a) Aerial and satellite tracking and surveys.

(b) Radio and sonic tagging, including transponders.

(c) Listening by means of hydrophone arrays.

(5) Estimates of stock sizes and sustainable yields are now being made by a greater variety and more refined methods. As a result, scientists are more confident of their conclusions. Nevertheless, there are sometimes divergent estimates, and we cannot yet assign useful limits of error to any estimates.

International regulations tend to be based on midrange estimates, and we cannot at this time suggest a better policy. It is essential, however, that some calculation be made of the likely consequences of acting upon midrange or other estimates if the true values are less or more favorable.

(6) Effective studies of cetaceans will be costly and must usually be on a scale requiring international cooperation. There are some marine mammal research projects under the International Biological Program, and a coordinated study of the exploitable biological production of Antarctic circumpolar waters is being considered as a major project of the International Decade of Ocean Exploration, which has been declared for the 1970's. Arrangements should now be made for scientists to draw up and for governments to fund in the framework of the Decade a comprehensive program of cetacean research, with emphasis on the ecology, behavior, and dynamics of the exploitable species and stocks, and on the monitoring of changes in protected stocks. Such a program will require funding greatly exceeding present levels and it is urged that this be provided. Present expenditures are quite inadequate for the declared aims of existing programs. They also represent an unacceptably small proportion of the income from the exploitation of cetacean stocks, even in their present depleted state.

(7) A disturbing feature of the present situation is the almost complete lack of scientists working full-time on the population dynamics of whales. Assessments of the major whale stocks have been made by scientists working on whales in the time that could be spared from their other main duties. There is great need within a general expansion of whale research for full-time studies of the population dynamics of whales, as well as for whale biologists to become familiar with the concepts of population dynamics, and for population dynamics experts to study whales in more depth.

(8) Implementation of regulations. Though the implementation of regulations will have to take into account economic, social, and political factors, as well as the scientific evidence, the meeting has come to the following general conclusions:

(1) The entire species and all its component stocks should be considered. The subgroups of a species are very likely part of the evolutionary strategy by which the species took its place in the modern world and thus the geographic pattern of distribution and abundance must be looked on as part of the health of the species as a whole. This implies that it is not sufficient to set a single species limit, but rather that limits should be set by stocks.

(2)   Management of whale stocks of the world must insure that each stock of each species is maintained at or close to the optimum level. This would be achieved by setting a quota separately for each stock and is clearly not insured by the current blue whale unit system.

(3)   It should not be the strategy of the International Whaling Commission merely to maintain overexploited stocks, but rather to allow the whale populations sufficient surplus reproductive potential to regain levels at which the populations can provide the maximum sustainable yield.

(4)   Estimation of what constitutes a prudent management practice should include consideration of the possible effects of population catastrophes, such as failure of local food supplies, epidemic diseases, etc.

(5)   It should be the role of the groups, both national and international, to foster and encourage study in the research areas discussed. A new agency should be created, or an existing one mandated, to coordinate research efforts, including the job of applying to member governments for support of specific projects.

Recommendations:

(1)   The international observer scheme should be instituted at once.

(2)   The blue whale unit should be rescinded and replaced by a quota scheme that provides management of each individual species and stock.

(3)   Quotas for those species estimated to be at or above their optimum levels should be set at values which will not reduce the stocks below those levels; quotas for species below their optimum levels should be brought to values that will insure rebuilding of the stocks. The consequences of acting upon possibly erroneous midrange estimates of sustainable yields should be studied and appropriate action taken.

(4)   A quota should be set for minke whaling in the Antarctic.

(5)   There appears to be a danger to certain isolated whale stocks which could be destroyed in a very short time. There could be a particular risk from the irresponsible use of existing surplus whaling

equipment, especially in countries not party to the International Whaling Commission. It is recommended that all nonmember nations engaged in or entering whaling operations immediately join the IWC and adhere to its regulations, and that member countries of the IWC should take all possible steps to bring this about and seek universal adherence to its recommendations.

(6) Regulations should be implemented as soon as possible to insure appropriate levels of protection for each sex of sperm whale.

(7) A comprehensive long-term program of cetacean research, involving a very substantial increase in the number of scientists engaged in the field and in available research facilities, including special ships, should be planned and implemented as soon as possible.

Chairman: Sidney J. Holt

Members: K. Allen
L. Boerema
D. Chapman
T. Doi
M. Dunbar
Y. Fukuda
J. Gulland
R. Laws
J. McHugh
S. McVay
M. Nishiwaki
K. Norris
R. Payne

12 June 1971

Participants

Anelio Aguayo L.
Departamento de Oceanología
Universidad de Chile
Viña del Mar, Chile

K. Radway Allen
Fisheries Research Board of Canada
Nanaimo, British Columbia, Canada

J. L. Bannister
Western Australian Museum
Perth, Western Australia

George A. Bartholomew, Jr.
University of California
Los Angeles, California

Peter B. Best
Division of Sea Fisheries
Cape Town, South Africa

L. K. Boerema
United Nations Food and Agriculture Organization
Rome, Italy

Sidney G. Brown
Whale Research Unit (IOS)
London, England

Douglas G. Chapman
University of Washington
Seattle, Washington

William H. Dawbin
University of Sydney
Sydney, N.S.W., Australia

Takeyuki Doi
Tokai Regional Fisheries Research Laboratory
Tokyo, Japan

Max J. Dunbar
McGill University
Montreal, Quebec, Canada

William E. Evans
Naval Undersea Research and Development Center
San Diego, California

Yoshio Fukuda
Far Seas Fisheries Research Laboratory
Shimizu, Shizuoka Ken, Japan

Ray Gambell
Whale Research Unit (IOS)
London, England

J. A. Gulland
United Nations Food and Agriculture Organization
Rome, Italy

Sidney J. Holt
Office of Oceanography
UNESCO
Paris, France

Tadayoshi Ichihara
Far Seas Fisheries Research Laboratory
Shimizu, Shizuoka Ken, Japan

Åge Jonsgård
Institutt for Marin Biologi, A og C
Oslo, Norway

Karl W. Kenyon
Department of the Interior
Fish and Wildlife Service
Seattle, Washington

Richard M. Laws
British Antarctic Survey
Monks Wood Experimental Station
Abbots Ripton
Huntington, England

J. L. McHugh
State University of New York
Stony Brook, New York

Scott McVay
Princeton University
Princeton, New Jersey

Edward D. Mitchell
Arctic Biological Station
Fisheries Research Board of Canada
Ste. Anne de Bellevue, Quebec,
   Canada

Masaharu Nishiwaki
Ocean Research Institute
University of Tokyo
Nakano, Tokyo, Japan

Kenneth S. Norris
Oceanic Institute
Waimanalo, Oahu, Hawaii

Seiji Ohsumi
Far Seas Fisheries Research
   Laboratory
Shimizu, Shizuoka Ken, Japan

Hideo Omura
Whales Research Institute
Tokyo, Japan

Roger S. Payne
Rockefeller University
New York, New York

G. Carleton Ray
Johns Hopkins University
Baltimore, Maryland

Dale W. Rice
National Marine Fisheries Service
La Jolla, California

Victor B. Scheffer
Bellevue, Washington

William E. Schevill
Woods Hole Oceanographic Insti-
   tution
Woods Hole, Massachusetts

Willem L. van Utrecht
Universiteit van Amsterdam
Amsterdam, Netherlands

Raúl Vaz-Ferreira
Universidad de la República
Montevideo, Uruguay

Observers

Department of the Interior
Washington, D.C.
   Curtis Bohlen
   Clyde Jones
   Frank Whitmore
   Earl Baysinger
   John Sayre
   Robert L. Brownell

Smithsonian Institution
Washington, D.C.
   Charles O. Handley
   Michael Huxley

New York Zoological Society
Bronx, New York
   James A. Oliver

U.S. International Biological
   Program
   William Milstead

Department of Commerce (NOAA)
Washington, D.C.
   George Y. Harry, Jr.

National Science Foundation
Washington, D.C.
   George Llano

Environmental Defense Fund
   Roderick A. Cameron

World Wildlife Fund
   Steven Seater

Council on Environmental Quality
Washington, D.C.
   Lee Talbot

International Union for the Con-
   servation of Nature
   Colin W. Holloway
   Harold J. Coolidge

University of California, Berkeley   Marine Technological Society
  Edwin R. Lewis   Robert Niblock

Conference Management

Marine Mammal Council
U.S. International Biological Program
  G. Carleton Ray, program director and session chairman
  William E. Schevill, associate program director and session chairman
  Kenneth S. Norris, session chairman
  Suzanne M. Contos, executive secretary

PART TWO

Current Status

CHAPTER 1

DISTRIBUTION AND ABUNDANCE OF WHALES IN RELATION TO BASIC PRODUCTIVITY

J. A. Gulland[1]

Management of marine resources has become an important problem
facing the international community. A special conference on the Law
of the Sea is being convened probably in 1974 by the United Nations,
in which this will be an important topic. Among marine resources,
whales have become the classic example of the difficulties involved
in achieving proper management and the losses arising from a failure
of management. They are, however, not unique in this regard. The
advance of modern technology has resulted in an increasing variety of
animals being harvested from the sea. Few stocks are being well
managed, even though different conditions, especially with regard to
the biology of the animals concerned, have resulted in the failure to
achieve good management in the case of most fish stocks, which has
less serious consequences than in the case of the whales. As an industry, whaling should be considered as one rather specialized way of
using the vast, but not limitless, primary production of the oceans
for the benefit of mankind. The present paper is a brief attempt to
relate the distribution and abundance of whales to the primary production in the oceans, with particular reference to the efficiency of
the whale stocks in harvesting this production (and making it more
readily available to man).

For the purpose of this paper, only the larger whales (rorquals
and sperm whales) will be considered. The most important differences
between these two groups in the present context is their food. Sperm
whales are top predators; the large squid on which they feed are themselves probably second or third level predators, so that the sperm

---

[1]The views expressed here are those of the author and not necessarily those of FAO.

whale is several steps removed from primary production. Relative to
this primary production, and the intermediate stages, the production
of sperm whales must therefore be low.

The baleen whales, on the other hand, are trophically much nearer
to the primary production; the typical food chain of Antarctic rorquals
is diatoms to euphausids to whales. This energy transfer implies that
the production of Antarctic baleen whales can be high. In considering
the links between whales and their food, a careful distinction must
be made between production and biomass. To the whale looking for a
meal or to the individual fisherman wanting a good catch the production
——the rate at which plants or animals are produced——is of less significance than the standing crop or biomass——the amount present at a
given time. Though high standing stocks are associated with high production, the relation is not necessarily a directly proportional one.
The average length of life of the animals or plants concerned is also
important.

In the warmer areas of high production, e.g., the upwelling areas
off Peru or California, the herbivores are short-lived; Cushing (1969)
states that the copepods, which are a major element of the zooplankton,
have up to ten generations per year, while in the Antarctic, krill
(*Euphausia superba*) lives several years (Marr 1962). The standing
stock of zooplankton off California will therefore be the production
of only a few weeks, but in the Antarctic the standing stock may be
the production of a year or more. For a given annual production, the
zooplankton standing stock in the Antarctic will therefore be much
higher than off California. This makes the Antarctic particularly
attractive to the rorquals.

Less is known about the average length of life of the squid on
which the sperm whales feed, but they are large animals and might be
expected to be moderately long-lived——probably a matter of years
rather than months even in the tropical areas. The biomass, therefore,
will be equal to the production of a year or more. Possibly also
sperm whales are less dependent on high biomass than rorquals because
they pursue individual animals and may therefore be able to harvest
the available food over a much wider area than do the baleen whales.

On this basis one might expect the rorquals to be confined to areas of high zooplankton biomass, but to be abundant in those areas, and sperm whales to be more widespread, though at a lower density.

Distribution

Primary and Secondary Production

The pattern of distribution of primary production in the oceans has been fairly well mapped out, at least in broad general terms, in the hundred years since the Challenger expedition (a convenient date for the start of oceanography as a science) and especially in the period since the development of the carbon-14 technique by Steemann Nielsen (1954).

Over much of the open oceans of the tropics and subtropics, both production and standing stock at all trophic levels are low. Much of the available nutrient material is locked up in living organisms. There is a fairly steady balance among the trophic levels: plants, herbivores, and carnivores. Any tendency of plants to increase is limited by grazing by herbivores, which in turn are controlled by predation by larger animals. In temperate and subarctic waters the balance is less well-maintained. At the end of winter, the deep mixing of the water by winter storms has added to the nutrients in the surface layers, and the increasing sunlight in spring triggers off a phytoplankton bloom which cannot be controlled by the herbivorous zooplankton. The latter in turn has an outburst a few weeks later. These waters (for instance, the North Atlantic in a band from New England south of Greenland to Iceland and Norway, and much of the Antarctic Ocean) are therefore regions of high primary production.

In the upwelling areas there is a somewhat similar, but more continuous, structure of unbalanced production, i.e., production in which the peaks of plant and herbivore production do not coincide, although the separation in the upwelling areas is a matter of space rather than time. In these areas—off Peru, northwest and southwest Africa, and along the equator—divergent current systems, or steady offshore winds (the same wind systems that produce the coastal deserts of Peru,

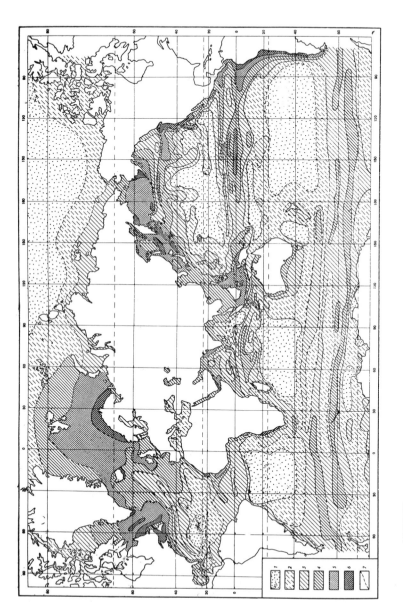

FIGURE 1-1. Distribution of zooplankton biomass in the world ocean, from Bogorov, Vinogradov, Voronina, Kanaeva, and Suetova (1968).

Symbols: 1 – < 25 mg/m^3 4 – 101–200 mg/m^3
           2 – 25–50 mg/m^3 5 – 201–500 mg/m^3
           3 – 51–100 mg/m^3 6 – > 500 mg/m^3
                                   7 – Conjectural boundaries

Mauritania, and the Kalahari), result in cool and nutrient-rich water rising to the surface. As it nears the surface, there is a phytoplankton bloom, similar to the spring bloom in temperate or subpolar waters, which in turn supports a later zooplankton outburst, generally somewhat downstream of the initial upwelling. These upwelling areas are among the most productive seas in the world and support some of the richest fisheries, notably that for the Peruvian anchoveta, which currently produces some 20% of the total world fish catch.

The distribution of the standing crop of zooplankton has been studied in most parts of the world. There are some difficulties of interpretation, because the more active animals, such as the euphausids, can escape to a greater or lesser extent from the nets usually employed. Thus summary distribution charts, for example produced by Bogorov, Vinogradov, Voronina, Kanaeva, and Suetova (1968) (Fig. 1-1) may underestimate the density of zooplankton in areas where euphausids predominate. Their chart, therefore, does not show the Antarctic as being particularly rich, but the other highly productive areas such as the North Pacific, the North Atlantic, and the waters off Peru are clearly illustrated.

General Distribution of Sperm Whales

The classical sperm whale fishery is that from New England,

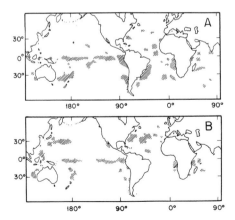

FIGURE 1-2. Major areas of nineteenth-century sperm whaling, from Cushing (1969) after Townsend (1935). A. October through March. B. April through September.

described so well and in rich detail by Herman Melville. Logbooks of
these whaling ships give the position of many of the whales caught;
these records have been summarized by Townsend (1935). The striking
correlation between the distribution of sperm whale catches and the
major upwelling areas has been noted by several authors. Cushing
(1969) notes that the main upwelling areas—California, Peru, Benguela,
and Canary, and particularly the equatorial system in the Pacific—
show up well as areas of high sperm whale catches (Fig. 1-2). The
South Arabian upwelling area is missing from the catch data. On the
other hand, there are some catches away from known upwelling areas—
in the central and western North Atlantic in the northern summer, and
rather similar areas north and west of New Zealand and west of the
Cape of Good Hope in the northern winter. The first of these may be
as much due to the number of whaling ships passing through the area
on the way to and from their home ports as to a real concentration of
whales.

The distribution of present-day whaling gives a less clear picture of sperm whale distribution. There is no pelagic operation
confined entirely to sperm whales, and the position of shore stations
is dependent upon a number of social and economic factors as well as
upon the distribution of whales. Coastal sperm whaling is now
restricted to half a dozen places: Japan, Chile-Peru, Western Australia, South Africa, and Spain-Portugal. These shore stations
account for only about a quarter of the total catches. The rest is
taken by pelagic expeditions, in three distinct operations: in the
Antarctic during, or immediately before or after, the rorqual season;
in temperate or tropical waters by Antarctic expeditions on passage;
and in the North Pacific. The second of these probably covers areas
similar to those of the New England fishery and takes both males and
females.

Possibly as a result of a more strict enforcement of size limits
in temperate and tropical waters, these catches have declined in the
last few years. The North Pacific, and more especially the Antarctic,
pelagic operations take almost entirely large males. These catches
are outside the distribution of New England catches shown by Townsend;

probably the general weather conditions were unfavorable to open-boat
whaling in these areas.

There is a very striking difference in distribution between the
large males, which occur in temperate or subpolar waters, and the rest
of the population, which is confined almost entirely to tropical and
subtropical waters. The reasons for the difference do not appear to
be known. The food seems to be similar for the two groups of animals.
Both eat squid and if the big sperm whales eat rather larger squid
there are no data on the distribution of different sizes of squid that
would suggest that they would have to enter the colder areas to get
the big squid. The separation between the main sperm whale groups,
mainly of females and young, and the solitary males is clearly a social
phenomenon, but it is not clear why this separation should be a matter
of up to 20 degrees of latitude.

In any case the main concentration of the population, and the
location of the main production, is where the females and young, and
also the harem bulls, are found, which is in the upwelling areas. On
the other hand, the solitary males represent the larger part of the
harvestable surplus. Thus, the main areas of production (in the subtropics) are different from the main harvesting areas (temperate and
subpolar) for sperm whales.

Relative Abundance of Sperm Whales in Different Areas

Few estimates are available on the abundance of sperm whales in
different areas, though some (noted below) have been reported to the
International Whaling Commission, especially by the IWC-FAO working
group on stock assessment (Anon. 1969b). The common indices of relative abundance—particularly indices of catch per unit effort, such
as catch per catcher day's work, number sighted per hour flown by
aircraft, etc.—are not easily comparable between areas, because of
the different measures used, or the difference in size and capacity
of the ships concerned, or the skill of gunners. Another difficulty
is that not enough is known about the stock structure of sperm whales,
and especially the relations between the males in high latitudes and
the females and young males in warmer water. Thus, even when abundance

estimates have been obtained, the limits of the area to which they
refer are far from clear.

With these reservations, the estimates given by the IWC-FAO working group which met in Rome in March 1968 and from more recent sources
are given in Table 1-1. Doi and Ohsumi (1970) believed that the mature

TABLE 1-1. Estimates of abundance of sperm whales.

Area	Period	Number of whales		Source
		Males only	Total	
North Pacific	1969	77,700	–	} Doi and
	Initial stock	129,400	–	} Ohsumi 1970
Western Australia	1962-1966	28,000-44,000	–	}
Durban, South Africa	1962-1965	–	22,000-56,000	Anon. 1969b
Chile-Peru	1957-1961	–	45,000	}

males in subarctic waters of the North Pacific, estimated in the first
line of Table 1-1, were about two-thirds of the initial male stock.
If the mature females number about the same, the total of mature animals in the North Pacific stock (or stocks) must initially have been
about $2 \times \frac{3}{2} \times 130,000$, or about 400,000.

These figures would suggest clear differences in the size of the
stocks considered. The North Pacific stock would appear to be much
the biggest (a little less than half a million animals), the Western
Australian stock being about half to a quarter of this size (100,000
to 150,000), and the stocks off Durban and Chile-Peru both about half
that off Western Australia. No estimates have been made of the magnitude of the stocks exploited in the Antarctic. Conventional analysis
based on catch-per-unit-of-effort data is impossible because of the
lack of suitable effort data. The total catches from the beginning
of Antarctic whaling have been close to 100,000 animals, almost
entirely males, which must have come from a total stock at least comparable with the North Pacific stocks. The distribution of these
catches among the different Antarctic whaling areas is shown in Table
1-2. Since the distribution of whaling activities is determined by

TABLE 1-2. Numbers of whales caught in the various areas of the
Antarctic 1931 to 1938 (from International Whaling Statistics), except
that in the third line the blue and fin catches are combined in blue
whale units, not individual whales.

Area	II $60^{\circ}W-0^{\circ}$	III $0^{\circ}-70^{\circ}E$	IV $70^{\circ}E-$ $130^{\circ}E$	V $130^{\circ}E-$ $170^{\circ}W$	VI $170^{\circ}W-$ $120^{\circ}W$	I $120^{\circ}W-$ $60^{\circ}W$	Total
Catches							
Blue	38,400	77,000	56,800	11,700	4,500	900	189,300
Fin	159,600	193,500	77,000	35,900	30,300	16,900	513,200
B.w.u.	118,200	173,750	95,300	29,650	19,650	9,350	445,900
Sperm	21,400	32,500	20,800	19,000	3,000	2,400	99,100
Sperm/b.w.u.	0.18	0.19	0.22	0.64	0.15	0.26	0.22

the availability of baleen whales, the total in blue whale units is
shown as well as the catches of blue and fin whales.

This table shows that, not surprisingly, the biggest sperm whale
catches have come from the areas (II, III, and IV) in which most
whaling as judged by baleen catches has been carried out. The ratio
of sperm to baleen (b.w.u.) catches does not vary very much, except
for a high value in Area V, south of eastern Australia and New Zealand. If the large male sperm whales in the different areas of the
Antarctic come from populations immediately to the north, this would
suggest a relative high concentration of sperm whales north of Area V,
i.e., in the central and western Pacific, between Hawaii and the
Philippines. This would be in general agreement with the high primary
production in this area.

Distribution of Baleen Whales

Compared with our knowledge on the abundance and distribution of
sperm whales, that on baleen whales is reasonably good—though some
of this better knowlege is due to the more serious depletion of the
initial baleen stocks. Because so many of the stocks have been reduced
to a very low level, the actual numbers caught give a good index of the

initial abundance of the different stocks.

The actual number of baleen whales taken in the Antarctic between
1931 (when the collection of good statistics on area of capture was
started) and 1968 was 825,300. The general distribution of the whaling grounds, and the boundaries of the statistical areas are shown in
Figure 5 of Mackintosh (1965, p. 52). The estimates of the initial
exploitable stocks in the whole Antarctic are 210,000 blue whales
(Anon. 1964), and between 120,000 and 180,000 sei whales (Anon. 1969a,
Doi and Ohsumi 1969). The initial stock of fin whales has not been so
well estimated, but was probably twice the average number alive in the
period 1950/51 to 1956/57 (Anon. 1964). The most recent estimate of
the abundance near to that period is 171,800 in 1958 (Anon. 1971),
which would suggest an initial exploitable population of around
450,000 animals. Adding in a few tens of thousands of humpback whales
gives a total initial population of around 850,000 animals, which happens to be very nearly equal to the actual numbers caught. Assuming
the arithmetical coincidence also holds for individual areas, the
distribution of catches according to statistical areas can be taken
as a fair approximation to the distribution of initial stock. These
are set out in Table 1-3. Because the statistical areas differ in

TABLE 1-3. Distribution of total number of baleen whales caught in
different areas of the Antarctic between 1931 and 1968 (from International Whaling Statistics), and density of whales estimated as
thousands of whales caught per degree of longitude.

Area	Total catch	Whales per degree of longitude (thousands)
II $(60°W-0°)$	240,611	4.01
III $(0°-70°E)$	296,256	4.23
IV $(70°-130°E)$	156,396	2.61
V $(130°E-170°W)$	64,542	1.08
VI $(170°-120°W)$	44,472	0.89
I $(120°-60°W)$	23,023	0.38
Total	825,300	2.29

extent, the numbers have been expressed in the last column in a more comparable form, as numbers of whales per degree of longitude. While it would be even better to express the initial density of whales as numbers per unit area, it is difficult with animals making long north-south migrations to determine what area should be chosen. Numbers per degree are less likely to be misleading, as well as being easier to calculate.

This table shows a very great difference between areas. The region to the south of the Atlantic (Areas II and III) appears to be up to ten times as productive as some of the other areas, especially Area I (south of the west coast of South America). These figures almost certainly exaggerate the real differences, since whaling activities have also been concentrated in the Atlantic areas. Indeed, for a period up to the 1955/56 season, factory ship operations for baleen whales were prohibited in the so-called sanctuary area, between 70° and 160°W, or most of Areas I and VI.

On the other hand, it appears from marking data (Brown 1962) that there is not much movement of whales between areas, so that if Areas I and VI had been as productive as the other areas there should have been very large catches after the sanctuary was opened. This was not the case. In fact the sanctuary was proposed as a known area of low abundance which could be closed without great disruption of the whaling industry, but which at the same time would provide some measure of assurance of the continued existence of the various species of whales. The setting up of the sanctuary was perhaps an acknowledgment by the industry at that time (immediately after World War II) of its wider responsibility for the continuation of the species.

A comprehensive comparison of this distribution of whales round the Antarctic with that of primary production is difficult because of the sparse data on the latter from much of the Antarctic, particularly the area south of the Indian Ocean. The best data, based on carbon-14 uptake, have been reviewed by El-Sayed (1968, 1970). These data are entirely confined to a segment of the Antarctic running westward from 10°W south of South America and New Zealand to 140°E, south of Tasmania. In this segment the differences in productivity are most striking. Most of the area south of the open Pacific (Areas VI and I) is very unproductive, with a production in terms of carbon fixation per unit volume at the surface of under 10 mg $C/m^3/hr$, or, integrated over the water column, under 30 mg $C/m^2/hr$. On the other hand, very high values occur close to the Antarctic Peninsula, and high values (over 30 mg $C/m^3/hr$ or 60 mg $C/m^2/hr$) also occur in the Southwest Atlantic, from Argentina to South Georgia and South Orkney (the western part of Area II) and between New Zealand and Tasmania and Antarctica (Area V).

The boundaries of high and low production areas do not agree too well with the boundaries of the whaling areas, and in particular a small part of the highly productive area round the Antarctic Peninsula lies west of $60^{o}W$, in Area I. Otherwise the primary production in Area I is poor, and ignoring this, the mean production per unit surface area in both Area VI and Area I can be estimated, from Figure 8 of El-Sayed (1970), as about 20 mg $C/m^2/hr$. Comparable figures for Areas II and V are 30 mg C and 55 mg C respectively. The figure for Area II is possibly misleading. As the charts in El-Sayed (1968) show, the samples in this area have been concentrated in the Weddell Sea, especially in the southern part close to the ice edge. The samples in the other areas (V, VI, and I) are more evenly spread through the open ocean on both sides of the Antarctic Convergence. The few samples in the open ocean in Area II tend to show high production, comparable to Area V.

While no detailed studies of production in the remaining areas (III and IV) appear to have been carried out using the carbon-14 technique, some generalized picture of the production in these areas has been given. For instance, Figure 53 of Moiseev (1969) shows a tongue of moderately high production (probably over 150 g $C/m^2/yr$) projecting eastward from South Georgia southward of South Africa and the Indian Ocean (see Fig. 1-1). This would suggest that Areas III and IV are intermediate in production between the rich areas (II and V), and the poor areas (I and VI).

The distribution of whales, or at least of whale catches, agrees only in part with that of primary production. There have been more whales caught in Area III, and fewer in Area V, than the data on primary production would suggest. The agreement with zooplankton distribution is much better, especially with that of krill, *Euphausia superba*. This distribution has been studied in detail, especially by Marr (1962). As his figure 135 (p. 394) shows, most of the largest catches of krill in plankton nets were taken in the area round South Georgia. As is inevitable when studying such a large area, the sampling is not uniform over the whole region, and it appears that the abundance of krill is very much higher in Area II than elsewhere, with Areas III and IV probably having rather more than the other areas. While this is more consistent with the distribution of whales, there still appear to have been more whales in Area III than can be immediately accounted for by either primary production or zooplankton.

The other major area of recent baleen whaling is the North Pacific. A number of estimates of the current and initial abundance of the major stocks have been made by both Canadian and Japanese scientists. Among the most recent are those by Allen (1969) and Doi and Ohsumi (1970), which give the estimates of the initial population given in Table 1-4.

TABLE 1-4. Estimates of initial populations of rorquals in the North Pacific from data of (a) Doi and Ohsumi 1970, (b) Allen 1969, and (c) Pike 1966.

Fin whales	40,000 (a)	18,000 (b)
Sei whales		
Pelagic stock	45,400-69,200 (a)	16,000 (b)
Japanese coastal stock	14,000-15,500 (a)	
Blue whales	1,100-2,500 (c)	

Adding a few humpback whales to these figures gives estimates of the total rorquals of about 120,000 using the Japanese figures for fin and sei whales, or rather less than 50,000 using Allen's data. Probably the true number was not in excess of 100,000. The North Pacific at around 50°N extends over about 80° of longitude, though narrowing rapidly further north to about 25° along 60°N, and to only a few miles at the Bering Strait (66°N). It is difficult to make a direct

comparison of production per unit area with the Antarctic, where there
is little land in the latitudes where most whaling is done. If the
80° of longitude is taken as a fair measure of the extent of the area
inhabited by the stocks throughout the year, then for an initial population of 100,000 animals the number of whales per degree will be 1.2
thousand. This value is almost half the average value for the Antarctic, though a higher figure would of course be derived if a narrower
width were used.

Some data are also available from the North Atlantic. Allen
(1970) estimated that the number of fin whales in the west Atlantic
between Cape Cod and 57°N was about 7,000 animals, and that the initial
population of blue whales off Newfoundland and Labrador did not greatly
exceed 1,100. The humpback population was probably about the same.
No estimate was made for sei whales, though catches had been increasing.
The fin whale population had been exploited before the period analyzed
by Allen, but probably had not been greatly decreased below its initial
value. If the sei whale population is of about the same magnitude as
the fin whale stock, the initial population of all rorquals off the
North American east coast was about 20,000 animals (perhaps 8,000 to
9,000 each of fin and sei whales and 3,000 others). Possibly this
stock extends eastward to the longitude of Cape Farewell, about fifteen or twenty degrees of latitude. If so the number of whale per
degree—1.0 to 1.3 thousand—is close to that estimated for the North
Pacific.

Relative Distribution and Abundance of Different Species of Rorquals

The various species of large rorquals differ in their habits and
food preferences, and these variations might be expected to be reflected in their distribution and abundance. The humpback whale is
somewhat of a special case, having a local and coastal distribution,
resulting in a rather low global abundance. The blue, fin, and sei
whale tend to be distributed successively in order of increasing distances from the pole; this zonation enabled the Antarctic industry
to concentrate its activities in turn on the three species, moving
successively further northward as the major interest changed from

blue whales to fin, and from fin to sei. At first the Antarctic whaling industries concentrated their attention on the species giving the
biggest yield of oil per animal—the blue whale. As the stocks of
this species were depleted, attention was switched to fin whales. The
switch to sei whales was encouraged by the ensuing depletion of the
fin whales; in addition, in the second half of the 1960's the greater
interest in meat rather than oil, as well as the blue-whale unit system and the allocation of quotas encouraged the concentration of whaling on sei whales rather than fin whales. The food habits of the
three species—blue, fin, and sei—are probably correlated with their
latitudinal distribution. The blue whale feeds mainly on euphausids,
the fin whale on euphausids and copepods, and the sei whale mainly on
copepods (Nemoto 1970). The data of Marr (1962) already referred to
show that in the Antarctic *Euphausia superba* is concentrated mainly
close to the ice edge, except for a concentration in the area of
South Georgia.

The relative abundance of blue and fin whales in the different
areas of the Antarctic can be gauged by the ratios of the catches.
Although there was a gradual change in attention from blue to fin
whales, for a long period up till the 1960's both species were caught
in significant numbers. The switch to sei whales was more abrupt,
and the detailed catch records show that those expeditions which did
go after sei whales took almost entirely sei whales. Thus, the relative catches of sei and other whales reflect catcher preference rather
more than relative abundance. The blue and fin catches are best
treated separately for different periods; the data are set out in
Table 1-5, expressed as the ratio of fin:blue whales.

The figures are reasonably consistent. First, the ratio steadily increases in all areas during the period; secondly, the ratio
decreases from Area II through Area III to Area IV, showing that there
were relatively more blue whales in the latter area. The low ratio
in Areas V, VI, and I in the period 1938 to 1953, for which separate
catch data are not available in the early years, may be due to the
later start of intense whaling there, and hence to the lesser depletion of the blue whale stock. This cannot apply to Areas III and IV,

TABLE 1-5. The ratio of catches of fin to blue whales in the Antarctic (data from International Whaling Statistics).

Period	Ratio in Area—			
	II	III	IV	V, VI, and I
1933-1938	1.64	0.81	0.48	-
1938-1953	3.19	2.48	1.65	1.76
1953-1968	17.78	5.46	2.91	12.83
Whole period 1933-1968	4.16	2.52	1.37	-

in which large catches were taken in the 1930's.

Over the Antarctic as a whole the ratio of catches has been almost 2.71:1—the ratio of initial stocks was probably lower, around 2.14:1—the difference being due to the relatively greater catches taken from the fin whale stock. While the blue whale stock was reduced in a single decade (1930 to 1940) to a level of low net productivity, fin whale exploitation was more gradual, partly due to the regulations of the International Whaling Commission in the postwar years. Thus for a considerable period—probably from almost 1940 to 1955—the surplus of births over natural deaths in the fin whale population was appreciable.

A very different fin:blue ratio seems to occur in the northern hemisphere, around 20:1 in the North Pacific and 5:1 in the Northwest Atlantic. The low abundance of blue whales in the North Pacific may be accounted for by the topography of the area, with much less ocean at the higher latitudes (60° to 70°) favored by blue whales than further south. The large numbers of blue whales that used to occur in the Antarctic may be explained by the relatively high population of euphausids in the Antarctic zooplankton.

The ratio of sei whales to other large baleen whales also shows considerable regional variation. For the Antarctic as a whole the original stocks of sei whales, excluding young animals, totaled around 130,000 animals (FAO 1970), compared with a combined stock of 560,000 blue and fin whales—a ratio of 0.23:1. The ratio by areas (in terms of original sei whale stock as given in the FAO report compared with blue and fin whale catches given in the International Whaling Statistics) varied from 0.09 in Area III and 0.10 in Area IV, to 0.4 to 0.6 in Areas V, VI, and I. The higher values in the latter areas may be partly due to the low catches of blue and fin whales in these Areas, rather than low original stocks. This explanation would, however, not hold for Area II, where there were relatively twice as many sei whales (a ratio of 0.23:1) compared with Areas III and IV.

Outside the Antarctic, sei whales appear to be relatively more abundant; in the North Pacific there seem to have been roughly equal quantities of fin and sei whales (Table 1-4), and the same may also be true of the Northwest Atlantic. The reason for the greater numbers of the generally less polar sei whales in these regions, and in Area II of the Antarctic, may be the location of the regions of higher primary productivity, which tend to be more in temperate, rather than subarctic (or subantarctic), waters.

The Magnitude of Production and Potential Harvest

The combined maximum sustainable yield of Antarctic whales is about 6,000 blue whales, 20,000 fin, 5,000 sei, and perhaps 1,000 humpback (Anon. 1964, and later reports of the IWC). Converting these to weight, using the data of Crisp (1962), gives a total of about 1.7 million tons. Previous sections have discussed the distribution of these resources in relation to primary production, but it is also of interest to attempt to compare the magnitude of the potential whale harvest with the production of plants and zooplankton.

The first step is to examine the amount of food consumed by whales. The annual consumption of krill by whales is not quite clear. Moiseev (1970) gives a figure of 140 million tons for the total food consumption of a population of whales in the Antarctic which is probably not very different from the population giving the maximum sustainable yield. Mackintosh (1970) gives a wider range of 33 to 330 million tons, with a more probable range of 100 to 150 million tons. This ratio of 1.7:100 or 150 would seem to imply that the whale population

is a very inefficient converter of the food it eats, with an efficiency
of only 1 to 2%. This compares with oft-quoted values of the efficiency
of conversion of 10 to 15%, as in Schaefer (1965). The figure of 1 to
2% is actually an underestimate of the ecological efficiency, defined
as the ratio of the food eaten by one trophic level to that consumed
by the next level. To the catch by man should be added the quantity
of whales dying of natural causes, and consumed by predators or bacteria. When the population is being exploited at the sustainable rate,
the latter quantity may be equal to man's harvest, giving an overall
efficiency of 2 to 4%.

The efficiency of food utilization by a whale, considered as the
ratio of growth to food consumed, changes through its life. While precise data are scanty, the young whale grows extremely fast, and there
seems little doubt that it is an efficient converter of the food it
gets. Perhaps in the first year of life the efficiency may be 20 to
30%. Later in life it is much lower. An adult whale may consume annually several times its own weight in food (Mackintosh, 1970, suggests
between 1.5 and 15 times), but virtually all this is used for maintenance and very little for growth. In the larger whales which are
growing very slowly, if at all, the efficiency of conversion from food
to growth may be only 1 to 2% or lower. Since in a balanced population being rationally exploited most of the whales will be mature, the
efficiency of the population as a whole must be low.

Though whales are the most obvious consumers of krill and other
zooplankton, it is not clear that the consumption by whales accounts
for the majority of the annual production of krill, let alone of zooplankton as a whole. Gulland (1970) gave very rough figures of zooplankton production of 150 million tons based on scattered observations of standing stock, and of 1,000 million tons, based on the
assumption of a 10% efficiency of utilization of a primary production
of 10,000 million tons. The former figure is likely to be low because
(a) many of the zooplankton animals may escape from the nets used, and
(b) the average life span of the smaller animals, other than euphausids,
may be less than the one year assumed by Gulland. Consumption by whales
therefore probably accounts for only a small part of the total animal

production, but may, if krill is only a minor fraction of the total zooplankton, be a significant proportion of the production of the preferred food.

These data on food consumption and food availability are in principle relevant to the study of the factors determining the absolute abundance of whales. Why should the original numbers of blue whales in the Antarctic have been 200,000 rather than 20,000 or 2,000,000? The proximate cause of the stability of whale populations at the upper levels of abundance—compared with the 5 to 10% increase at lower levels, such as those occurring at present—seems to be a delay in the age at maturity, and a decrease in pregnancy rate (Gulland 1971a, Laws 1962). If indeed whales could account for a large proportion of the total krill production, these changes might in turn be due to changes in food supply, rather than, say, social pressures, but direct evidence is absent. For the present, therefore, these can be no more than interesting speculations, and the detailed understanding of the factors determining whale abundance—other than predation by man—awaits further study.

Rational Utilization and Conservation

The immediate practical importance of the studies discussed here is to assist in the rational utilization of the rich resources of the ocean. While it appears that in the Antarctic whales are only a small part of the total resources, and the same is probably true of other areas, the immediate problem of conservation in the Antarctic and elsewhere is the proper management of the whale stocks.

Whale conservation has two major objectives: to preserve the individual species of whales, and to maintain a high yield of meat and oil from what has been one of the most important renewable living resources of the oceans. Everyone would agree with the vital importance of the first objective. There may be less agreement on its importance in determining whaling regulations. In practice, it may be extremely difficult to exterminate a species of whale.

On land, a determined hunter can search out and kill the last few surviving buffaloes, and will do so if not prevented and if he places sufficient value, as trophies, on the few survivors. Searching for a few whales in the open ocean is a much more difficult business, and would be entirely impracticable as an economic operation. Captain Ahab may have found the white whale again, but in reality the chances of a ship finding an individual whale are extremely small. There are exceptions to this generalization, however. The gray whale is highly vulnerable to exploitation (or pollution) in its breeding lagoons, and indeed it is conceivable that the Atlantic gray whale was exterminated by man in prehistoric times. The humpback whale likewise is unusually vulnerable because of its very patchy inshore distribution.

Another exception is when a rare species of whale continues to be hunted by an industry whose economic base is some other species with a similar distribution. Thus the Scientific Committee of the IWC expressed serious concern at the possible fate of the blue whale in the 1960's if it continued to be caught by an industry which depended for its existence on fin whales (and later also on sei whales). As will be explained, this concern is now less serious.

From calculations of birth and mortality rates it appears that a moderate-sized population of baleen whales can sustain a harvest on the order of 5 to 10% per year (Anon. 1964). If more than this sustainable rate is taken, the stock will decrease; if less, it will increase towards the equilibrium level under natural unexploited conditions. As the stock approaches the equilibrium level, the sustainable rate of harvest expressed as the percentage of the stock that can be harvested each year decreases, but the sustainable rate appears to be reasonably constant at moderate to low levels of stock, or less than about half to a third of the unexploited stock.

Around 1960 about 30% of the fin whale stock was being caught each season. As a result, the stock was decreasing by some 20% or more per year. The situation regarding blue whales was somewhat similar. On the one hand, the fleets were concentrated in the areas mainly inhabited by fin whales rather than by blue whales; on the other hand, if a catcher saw a fin whale and a blue whale at the same time, it would almost certainly chase the bigger animal. The blue whale also had, before the complete closing of blue whale catching,

greater protection by the early closing of the open season for blue
whales. During the later part of the season only fin and sei whales
could be caught. There is evidence that the closed season for blue
whales was not entirely rigorously observed, but the extent and impact
of the breaches should not be exaggerated. Overall the regulations
recommended by the IWC regarding both the quantities taken and the
species and size of whales killed have been observed by the great
majority of expeditions and by individual gunners.

Even discounting the complete legal protection given to blue and
humpback whales, the general level of present-day whaling presents a
greatly reduced threat to their continued existence. In the Antarctic
the number of expeditions has been greatly reduced since 1960. Instead
of some 30% of the Antarctic fin whale stock being taken each year,
now only about 5% is caught, according to recent reports of the IWC
Scientific Committee. There is considerable argument as to whether
the recent catches of sei and fin whales are slightly below or highly
above the combined sustainable yield of these stocks, but the difference is in any case not great, so at worst the present catches are
only reducing those stocks slowly. It may be noted here that since
1965 the expression of the Antarctic quota in blue whale units rather
than in terms of quotas for each species has allowed the whaling expeditions to concentrate (until very recently) on the underexploited
sei whales, and thus take considerably less than the sustainable
yield of the heavily exploited fin whales. This is not to say that
the management measures should not, under optimum conditions, be
applied to each stock (not species) separately; in fact, there has been
an undesirable imbalance between the extent of exploitation of the
different stocks of sei whale in the Antarctic.

The present number of expeditions and catchers in the Antarctic
therefore harvest fin and sei whales at a rate, expressed as the
percentage of the stock killed each year, which is not much different
from the rate those stocks can withstand. The large rorquals have
very similar vital parameters (a reproductive rate of about one young
every other year, and natural mortality rates of a few percent), which
suggests that the sustainable harvesting rate for blue whales is

similar to that for fin and sei whales. The present fleets therefore,
if unrestricted and harvesting blue whales at the same rate as they
do fin whales and sei whales, would catch blue whales at about the
rate (some 5 to 10% per year) which the stock can withstand. In fact,
so long as the main commercial interest is in sei whales, the expeditions will tend to be distributed north of the main concentrations of
blue whales. Then they will probably see within catching range less
than the 5 to 10% of the stock, which is the sustainable yield. Of
course from the commercial viewpoint the need is for a considerable
period of complete and properly enforced protection of blue whales,
and also for a shorter or less complete period of protection for fin
whales, so that the stocks can be rebuilt in the shortest term to a
level where they can produce a high sustained catch. It seems that
even though the present prohibition on blue whale catches may not be
fully observed, the present catches probably do not pose a serious
threat to the survival of blue whales in the Antarctic. The situation regarding blue whales and also right and humpback whales would
be still more comforting if the prohibition was seen to be properly
enforced, e.g. through the international inspection scheme.

While the dangers to the survival of various whale species may
not now be very serious, the proper use of the living resources of
the oceans, of which whales are a part, is certainly a vitally important question. In this respect whales cannot be treated in isolation.
In the oceans generally it is becoming clear that the stocks of the
more familiar large animals such as cod or sea bream, or easily caught
shoaling fish such as herring and anchovies, cannot continue to supply the increasing world demand for fish much longer. The potential
catch of these more familiar types of fish has been estimated as 100
million tons (Gulland 1971b). The world marine fish catch was 56
million tons in 1969, and it has been doubling every decade.

Increased catches will require harvesting the smaller or less
familiar animals, some of which, squid, for example, are already
caught in large quantities in several parts of the world. The biggest increases will require the harvesting of the small fish and other
animals on which the more familiar animals feed. The potential is

vast, if small enough animals can be taken; the annual production of herbivorous zooplankton is probably at least 20×10^9 tons (Schaefer 1965)—four hundred times the present world fish catch. Economic factors rule out harvest of most of these in the foreseeable future, but there are some whose size or shoaling behavior makes them potentially attractive. Among these the Antarctic krill is often mentioned. Trials of krill fishing have been fairly promising, and satisfactory methods of processing the krill into a form of shrimp paste have been developed by Russian scientists. Some products derived from this paste have become very acceptable on the Russian market, but the catching side still presents difficulties. Not enough is known about the local distribution to maintain catches at a sufficiently high sustained level to make krill fishing an economic prospect at the present time. However, the gaps which must be crossed before commercial exploitation begins are not large, and it is reasonable to expect krill fishing to be in operation before the end of this century—sooner than the blue whale stocks could be rebuilt to their optimum level.

The question of whether to harvest krill or whales may then become a pressing one. The choice has been compared to the choice between grass and beef-cattle (and unless the population explosion can be controlled it is a real question how long rich people can be allowed to go on eating beef). In terms of their positions in the food chain, the proper analogy is the choice between wolves or lions and cattle.

If the problem facing the world were one of more food at all costs, and the choice were between 1.7 million tons of whales, and 100 to 150 million tons of krill, there would be no doubt that preference would have to be given to harvesting krill. The decision is not likely ever to be quite so sharp. Total weight is never likely to be the objective, and the unusable fraction of the whale (mainly the bones) will be less than that of krill, of which the large proportion of the total weight made up of the exoskeleton makes processing difficult. Also whales can be and are eaten directly by man, while krill may have to be used indirectly by feeding meal to chickens or cultured fish. Kept under conditions that are optimum from the farmers' point of view, and harvested as soon as the main growth is over, domestic

animals are efficient converters. These differences can balance only
part of the 50 to 100 fold difference between the potential harvest
of krill and whales, and more food for man would be obtained by harvesting krill than by catching whales.

It is not at all certain that the catching of whales and the
catching of krill are mutually exclusive. The interaction of predators
(whales) and prey (krill) can be complex, and unless the particular
predator concerned is the main controlling agent on the abundance of
the prey, the removal of that item of predation need not result in any
increase of the prey. For example, there may be an increase in disease, or of some more drastic predator. Though it has often been
stated that the reduction of whales must have resulted in an increase
of krill (and sometimes also of other animals that feed on krill), the
evidence for this is poor. The argument in favor of a potential krill
harvest of at least 50 million tons is not that there has been any
increase of krill, but that by substituting directly for predation by
whales (and the substitution might have to include a proper balance
in the time and place of harvest, and the sizes of animals taken), the
Antarctic ecosystem can be maintained in the position it was in around
1900, except for man taking the part of the whales. In fact, it may
well be possible to take both 100 million tons of krill and 1.7 million tons of whales. Estimates of the possible total krill production
range up to 500 million tons (Gulland 1970). While at some time in
the future it may be necessary to give most attention to the rational
exploitation of krill, this seems no reason for not, at the present
time, attempting proper management of the whale stocks.

REFERENCES

Allen, K. R. 1969. Further estimates of whale population and sus-
        tainable yields in North Pacific areas. Rpt. IWC, 19:120-122.
_____ 1970. A note on baleen whale stocks of the north west
        Atlantic. Rpt. IWC, 20:112-113.
Anonymous. 1964. Final report of the special committee of Three
        Scientists. Rpt. IWC, 14:39-92.
_____ 1969a. Report on the effects on the baleen whale stocks of
        pelagic operations in the Antarctic during the 1967/68 season,
        and on the present status of those stocks. (Prepared by Fishery

Resources and Exploitation Division, Fish Stock Evaluation Board,
FAO.) Rpt. IWC, 19:29-38.
_____ 1969b. Report of the IWC-FAO Working Group on sperm whale
stock assessment. Rpt. IWC, 19:39-83.
_____ 1971. Report of the special meeting on Antarctic fin whale
stock assessment, Honolulu, Hawaii, 13th-25th March, 1970. Rpt.
IWC, 21:34-39.
Bogorov, V. G., Vinogradov, M. E., Voronina, N. M., Kanaeva, I. P.,
and Suetova, I. A. 1968. Raspredelenie biomass zooplanktona v
poverkhnostnom sloe mirovogo okeana. Dokl. Akad. Nauk SSSR.
182, 5:1205-1207.
Brown, S. G. 1962. The movements of fin and blue whales within the
Antarctic zone. Discovery Rpt. 33:1-54.
Crisp, D. T. 1962. The tonnages of whales taken by Antarctic pelagic
operations during twenty seasons and an examination of the blue
whale unit. Norsk Hvalf.-Tid., 51:389-393.
Cushing, D. H. 1969. Upwelling and fish production. FAO Fish. tech.
pap. 84.
Doi, T., and Ohsumi, S. 1969. The present state of sei whale popula-
tion in the Antarctic. Rpt. IWC, 19:118-120.
_____ _____ 1970. Sixth memorandum on results of Japanese stock
assessment of whales in the North Pacific. Rpt. IWC, 20:97-111.
El-Sayed, S. Z. 1968. Primary productivity of the Antarctic and
Subantarctic. Antarctic Map Folio Ser., American Geogr. Soc.,
Folio 10.
_____ 1970. On the productivity of the Southern Ocean. In Antarctic
Ecology, ed. M. W. Holdgate. London, Academic Press. Vol. 1,
pp. 119-135.
Gulland, J. A. 1970. The development of the resources of the Antarc-
tic Seas. In Antarctic Ecology, ed. M. W. Holdgate. London,
Academic Press. Vol. 1, pp. 217-223.
_____ 1971a. The effect of exploitation on the numbers of marine
animals. Proc. Advanced Study Inst. on "Dynamics of Numbers in
Populations," Oosterbeek, 7-18 September 1970. Eds. P. J. den
Boer and G. R. Gradwell. Wageningen, Centre for Agricultural
Publishing and Documentation. Pp. 450-468.
_____ 1971b. The fish resources of the ocean. London, Fishing News
(Books) Ltd.
Laws, R. M. 1962. Some effects of whaling on the southern stocks of
baleen whales. In The exploitation of natural animal populations,
eds. E. D. LeCren and M. W. Holdgate. Blackwell Sci. Publ., Ltd.,
and Wiley, N.Y. Pp. 137-158.
Mackintosh, N. A. 1965. The stocks of whales. London, Fishing News
(Books) Ltd.
_____ 1970. Whales and krill in the twentieth century. In Antarctic Ecology, ed. M. W. Holdgate. London, Academic Press. Vol.
1, pp. 195-212.
Marr, J. W. S. 1962. The natural history and geography of the Antarctic krill (Euphausia superba Dana). Discovery Rpt. 32:33-464.
Moiseev, P. A. 1969. Living resources of the world ocean (in Russian). Moskva, Pishchevaia promyshlennost'.

_____ 1970. Some aspects of the commercial use of the krill re-
   sources of the Antarctic Seas. *In* Antarctic Ecology, ed. M. W.
   Holdgate. London, Academic Press. Vol. 1, pp. 213-216.
Nemoto, N. 1970. Feeding patterns of baleen whales in the ocean.
   In Marine Food Chains, ed. J. Steele. Edinburgh, Oliver and
   Boyd. Pp. 241-252.
Pike, G. C. 1966. Report of fifth meeting of working group on North
   Pacific whale stocks. Rpt. IWC, 16:59-62.
Schaefer, M. B. 1965. The potential harvest of the sea. Trans. Am.
   Fish. Soc., 94:123-128.
Steemann Nielsen, E. 1954. On organic production in the oceans. J.
   Cons. Perm. Explor. Mer 19:309-328.
Townsend, C. H. 1935. The distribution of certain whales as shown
   by the log book records of American whale ships. Zoologica, N.Y.
   19:1-50.

CHAPTER 2

STATUS OF THE WHALE POPULATIONS OFF THE WEST COAST

OF SOUTH AFRICA, AND CURRENT RESEARCH

Peter B. Best

Since 1936 whaling operations off the west coast of South Africa
have been conducted from a single land station at Donkergat in Saldanha Bay (33°S, 18°E). The station has remained the only land-based
operation in the southeast Atlantic, apart from a brief spell of humpback whaling off the Congo from 1949 to 1952. From 1962 to 1967, when
the Saldanha Bay station closed down, the author carried out biological observations on the whales landed there.

Since operations from this station extended only to 150 miles
from the coast, conclusions drawn from such a restricted area might
not be considered applicable to the ocean region as a whole. Nevertheless, the wide-ranging habits of whales, and the faithful reflection of fluctuations in their abundance seen elsewhere in South African catches, reduce the strength of these reservations. The analysis
of records from land-based operations also has some advantages. Sampling of the population occurs at the same time and in the same place
each year, thus avoiding corrections that have to be made for the
disposition of pelagic fleets. Effort is relatively easily calculated, and the low level of catches means there is hardly any species
selection. In the case of South African land stations there is the
additional advantage that spotter aircraft have been in use for
several years, providing an independent estimate of whale abundance.

An analysis of the status of whale stocks in the temperate southeast Atlantic must obviously take the catches of the Antarctic fleets
into consideration. For this reason, and because better figures for
stock assessment are usually available for the Antarctic, frequent
references have been made to stock sizes and sustainable yields

calculated for the relevant areas of the Antarctic by other workers.
Where possible, local data have also been included, but since the
Saldanha Bay station closed down in 1967, no more recent evidence is
available.

Minke whales have become an integral part of the catch at Durban
since 1968, and a program of research has been directed towards this
species. The discovery of a sizable population of right whales along
the southern coast of South Africa has also led to an annual census
being undertaken from the air. Although both these populations are
outside the geographical limits set by the title of this paper, they
have been included for general interest.

I am indebted to Mr. S. G. Brown of the Whale Research Unit
(I.O.S.), London, for permission to quote from an unpublished paper.

Status of Populations

Fin Whales, Balaenoptera physalus *(Linné 1758)*

Stock identity. Two returned whale marks illustrate a northward
migration of fin whales from Area III in the Antarctic to the west
coast of South Africa (Brown 1962a). The relationship of fin whales
on the east coast of Africa to those in Area III is shown by seven
mark recoveries at Durban (Brown 1970MS). The validity of the areas
within the Antarctic, originally based on the distribution of the whaling fleet, seems to be confirmed by the small amount of movement of
whales between areas (Brown 1962b). The analysis of fin whale blood
types by Fujino (1964) indicates four different breeding populations
in Areas II, III, and IV, an "Atlantic" stock in Area II, a "West
Indian Ocean" stock in higher latitudes and a "Lower Latitudinal"
stock in lower latitudes of Area III, and an "East Indian Ocean" stock
in Area IV. Fin whales from South African waters would seem more
likely (on geographical grounds) to be associated with the "West
Indian Ocean" rather than "Lower Latitudinal" stock, and the fact that
the "Atlantic" stock is genetically distinct from both makes it unlikely that animals from Area II regularly visit South African waters.

Trends in abundance. Catch returns for land stations in the
Cape Province are listed in Table 2-1 for the years 1917 to 1967,

extracted from the International Whaling Statistics, except that data
for the 1923 season are taken from Harmer (1928), as the returns for
all species given in this season by the International Whaling Statistics seem erroneous. The length of the whaling season each year is
not known, but as a rough measure of abundance the catch per boat per
season has been calculated.

TABLE 2-1. Fin whale catch returns for land stations in Cape Province,
1917-1967.

Year	Catch	Catch/boat	Year	Catch	Catch/boat
1917	342	42.8	1948	134	22.3
1918	200	22.2	1949	188	31.3
1919	219	18.3	1950	242	34.6
1920	228	22.8	1951	218	31.1
1921	139	19.9	1952	303	25.3
1922	288	32.0	1953	126	21.0
1923	460	41.8	------	------	------
1924	572	40.9	1957	143	35.8
1925	698	46.5	1958	269	53.8
1926	798	42.0	1959	357	71.4
1927	761	38.1	1960	255	42.5
1928	436	21.8	1961	201	33.5
1929	411	20.6	1962	52	10.4
1930	554	23.1	1963	56	9.3
------	------	------	1964	29	5.8
1936	566	40.4	1965	19	3.8
1937	398	30.6	1966	32	6.4
------	------	------	1967	27	5.4
1947	44	8.8			

| | (May-October) | | | (May-October) | |
Year	Catch per 10^5 c.t.d.	Whales seen per 100 hrs	Year	Catch per 10^5 c.t.d.	Whales seen per 100 hrs
1958	79.1	-	1963	20.2	3
1959	94.5	-	1964	10.0	1
1960	71.0	-	1965	6.8	4
1961	70.3	24	1966	11.0	5
1962	18.5	11	1967	12.3	2

The catch per boat per season shows little sign of a significant decrease until 1962. A more detailed analysis of fin whale abundance for the seasons 1958 to 1967 indicates that the catch per 10^5 catcher-ton-days (c.t.d.) averaged 78.7 for 1958-1961 but fell to an average of 10.0 for 1964-1967, a decrease in availability of about 88%. The number of fin whales sighted from the spotter aircraft averaged 3 per 100 hours for 1964-1967, a drop of 87.5% from 1961.

The catch per unit effort (c.p.u.e.) for Antarctic fin whales prior to 1954 is considered to be an unrealistic measure of their abundance because of the greater emphasis placed on taking blue whales, and after the 1965/65 season the shift in emphasis to sei whales also introduced a bias. Chapman (1971, p. 72) has calculated c.p.u.e. figures (tonnage-corrected) from 1961/62 to 1969/70 for zones A and B of Area III; these zones have never yielded extensive catches of sei, and effort can be regarded as concentrated on fin whales. Separation of Area III effort and catches by zone prior to 1961/62 is not available to the author, but as the majority (95% in 1958/59) of catching occurred in zones A and B in these earlier seasons, the total c.p.u.e. for Area III can probably be considered a good measure of fin whale abundance. These figures for the seasons 1957/58 to 1960/61 (tonnage-corrected) have been added to Chapman's for subsequent seasons in Table 2-2.

TABLE 2-2. Catch of fin whales per unit effort (per thousand catcher-ton-days) in Area III, 1957-1970.

Season	Catch per 10^3 catcher-ton-days	Season	Catch per 10^3 catcher-ton-days
1957/58	2.91	1964/65	1.07
1958/59	2.88	1965/66	0.49
1959/60	2.02	1966/67	0.47
1960/61	1.94	1967/68	1.00
1961/62	0.84	1968/69	0.52
1962/63	1.55	1969/70	0.50
1963/64	1.31		

The average c.p.u.e. for the seasons 1957/58 to 1960/61 is 2.44 whales per 10^3 catcher-ton-days, and the average for the seasons

1964/65 to 1966/67 is 0.68, a decrease of 72%. This is somewhat less than the drop in c.p.u.e. recorded for the Cape Province. It is possible, however, that there was a shift in emphasis towards catching sei whales off the west coast of South Africa as the fin stocks declined, which accentuated the apparent drop in fin whale availability; some evidence to this effect is given by Best and Gambell (1968).

Stock size. The original stock size for the Antarctic prior to 1935 is estimated to have been between 300,000 and 400,000 whales (Doi, Ohsumi, Nasu, and Shimadzu 1970). The proportion of fin whales occurring in Area III in 1957/58 was about 40% (Chapman 1971), so the original Area III population may have consisted of 120,000 to 160,000 fin whales.

The present (1969/70) population size for Area III is given by Chapman (1971) as about 29,000. It has apparently been steady at this level since 1965/66. The fin-whale population in this area in 1957/58 is given as 66,500. The reduction since then has therefore been of the order of 56%.

Sustainable yield. The maximum sustainable yield (m.s.y.) for the Antarctic fin-whale stocks was probably 9,000 to 12,000 whales (Anon. 1971). If 40% of animals occurred in Area III, the m.s.y. for this area would be 3,600 to 4,800 whales.

Considerable controversy exists over the present sustainable yield of Antarctic fin whales, according to the net rate of recruitment assumed. If this is taken to be 0.04, the most reasonable estimate, then the present sustainable yield for Area III would be about 1,000 whales (Chapman 1971). According to Allen (1971), however, the net recruitment rate may be as low as 0.01, whereas Doi, Ohsumi, and Shimadzu (1971) estimate it to be as high as 0.057.

Level of present catches. From Table 2-3 it may be seen that the

TABLE 2-3. Catch of fin whales, 1965-1970.

Season	Antarctic Area III	Durban (1966, etc.)	Donkergat (1966, etc.)	Total
1965/66	1,008	199	32	1,239
1966/67	1,554	124	27	1,705
1967/68	780	62	–	842
1968/69	552	150	–	702
1969/70	1,546	47	–	1,593

average catch in Area III during the last three seasons (including that
at Durban) has been 1,046 whales, at or slightly above the present sustainable yield (if $r - m = 0.04$).

Blue whales, Balaenoptera musculus *(Linné 1758)*

   Stock identity. No direct evidence is available, though from
analogy with the fin whale it can be assumed that there is a close
relationship between Area III stocks in the Antarctic and South African
stocks. While there are no authentic records of pygmy blue whales off
the west coast of South Africa, there is a record from Durban (Gambell
1964). Pygmy blue whales *(B. m. brevicauda* Zemsky and Boronin 1964)
are almost completely confined to Area III, being found from $0°$ to $80°E$
and north of $54°S$ (Ichihara and Doi 1964).

   Trends in abundance. Catch returns for land stations in the Cape
Province are listed in Table 2-4, including the catch per boat per
season as a rough index of abundance.

TABLE 2-4. Blue whale catch returns for land stations in Cape Province
1917-1966.

Year	Catch	Catch/boat	Year	Catch	Catch/boat
1917	337	42.1	1947	1	0.2
1918	127	14.1	1948	4	0.7
1919	108	9.0	1949	5	0.8
1920	144	14.4	1950	4	0.6
1921	125	17.9	1951	4	0.6
1922	599	66.6	1952	11	0.9
1923	669	60.8	1953	4	0.7
1924	503	35.9	-----	-----	-----
1925	784	52.3	1957	4	1.0
1926	1,000	52.6	1958	4	0.8
1927	1,020	51.0	1959	0	0
1928	554	27.7	1960	1	0.2
1929	316	15.8	1961	3	0.5
1930	468	19.5	1962	1	0.2
-----	-----	-----	1963	2	0.3
1936	79	5.6	1964	1	0.2
1937	57	4.4	1965	3	0.6
-----	-----	-----	1966	0	0

In the 1920's, when exploitation of blue whale stocks was beginning to escalate, the catch per boat per season averaged about 40
whales, but from the 1930's onwards there was a sharp reduction in
catches. By the 1960's the catch per boat had fallen to around 0.3,
or 1/130 of the original level. These figures ignore the effect of
increasing catcher size and efficiency over the years.

   Stock size. The Antarctic blue whale population in 1962/63 was ss
estimated to be less than 1,000—possibly 600—(Anon. 1967 p. 28).
This did not include pygmy blue whales. More recent estimates based
on sightings data have been made by Doi, Ohsumi, and Shimadzu (1971),
in which the blue whale population for the Antarctic varied from
3,180 to 10,940 in the years 1965/66 to 1969/70, with a mean of
6,400. These figures include an unknown number of pygmy blue whales.

   Calculations of the average stock size in Area III for 1947 to
1951, and a knowledge of the total population size in 1953/54 (Anon.
1967), enable one to estimate that blue whales in Area III probably
comprised about 22% of the Antarctic stocks. Hence the Area III stock
size for 1962/63 can be calculated as less than 220—possibly 130.

   According to the Special Committee of Three Scientists (Anon.
1964), the Antarctic blue whale population level at maximum sustainable
yield is about 100,000 to 125,000 whales. This is 100 to 200 times
greater than the present population.

   The original stock size of pygmy blue whales in the Antarctic
was probably 10,000 (Ichihara and Doi 1964).

   Sustainable yield. The sustainable yield for the Antarctic in
1962/63 was estimated to be less than 200, while the maximum sustainable yield was estimated to be 6,000 whales (Anon. 1964). Separate estimates for each area are not available, but a rough approximation of the maximum and "present" (1962/63) sustainable yields for
Area III can be made using the proportion of the population estimated
to be in Area III (see above). This provides estimates for the m.s.y.
of 1,320 and for the present sustainable yield of less than 42 whales.

   The maximum sustainable yield of pygmy blue whales is about 200
(Ichihara and Doi 1964).

   Level of present catches. Blue whales have been completely protected in the southern hemisphere since 1966.

Humpback Whales, Megaptera novaeangliae *(Borowski 1781)*

Stock identity. Mackintosh (1942) has summarized the evidence
showing that the Antarctic population of humpbacks is divided into five
more or less self-contained stocks. One of these is located between
10° and 40°E (in Area III), and although there is no direct evidence
of a link between this population and that off the west coast of South
Africa, two mark returns (from Madagascar) illustrate its link with
the east coast. Mackintosh has suggested that another group of humpbacks in the Antarctic between about 10° and 35°W (in Area II) might
be associated with the west coast of South Africa, though again there
is no direct evidence of this. The fact that catches of humpbacks off
Angola declined simultaneously with catches at South Georgia from
1913 onwards, while humpbacks continued to be caught in large numbers
off Natal and Madagascar in the 1930's, suggests that there may indeed
be a link between the Area II stock and that off the Cape Province.

Mackintosh (1965) has suggested that the Area II stock might be
further subdivided into two populations, IIa and IIb, wintering respectively in the southwest and southeast Atlantic.

Trends in abundance. The annual catch of humpback whales off the
Cape Province and the catch per boat per season are shown in Table 2-5
for the period 1917 to 1963. During this period the species never
featured very prominently in the catch, and the catch per boat per
season only varied from 0.4 to 4.3 throughout. As mentioned above,
these small catches were probably a reflection of the decline in stocks
off South Georgia from 1913 onwards.

Catch-per-unit-effort figures for Area II or Area III are not
available for this species. However, trends in the catch per boat per
season for South Georgia are interesting. From 1909/10 to 1911/12,
the catch per boat averaged 242 whales, but fell in 1912/13 to 91
whales, from 1913/14 to 1916/17 averaged 31 whales, and from 1917/18
onwards never exceeded 14 whales per boat per season. In the 1930's
the catching of humpbacks at South Georgia was only permitted under
special conditions, and from 1955 the catching of humpbacks in Area
II was completely prohibited.

TABLE 2-5. Humpback whale catch returns off Cape Province, 1917-1963.

Year	Catch	Catch/boat	Year	Catch	Catch/boat
1917	7	0.9	----	----	----
1918	19	2.1	1947	5	1.0
1919	14	1.2	1948	14	2.3
1920	20	2.0	1949	15	2.5
1921	30	4.3	1950	7	1.0
1922	13	1.4	1951	9	1.3
1923	13	1.2	1952	15	1.3
1924	19	1.4	1953	9	1.5
1925	9	0.6	----	----	----
1926	19	1.0	1957	3	0.8
1927	12	0.6	1958	2	0.4
1928	21	1.1	1959	7	1.4
1929	40	2.0	1960	4	0.7
1930	30	1.3	1961	4	0.7
----	----	----	1962	9	1.8
1936	27	1.9	1963	3	0.5
1937	28	2.2			

Stock size. No separate estimates of the stock size of either
Area II or Area III humpbacks are available. Using sightings data
provided by Mackintosh and Brown (1956) for the period 1933 to 1939,
however, it can be estimated that the Antarctic population totaled
22,000 to 34,000 humpbacks in the 1930's, of which apparently 10% were
in Area II and 20% in Area III (Anon. 1964, p. 69). Estimates of
2,200 to 3,400 whales for Area II and 4,400 to 6,800 whales for Area
III can thus be obtained for this period.

According to an analysis of sightings, the humpback population
in the Antarctic ranged from 790 to 3,970 in the years 1965/66 to
1969/70, with a mean of 1,700 (Doi, Ohsumi, and Shimadzu 1971).
Applying the same proportions to these population figures means that
the Area II population may now consist of 170 and the Area III population of 340 humpbacks.

Sustainable yield. There are no estimates available for the sustainable yield of Area II or III humpback stocks, but it must be very

close to zero at present if the population figures calculated above
are of the right order.

Level of present catches. The species has been protected in the
Southern Hemisphere since 1963.

Sei Whales, Balaenoptera borealis *Lesson 1828*

Stock identity. One whale mark has been recovered off the west
coast of South Africa that was fired into a sei whale close to Tristan
da Cunha; the details are given below.

Mark no.	Date fired	Date recovered	Position marked	Position recovered	Length, sex
USSR 650151	17.XI.1965	10.IX.1966	$36^{\circ}19'S$, $12^{\circ}57'W$	$34^{\circ}38'S$, $17^{\circ}20'E$	43' male

This suggests a possible relationship between sei whales in Area IIE
and the west coast of South Africa. On geographical grounds, and
from analogy with fin whales, however, it is to be expected that the
west coast population will also be linked with that in Area III. Two
marks fired off Durban have been recovered in the Antarctic, one in
the eastern half of Area III and the other (from a refrigerator vessel) on the western edge of Area IV (Brown 1971).

These recoveries all indicate that sei whales may disperse considerably in a longitudinal direction during their annual migrations,
thereby complicating the identification of stocks. According to an
analysis of the density distribution of sei whales in the Antarctic
by Doi, Ohsumi, and Nemoto (1967), there is at least one stock between
$70^{\circ}W$ and $60^{\circ}E$ (i.e. in Areas II and III).

Trends in abundance. The annual catch of sei whales off the
Cape Province from 1917 to 1967 is listed in Table 2-6, including the
catch per boat per season. More detailed figures of the catch per
unit effort and the number of whales seen per 100 flying hours are
included for recent seasons.

Catches of sei whales escalated in years after World War II, but
seriously declined in 1966 and 1967. Recent trends in the availability of sei whales off South Africa have been analyzed by Best and
Gambell (1968). The decline in catch and aircraft sightings on the
west coast from 1963-65 to 1967 was estimated to be 72% and 89%

TABLE 2-6.   Sei whale catch returns off Cape Province, 1917-1967.

Year	Catch	Catch/boat	Year	Catch	Catch/boat
1917	35	4.4	1948	83	13.8
1918	95	10.6	1949	119	19.8
1919	190	15.8	1950	324	46.3
1920	127	12.7	1951	237	33.9
1921	34	4.9	1952	711	59.2
1922	79	8.8	1953	295	49.2
1923	128	11.6	----	----	----
1924	364	26.0	1957	263	65.8
1925	33	2.2	1958	405	81.0
1926	258	13.6	1959	525	105.0
1927	65	3.3	1960	498	83.0
1928	355	17.8	1961	228	38.0
1929	193	9.7	1962	388	77.6
1930	159	6.6	1963	721	120.2
----	----	----	1964	673	134.6
1936	214	15.3	1965	764	152.8
1937	49	3.8	1966	417	83.4
----	----	----	1967	152	30.4
1947	39	7.8			

Year	(May-October) Catch/ 10^5 c.t.d.	Whales seen/ 100 hrs	Year	(May-October) Catch/ 10^5 c.t.d.	Whales seen/ 100 hrs
1958	120.8	-	1963	260.2	82
1959	139.0	-	1964	224.3	49
1960	140.0	-	1965	273.5	82
1961	79.0	32	1966	143.6	53
1962	136.1	108	1967	69.4	8

respectively, with an average of 82%. Similar values were obtained
at Durban, where the decline in abundance has continued since 1967.

Changes in sei whale catch per unit effort in the Antarctic must
be viewed with caution because of changes in the preference for sei
and fin whales from year to year, and because the two species are differently distributed. The catch per unit effort of sei whales in

Series D in December and January has been chosen as being an area and
time of high density for which data are available for several seasons
(Chapman 1971). The apparent availability of sei whales in Series D
of Area II has decreased by 38% from 1965/66 to 1969/70; their availability in Area III has fluctuated, though the decrease from 1966/67 to
1969/70 is about 62% (see Table 2-7).

TABLE 2-7.   Catch of sei whales per unit effort (per thousand catcher-
ton-days).

| Season | Catch/10^3 c.t.d. | |
	Area II	Area III
1965/66	3.27	0.84
1966/67	2.20	2.86
1967/68	-	1.34
1968/69	-	1.03
1969/70	2.02	1.08

Neither of these trends is so large or so consistent as the changes
in c.p.u.e. seen off South Africa. It should be remembered that South
African whaling operations cover the same geographical region at the
same time each year, and there is hardly any diversification of effort
between species; furthermore, independent estimates of abundance are
available from spotter plane sightings. Chapman (1971) has indicated
that with the reduction of Antarctic catches that has taken place,
changes in the c.p.u.e. do not become so marked and are rather sensitive to random fluctuations in the distribution of whales, weather
conditions, etc. There may be some doubt, therefore, over the validity
of these figures as indices of sei whale abundance in the Antarctic.

Stock size. Estimates of the original stock size for Area II
vary from 45,400 to 55,000, and for Area III from 24,800 to 32,500
(Anon. 1970; Doi and Ohsumi 1969).

The stock in Area II was reduced by 31,000 whales from 1959/60
to 1969/70 (Chapman 1971). This is a reduction of 62% if a figure of
50,200 whales is taken as the mean of the proposed original stock
sizes. In Area III the reduction can be calculated as 58%. These
figures are considerably smaller than the observed changes in c.p.u.e.
and sightings off South Africa.

Sustainable yield. The maximum sustainable yield of Area II sei whale stocks is given as 1,700 (Anon. 1970) or 1,680 (Doi and Ohsumi 1969), and of Area III as 930 (Anon. 1970) or 1,000 (Doi and Ohsumi 1969).

Estimates of the sustainable yield in 1968/69 vary from 1,670 to 1,900 for Area II and from 850 to 930 for Area III (see Chapman 1971). These are close to the maximum sustainable yields. Chapman expresses reservations about the sustainable yield of Area II being as high as 1,900, and points out that estimates of the rate of sei-whale recruitment were derived partly by analogy from those of fin whales. As the latter are now considered to be somewhat lower than originally estimated, the sustainable yields calculated for sei whales may be too high.

Level of present catches. Table 2-8 shows the annual catches of sei whales in Areas II and III, combined with those of the land stations at Durban and Donkergat, for the seasons 1965/66 to 1969/70.

TABLE 2-8. Level of present catches of sei whales.

Season	Antarctic Area II	Donkergat (1966 etc.)	Total	Antarctic Area III	Durban (1966 etc.)	Total
1965/66	12,718	416	13,134	2,756	273	3,029
1966/67	1,540	152	1,692	6,865	66	6,931
1967/68	195	-	195	2,352	24	2,376
1968/69	188	-	188	1,771	40	1,811
1969/70	1,278	-	1,278	1,997	8	2,005

Catches in Area II over the last four seasons have been generally below the present sustainable yield as calculated above, while catches in Area III have consistently exceeded the sustainable yield many times over.

Bryde's whales, Balaenoptera edeni *Anderson 1879*

Stock identity. Off the west coast of South Africa there are two population groups of Bryde's whales (Best 1970d). One group is found inshore over the continental shelf (within 20 miles of the coast), where it is apparently resident throughout the year; these whales have no fixed breeding season and their baleen plates are similar to those

of sei whales in shape. The second group is usually found 50 miles
or more from the coast, is strongly seasonal in its appearance, has
a restricted breeding season in autumn, is two to three feet longer
than inshore whales on average and has relatively broader baleen plates.
These two groups must be considered separately for stock assessment
purposes.

To date no Bryde's whales have been recorded from the Antarctic
(south of 40°S), and catches at Durban have been limited to one or two
specimens per year (apparently a true reflection of their abundance).
The whaling station in Saldanha Bay was therefore the only factory
exploiting this resource.

Trends in abundance. The annual catch of Bryde's whales at Cape
Province whaling stations from 1917 to 1967 is listed in Table 2-9.

TABLE 2-9. Bryde's whale catch returns for land stations in Cape
Province, 1917-1967.

Year	Catch	Year	Catch	Year	Catch	Offshore group (Oct. c.p.u.e.)
1917	0	1929	29	1953	7	–
1918	0	1930	5	------------		–
1919	0	------------		1957	34	–
1920	0	1936	7	1958	26	–
1921	0	1937	36	1959	40	–
1922	0	------------		1960	9	–
1923	11	1947	55	1961	8	–
1924	52	1948	238	1962	64	–
1925	0	1949	139	1963	95	72 (7.8)[a]
1926	64	1950	100	1964	107	17 (0)
1927	28	1951	23	1965	180	97 (144.2)
1928	47	1952	0	1966	100	2 (1.8)
				1967	60	27 (15.4)

[a] Catchers instructed not to shoot Bryde's.

These statistics, especially for the early years, must be viewed with
caution, as the species may not have been distinguished from the sei

whale in some of the returns. The catch per boat has been omitted because valid effort figures for Bryde's whales are difficult to obtain; the inshore population was usually fished only when the weather outside was too bad. It is also impossible to separate inshore from offshore whales in catch returns prior to 1963.

The number of offshore-group whales caught each year since 1963 is given, plus, in parentheses, the catch per 10^5 catcher-ton-days for October, the month in which this group of Bryde's whales was usually most abundant. In most seasons more inshore than offshore whales were caught, except in 1963, when catching sei and Bryde's whales was permitted in March and April under special license. The great fluctuations in c.p.u.e. for offshore whales in October are partly a reflection of fluctuating interest in this species, but also possibly indicate that operations were only sampling the edge of a much larger population. Large concentrations of "sei whales" about 200 miles west of Cape Town in summer have been reported to the author by crew members of the catcher/factory ship "Run," and photographs of these animals show them to be Bryde's whales of the offshore group. The proximity of such concentrations to the Antarctic whaling grounds raises the interesting question of whether this population group has featured in recent Antarctic catches.

Stock size. A rough approximation of the population size of the offshore group can be made using the relative proportions of sei and Bryde's whales in October 1965. In this month 82 offshore Bryde's were shot, indicating that considerable attention was paid to them. October is also a month during which sei whale abundance is close to its maximum (Best 1967), and 165 sei whales were shot in October 1965. The c.p.u.e. in the Area II sei whale population in 1965/66 was 1.62 times as great as in 1969/70. If we accept a population size of 19,200 for this area in 1969/70 and assume the c.p.u.e. to be proportional to stock size, the sei whale population in Area II can be calculated as 31,100 for 1965/66. The offshore Bryde's whale stock size in 1965 can then be estimated as approximately 15,500.

Because of the low levels of catches on this stock, this estimate can be accepted as being close to the maximum stock size.

Sustainable yield. Adopting the crude estimate of population
size given above, and assuming that, as for the sei whale, the maximum
sustainable yield is reached when the stock is about half its original
size, the value for net recruitment for sei whales (0.075, Anon. 1968)
gives a maximum sustainable yield of about 580 whales for the offshore
population of Bryde's whales.

Level of present catches. Prior to 1967, when the last whaling
station in Cape Province closed down, the annual catch of Bryde's
whales never reached 250. Consequently the stocks of offshore Bryde's
whales should still be near their maximum size, though the operation
of catcher/factory ships in the South Atlantic in recent years may
have affected the condition of the stock.

Sperm Whales, Physeter catodon *Linné 1758*

Stock identity. There is very little information on stock identity in southern hemisphere sperm whales. It has been assumed, rather
as a matter of convenience for stock assessment purposes, that the
populations of sperm whales off both coasts of South Africa are
related to that in Area III of the Antarctic (Anon. 1963MS). A
Soviet whale mark fired into a sperm whale in the western half of Area
III and recovered at Durban tends to confirm this. The continuity of
stocks latitudinally is also shown by the recovery of Soviet marks at
Saldanha Bay that were fired in $5^{\circ}58'S$ and $21^{\circ}33'N$ respectively (Best
1969). On empirical grounds a southeast Atlantic stock has been proposed that is related to Area IIE and IIIW in the Antarctic (Best 1970a).

Trends in abundance. The annual catch of sperm whales from the
Cape Province from 1917 to 1967 is shown in Table 2-10, together with
the catch per boat per season.

The catch of sperm whales in this area increased rapidly after
the war, and particularly from 1957 onwards. Detailed c.p.u.e. and
aircraft sightings data for the period 1958 to 1967 show that although
sperm whale availability fluctuated strongly from year to year, there
was no indication of a decrease in overall abundance. A breakdown of
the catch by size and sex, however, has shown that medium-sized (12.2
to 13.7 m long) and large (longer than 13.7 m) males have declined in
abundance since 1960 (Best 1970a).

TABLE 2-10.   Sperm whale catch returns from Cape Province, 1917-1967.

Year	Catch	Catch/boat	Year	Catch	Catch/boat
1917	25	3.1	1948	105	17.5
1918	111	12.3	1949	209	34.8
1919	108	9.0	1950	161	23.0
1920	85	8.5	1951	211	30.1
1921	28	4.0	1952	331	27.6
1922	28	3.1	1953	252	42.0
1923	30	2.7	----	----	----
1924	35	2.5	1957	469	117.3
1925	60	4.0	1958	694	138.8
1926	95	5.0	1959	651	130.2
1927	155	7.8	1960	769	128.2
1928	225	11.3	1961	765	127.5
1929	221	11.1	1962	941	188.2
1930	125	5.2	1963	691	115.2
----	----	----	1964	728	145.6
1936	108	7.7	1965	792	158.4
1937	207	15.9	1966	684	136.8
----	----	----	1967	630	126.0
1947	48	9.6			

Year	(May-October) Catch/ 10^5 c.t.d.	Whales seen/ 100 hrs	Year	(May-October) Catch/ 10^5 c.t.d.	Whales seen/ 100 hrs
1958	188.4	-	1963	230.8	205.7
1959	151.0	-	1964	174.7	198.5
1960	203.0	-	1965	184.0	103.1
1961	263.7	213.6	1966	122.6	(incomplete data)
1962	271.3	158.0	1967	245.4	(incomplete data)

Catch-per-unit-effort figures for Areas II and III (tonnage-corrected) have been calculated from data supplied by the Bureau of International Whaling Statistics for the period outside the baleen whale season (November, December, March, April, and May). These figures are not an ideal measure of sperm whale abundance in the Antarctic, for it is difficult to judge how much attention has been paid to the species each season. Nevertheless, there seems to have been an overall decrease in the c.p.u.e. from 1961/62 to 1969/70 in both Areas II and III—especially in the last two seasons, when the c.p.u.e. has fallen to about half the level in 1961/62 to 1963/64 (see Table 2-11).

TABLE 2-11. Catch of sperm whales per unit effort (per thousand catcher-ton-days).

Season	Catch/ 10^3 c.t.d.	
	Area II	Area III
1961/62	1.301	1.123
1962/63	1.343	1.014
1963/64	1.351	1.291
1964/65	1.011	1.021
1965/66	0.960	1.017
1966/67	1.618	1.357
1967/68	1.075	1.040
1968/69	-	0.564
1969/70	0.520	0.845

As the majority of sperm whales in the Antarctic are medium-sized or large, this tendency for the c.p.u.e. to decrease agrees with the trend seen off the Cape Province. A possible shift in sperm whaling effort away from Areas II and III has already been noted and considered to be a sign that the stocks might be declining (Best 1970a).

Stock size. The stock of mature females (age 9 years or more) in the southeast Atlantic has been estimated as between 15,500 and 32,000 (Best 1970a). As the level of exploitation of the females is considered to have been low, this population size is probably close to the maximum. It is more difficult to estimate the size of the male population because of changes in their mortality rate with age.

The stock of mature females (age 10 years or more) belonging to the same stock as the males in Area III plus South Africa has been calculated as about 57,000 (Chapman and Boerema 1971). This is not directly comparable to the estimate given above because of the different stock limits adopted. However, if the stock of mature females on the east coast of South Africa is taken to be 18,000 to 30,000

(Gambell 1972), then the sum of both east and west coast stocks (33,500
to 62,000) is very close to Chapman and Boerema's figure.

Sustainable yield. Models developed at the 1968 Rome meeting
(Anon. 1969) can be used to calculate sustainable yields from the
population sizes given in the previous section.

The details behind these calculations are given more fully in my
second presentation (see Chapter 11), but the maximum sustainable yield
of an initial, unexploited population of 10,000 mature females is about
180 females. Assuming the southeast Atlantic stock to be close to the
unexploited level, the m.s.y. can be calculated as 280 to 580 females.

The maximum sustainable yield of males can be similarly calculated if it is assumed that the female population has been reduced to
the level giving the m.s.y.—that is, to about 60% of its original size.
Assuming a mean age at recruitment for males of 12 years, the m.s.y.
can be calculated as 220 to 240 males per 10,000 females in the original
stock, depending on the average size of a harem (10 to 16 females). The
m.s.y. for southeast Atlantic stock would therefore be 340 to 770 males.

The actual sustainable yield of mature males in the southeast
Atlantic in 1964/65 has been calculated as 290 to 440 (Best 1970a).
This took into account an apparent change in mortality rate with age,
attributed to heavier catches of older animals in the Antarctic.

It should be added that the sustainable yields calculated for a
heavily exploited population assume certain responses from the stock
that may or may not occur to the extent visualized.

Level of present catches. The annual catch of sperm whales in
Areas IIE and IIIW, both north and south of 40°S, has been extracted
from statistics supplied by the Bureau of International Whaling Statistics for the seasons from 1965/66 to 1969/70. This has been added to
the annual catch at Donkergat to give a figure for the total catch of
southeast Atlantic whales. Separate tables are given in Table 2-12.

The catch of females in the southeast Atlantic has generally
been below the calculated sustainable yield (Best 1970a). Catches
of males have, however, been high in relation to the sustainable
yield, exceeding 1,500 whales in 11 of the postwar years up to 1965/66,
and reaching a maximum of 2,300 in 1961/62. They continue to be above the
present sustainable yield as calculated by Best (1970a) for mature males.

TABLE 2-12.   Level of recent catches of sperm whales.

		Females		
Year	II E	III W	Donkergat	Total
1966	28	0	163	191
1967	15	20	247	282
1968	23	38	-	61
1969	21	64	-	85
1970	-	-	-	0
		Males		
Season	II E	III W	Donkergat	Total
1965/66	705	314	520	1,539
1966/67	514	325	383	1,222
1967/68	337	518	-	855
1968/69	273	823	-	1,096
1969/70	------no data------		-	no data

Minke Whales, Balaenoptera acutorostrata *Lacépède 1804*

Stock identity.   There are no data from marking or immunological
procedures.   Arsenev (1960) distinguished three population groups in
the Antarctic based on catch densities, one between 50°W and 40°E
(an "Atlantic" stock), one between 50° and 140°E (an "Indian" stock),
and a possible third stock around 160° to 170°E.   From an analysis of
sightings data, Ohsumi, Masaki, and Kawamura (1970) have shown that
minke whales tend to be distributed in the higher latitudes of the
Antarctic.   Insufficient data were available for Areas II and III in
higher latitudes, but a concentration of minke was found from about
70° to 130°E, and a smaller concentration possibly around 140° to
160°E (in approximately Areas IV and V).   These findings confirm the
presence of Arsenev's "Indian" stock, and possibly indicate that his
third group is equivalent to an Area V or southwestern Pacific stock.
    It is conceivable that populations of southern minke can be dis-
tinguished by differences in coloration.   The animals from Area IV
examined by Ohsumi, Masaki, and Kawamura all apparently lacked a
white band on the flipper; while of those landed at Durban, 45% had
a gray band on one or both flippers and a further 7.5% a white blaze

on the shoulder that extended on to the flipper. The latter animals also had almost completely white baleen, as in *acutorostrata*. It is possible, therefore, that the Durban catch is more closely related to the "Atlantic" population than the "Indian."

Trends in abundance. The exploitation of minke whales has been too recent an innovation off Durban (the first extensive catches were made in 1968), and the effort directed toward catching too irregular, for any meaningful trends in availability to emerge. See Table 2-13.

TABLE 2-13. Minke whales off Durban, 1968-1970.

Year	Catch	Catch/10^5 c.t.d.	Whales seen per 12,000 miles flown
1968	97	30.2	6.7
1969	112	26.0	10.9
1970	171	42.8	8.7

Antarctic catches also have fluctuated considerably. From 1955/56, when the species first began to occur regularly in the catch, the number of minke whales taken annually has fluctuated from 2 to 605, with a mean of 134. Catch-per-unit-effort figures are not available.

Stock size. Ohsumi, Masaki, and Kawamura (1970) have estimated from sightings that the Antarctic minke population is about 70,000. There are no separate estimates available by area.

Tentative estimates of the size of the Area III stock can be made by comparing the indices of availability of minke and fin whales off Durban from 1968 to 1970, using the stock sizes for Area III fin whales calculated by Chapman (1971) for each year. These are shown in Table 2-14. The mean population size for Area III minke whales for all three seasons is approximately 55,500. This estimate contrasts with the

TABLE 2-14. Tentative size estimates of Area III minke whale stock.

Year	Minke whale/fin whale availability		Size of Area III fin population	Size of Area III minke population
1968	catch:	1.6	29,400	47,040
	sightings:	1.3		38,220
1969	catch:	0.7	29,600	20,720
	sightings:	1.1		32,560
1970	catch:	3.6	29,000	104,400
	sightings:	3.1		89,900

figure of 70,000 just cited for the entire Antarctic, though the latter
was considered to be a possible underestimate.

Sustainable yield. Adopting Ohsumi, Masaki, and Kawamura's (1970)
figures for net recruitment of 0.13 when the population has. been reduced
by half (so giving the maximum sustainable yield), the m.s.y. of the
Area III population can be calculated as about 3,600. Their figure for
the m.s.y. of the entire Antarctic is 4,200.

Level of present catches. These have been given above for Durban
and for the entire Antarctic (there is no breakdown by area available
for the latter). Catches have been well below either estimate of the
maximum sustainable yield.

Right Whales, Eubalaena australis *(Desmoulins 1822)*

Stock identity. There is no direct evidence on the stock limits
of southern right whales. Their scarcity in tropical regions indicates a discontinuity between northern and southern populations, and
there may be morphological differences between northern and southern
whales (for example, the presence or absence of callosities on the
margin of the lower jaw—Best 1970b). The pattern of movement north
and south suggests that whales found in the Antarctic in summer and
early autumn are related to the population off South Africa in spring.
It has been assumed, from analogy with fin whales, that right whales
found along the southern coast of South Africa are referable to an
Area III stock in the Antarctic (Best 1970b).

Trends in abundance. Right whales have been protected since 1935,
so there are no recent figures for availability from catches. Information on the trends in abundance in the eighteenth and nineteenth
centuries shows that the big decline in the population off South Africa
took place as long ago as 1835 (Best 1970b). The prevalence of recent
sightings, when compared with the low level of catches off the Cape in
the years just prior to their protection, suggests that the population
has increased under protection.

Stock size. The size of the population off South Africa (an
"Area III" stock) in 1969 was estimated to be about 180 (Best 1970b).
The 1970 aerial census produced a total of 70 right whales, compared
to 60 in 1969. The area searched was somewhat larger in 1970, and in

a comparable stretch of coastline the increase from 1969 to 1970 was about 13%. This would mean a population of about 200 animals in 1970. Extrapolation of the "Area III" stock size to the rest of the Antarctic gave a population figure of about 925 for 1969, and so for 1970 the population might be 1,030.

Japanese sighting data indicate a population of 1,520 right whales for the Antarctic, this fugure being the mean calculated for the five seasons 1965/66 to 1969/70 (Doi, Ohsumi, and Shimadzu 1971). No estimates are available for the original stock size of southern right whales.

Sustainable yield. The present sustainable yield is probably close to zero.

Level of present catches. The species has been protected since 1935. Its vulnerability to any sort of exploitation should be stressed. A considerable population of right whales around Tristan da Cunha was recently decimated by the operations of a pelagic whaling fleet, and the presence of a fleet of catcher/factory ships in the South Atlantic registered in countries outside the International Whaling Commission poses a continual threat to the recovery of this species.

Current Research

West Coast

Since the only station on the west coast of South Africa closed down in 1967, research has been confined to writing up certain projects remaining from the original program. These include these three projects:

Biology of Bryde's whales. The presence of two allopatric forms of the species off the west coast, and some of the major differences between them, were summarized in a previous report (Best 1970d). With the help of Dr. H. Omura and Mr. R. L. Brownell, Jr., it has now been possible to study photographs and measurements of baleen plates from Bryde's whales off Brazil, from Mexico, and off the west coast of Kyushu, Japan. These show that both inshore and offshore forms occur off Brazil, and that inshore whales are found off Mexico and the west coast of Kyushu. A final report is being prepared.

Social groupings and segregation of sperm whales. An analysis of the composition of 27 sperm whale groupings sampled by the catchers

TABLE 2-15. Status of whale populations off the coasts of South Africa.

Species	Area	Stock size		Sustainable yield		Average catch for 1965/66 to 1969/70
		Maximum	Present	Maximum	Present	
Fin	Area III and South Africa	120,000 to 160,000	29,000	3,600 to 4,800	1,000	1,216
Blue	Area III and South Africa	–	130	1,320	< 42	(Protected since 1966)
Pygmy blue	Area III and South Africa	10,000	–	200	–	(Protected since 1966)
Hump- back	Area II and west coast of South Africa	–	170	–	–	(Protected since 1963)
	Area III and east coast of South Africa	–	340	–	–	(Protected since 1963)
Sei	Area II and west coast of South Africa	45,400 to 55,000	14,400 to 24,000	1,680 to 1,700	1,670 to 1,900	3,297
	Area III and east coast of South Africa	24,800 to 32,500	8,200 to 15,900	930 to 1,000	850 to 930	3,230
Bryde's	Offshore group off west coast of South Africa	15,500	15,500	580	> 580[a]	(None since 1967)
Sperm	Area IIE and IIIW and west coast of South Africa	♂ –	–	340 to 770	290 to 440	1,178
		♀ 15,500 to 32,000	15,500 to 32,000 (mature)	280 to 580	> 280 to 580[a]	124
Minke	Area III and east coast of South Africa	55,500	55,500	3,600	> 3,600[a]	127[b]
Right	Area III and south coast of South Africa	–	200	–	–	(Protected since 1935)

[a] Surplus still to be removed to reach population level giving m.s.y.

[b] Average Durban catch 1968 to 1970.

in 1963 was submitted to the special meeting on sperm whales in Honolulu (Best 1970c). These groupings were chosen because there was good
information on the composition of each before hunting started, and
animals shot from them could be identified on the flensing platform.
Findings show that the basic school is a mixed one of 20 to 30 animals,
80% of which are females of all ages and in all stages of the reproductive cycle. These schools seem to be fairly stable, though males begin
to leave their parent school around the onset of puberty, to form
bachelor schools. Harem-master bulls occur more frequently in mixed
schools in the breeding season than outside it, serving 10 to 15 females
each. Information on the percentage of each school shot shows that
mixed schools are subject to the least fishing (0.51), bachelor schools
of small males slightly more (0.77), whereas nearly all members of
bachelor schools of medium-sized and large males are killed (0.94 to
0.97). The change in the species of cyamid most predominant on males,
which occurs at a length of 39 to 40 ft (11.9 to 12.2 m), is taken to
represent the stage at which males first enter the Antarctic. An
analysis of the size composition of male sperm whales and their distribution is still being prepared as part of this report.

   Feeding of sperm whales. The identification of samples of squid
beaks from stomach contents is being undertaken at present (by Dr. M. R.
Clarke of the Institute of Oceanographic Sciences, U.K.) and must await
the description of several new species of squid. Analysis of the
results will be presented to show any seasonal differences in feeding
behavior, and any differences between size groups and sexes of sperm
whales. The noncephalopod content of the diet has already been analyzed and shows that females feed more frequently on bathypelagic
mysids *(Gnathophausia)*, and less frequently on crabs *(Geryon)* and
tunicates *(Pyrosoma)*, than males—amongst whom there is no difference
in these aspects of their feeding behavior.

East Coast

   Since 1967, research has been switched to the whaling station at
Durban. Long-term programs include the following:

   (1) Monitoring of changes in availability of all species through

catch per unit effort and aircraft sightings analysis.

(2) Collection of a mandibular tooth from all whales landed, plus information on the sexual condition of females.

(3) Regular preseason marking cruises (undertaken in conjunction with the Institute of Oceanographic Sciences, U.K.). To date 159 sperm, 6 sei, 1 fin, and 1 minke have been marked.

Special projects include the following:

Sperm whales. Several short-term investigations are being carried out to clarify specific problems. These include: (*a*) the attainment of sexual maturity in males through studying changes in the density of spermatozoa in the vas deferens, (*b*) the duration of lactation and the number of dentinal growth layers laid down per annum during the first years of life through the capture of calves, (*c*) differences in diving behavior (depth and duration) between sexes and size groups using asdic recordings, (*d*) the incidence of atherosclerosis in hearts of females and all male size groups, and (*e*) the efficiency of penetration of increased—charge marks by trial-firings on carcasses at the whaling station. Preliminary results of some of these projects will be mentioned in Chapter 11.

Minke whales. When this species first started to appear in the catch in any numbers (1968), a program of research into its biology was initiated. The first part consists of an investigation into the taxonomic position of the southern minke whale in relation to *B. acutorostrata*. The skulls of three animals have been examined and compared with the types of *B. huttoni* and *B. bonaerensis*, though more information is still needed from the latter. A comparison has also been made with skulls of *B. acutorostrata* in the British Museum and in the literature. To date only one significant difference has been found between skulls of the two forms. Data on the external appearance of the animals are also being collected, with emphasis on the coloration of the flippers and baleen (see section on stock identity of minke whales, above). No firm conclusions have yet been reached on the identity of the species involved. The second part of the program is an investigation into the biology of the minke whale, especially age determination, reproduction, feeding, and migration. The most striking feature of the species' biology is its high reproductive rate compared with other baleen whales (Ohsumi, Masaki, and Kawamura 1970). Although no pregnant females have yet been taken at Durban, apparently because of segregation of this class away from the area, the incidence of post-partum ovulations in lactating whales taken under special permit in 1970 was very high (six out of seven, or about 86%). None of 12 lactating females examined in 1970 was pregnant, but this may have been because the whale had only just ovulated or because the conceptum was too small to see with the naked eye. A further sample of 12 lactating animals will be taken in August and September 1971 to establish more accurately the incidence and success of postpartum ovulation.

Special Project on Right Whales

Each year since 1968 an aerial survey has been undertaken along the southern coast of South Africa in late September or early October (for details see Best 1970b), and a count made of all right whales seen. In 1970 the limits of the survey were extended up both east and west coasts as far as 32°S, but in fact only two (or 2.9%) of the whales found were outside the normal search limits at Muizenberg, False Bay, and Woody Cape, Algoa Bay. It is hoped to undertake a similar survey earlier in the year to determine whether the whales are distributed differently, and to see how many are already accompanied by calves. Two special flights were made in 1970, when groups of adults and an adult plus calf were each circled for an hour to determine the time spent on or at the surface relative to that spent below the surface, so that a factor to correct for whales missed during the census could be obtained. As expected, the cow and calf spent nearly all the time (96%) in plain view at the surface, whereas five unaccompanied adults spent less time (21 to 91%, average 50%) at the surface. More data are needed before a factor can be accurately determined. Future plans include the possibility of determining the length of animals by aerial photography and their sex by scuba divers, so that the composition of the population can be established, and of developing a means of marking individuals externally. Aerial surveys will continue on an annual basis.

REFERENCES

Allen, K. R. 1971. Notes on the assessment of Antarctic fin whale
    stocks. Rpt. IWC, 21:58-63.
Anonymous. 1963 (MS). Report of the meeting of the scientific sub-
    committee of the International Whaling Commission, Seattle,
    Washington, 18-22 November, 1963. (Cf. 1965, Rpt. IWC, 15:
    29-30.)
_____ 1964. Special Committee of Three Scientists. Final report.
    Rpt. IWC, 14:39-92.
_____ 1967. Report of IWC/FAO joint working party on whale stock
    assessment held from 26th January to 2nd February, 1966, in
    Seattle. Rpt. IWC, 17:27-47.
_____ 1968. Report on the effects on whale stocks of pelagic
    operations in the Antarctic during the 1966/67 season and on the
    present status of those stocks. (Prepared by Fish Stock Evalu-
    ation Branch, FAO.) Rpt. IWC, 18:23-47.
_____ 1969. Report of the IWC/FAO Working Group on sperm whale
    stock assessment. Rpt. IWC, 19:39-69.
_____ 1970. Report on the effects on the baleen whale stocks of
    pelagic operations in the Antarctic during the 1968/69 season,
    and on the present status of those stocks. (Prepared by Fish
    Stock Evaluation Branch, FAO.) Rpt. IWC, 20:21-32.
_____ 1971. Report of the special meeting on Antarctic fin whale
    stock assessment, Honolulu, Hawaii, 13th-25th March, 1970. Rpt.
    IWC, 21:34-39.
Arsenev, V. K. 1960. Distribution of *Balaenoptera acutorostrata*
    Lacep. in the Antarctic. Norsk Hvalf.-Tid., 49:380-382.
Best, P. B. 1967. Distribution and feeding habits of baleen whales
    off the Cape Province. Investl. Rpt. Div. Sea Fish. S. Afr.,
    57:1-44.
_____ 1969. The sperm whale (*Physeter catodon*) off the west coast
    of South Africa. 4. Distribution and movements. Investl. Rpt.
    Div. Sea Fish. S. Afr., 78:1-12.
_____ 1970a. The sperm whale (*Physeter catodon*) off the west coast
    of South Africa. 5. Age, growth and mortality. Investl. Rpt.
    Div. Sea Fish. S. Afr., 79:1-27.
_____ 1970b. Exploitation and recovery of right whales *Eubalaena
    australis* off the Cape Province. Investl. Rpt. Div. Sea Fish.
    S. Afr., 80:1-20.
_____ 1970c (MS). The sperm whale (*Physeter catodon*) off the west
    coast of South Africa. 6. Social groupings. (Paper submitted
    to the special meeting on sperm whale biology and stock assess-
    ments, Honolulu, 1970.)
_____ 1970d (MS). Two allopatric forms of Bryde's whale off the
    west coast of South Africa. (Paper submitted to the Scientific
    Committee, International Whaling Commission, 22nd meeting,
    1970.)
_____ and Gambell, R. 1968. The abundance of sei whales off South
    Africa. Norsk Hvalf.-Tid., 57:168-174.
Brown, S. G. 1962a. A note on migration in fin whales. Norsk Hvalf.-
    Tid., 51:13-16.

_____ 1962b. The movements of fin and blue whales within the Antarctic zone. Discovery Rpt. 33:1-54.
_____ 1970 (MS). A note on the migrations and movements of fin
whales in the Southern Hemisphere as revealed by whale mark
recoveries. (Paper submitted to the special meeting on Antarctic fin whale stock assessment, Honolulu, 1970.)
_____ 1971. Whale marking—progress report, 1970. Rpt. IWC,
21:51-55.
Chapman, D. G. 1971. Analysis of 1969/70 catch and effort data for
Antarctic baleen whale stocks. Rpt. IWC, 21:67-75.
_____ and Boerema, L. K. 1971. An example of estimation of stock
size of sperm whales in the Southern Hemisphere, based on mortality rates estimated from age distributions of adult males.
Rpt. IWC, 21:49-50.
Doi, T., and Ohsumi, S. 1969. The present state of sei whale population in the Antarctic. Rpt. IWC, 19:118-120.
_____ _____ Nasu, K., and Shimadzu, Y. 1970. Advanced assessment
of the fin whale stock in the Antarctic. Rpt. IWC, 20:60-87.
_____ _____ and Nemoto, T. 1967. Population assessment of sei
whales in the Antarctic. Norsk Hvalf.-Tid., 56:25-41.
_____ _____ and Shimadzu, Y. 1971. Status of stocks of baleen
whales in the Antarctic, 1970/71. Rpt. IWC, 21:90-99.
Fujino, K. 1964. Fin whale subpopulations in the Antarctic whaling
areas II, III and IV. Scient. Rpt. Whales Res. Inst., Tokyo,
18:1-27.
Gambell, R. 1964. A pygmy blue whale at Durban. Norsk Hvalf.-Tid.,
53:66-68.
_____ 1972. Sperm whales off Durban. Discovery Rpt. 35:199-358.
Harmer, S. F. 1928. Presidential address (history of whaling).
Proc. Linnean Soc. London, 140:51-95.
Ichihara, T., and Doi, T. 1964. Stock assessment of pigmy blue
whales in the Antarctic. Norsk Hvalf.-Tid., 53:145-167.
Mackintosh, N. A. 1942. The southern stocks of whalebone whales.
Discovery Rpt. 22:197-300.
_____ 1965. The stocks of whales. London, Fishing News (Books)
Ltd.
_____ and Brown, S. G. 1956. Preliminary estimates of the southern
populations of the larger baleen whales. Norsk Hvalf.-Tid.
45:469-480.
Ohsumi, S., Masaki, Y., and Kawamura, A. 1970. Stock of the Antarctic minke whale. Scient. Rpt. Whales Res. Inst., Tokyo, 22:75-125.

CHAPTER 3

A. THE FIN AND SEI WHALE STOCKS OFF DURBAN

Ray Gambell

The aircraft sighting data and catch records of fin and sei whales
(*Balaenoptera physalus* and *B. borealis*) off Durban since 1954 have been
analyzed on a unit-of-effort basis to determine the seasonal and annual
variations in the densities of these species.

The data available up to 1963 and the general methods employed in
their manipulation have been given by Bannister and Gambell (1965).
Records collected in subsequent seasons have substantiated the seasonal
frequency patterns they deduced. These patterns are illustrated in
Fig. 3-1 and indicate that fin whales are present off Durban from April
to October and most abundant in July; sei whales also appear in the
area from April to October, but are at their greatest concentrations
in September.

Taking the baleen whaling season as May to September, comparisons
of the annual densities of fin and sei whales since 1954 show marked
changes (Fig. 3-2). Fin whales, after an initial period of high abundance, declined rapidly in density after 1957 to a more stable position at a much lower level, although their concentration has continued
to fall slowly since 1963. Sei whales showed a definite increase in
abundance in the period 1959-1965, but their numbers have since dropped
to the lowest levels recorded during the period under review. The
similarities between the trends indicated by the density estimates for
both species derived from the independent catch and aircraft sightings
data strongly support the contention that real changes have occurred
in the numbers of fin and sei whales in the area each year.

The basic biology of the southern fin whale has been described by
Laws (1961), and that of the sei whale by Gambell (1968). Both species
have reproductive cycles geared to their annual migrations between
the Antarctic summer feeding grounds and the subtropical winter breeding

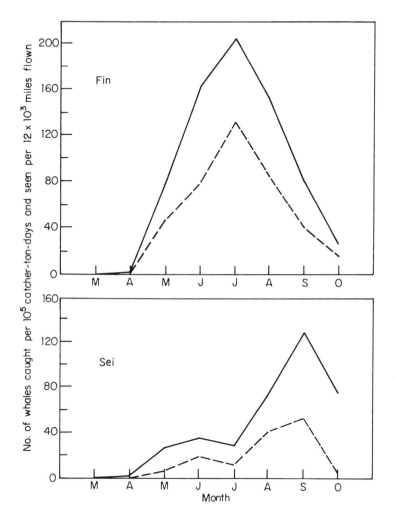

FIGURE 3-1. Average monthly catches (solid line) and sightings (broken line) per unit of effort of fin and sei whales off Durban, 1954 to 1963.

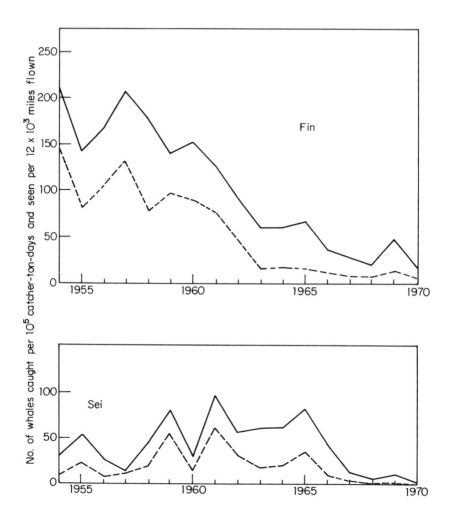

FIGURE 3-2.   Annual catches (solid line) and sightings (broken line) per unit of effort of fin and sei whales off Durban.

areas. Direct evidence of a connection between the fin whales off Durban and those in Antarctic Area III (0° to 70° E) is provided by a number of recovered whale marks (Brown 1964). Only one marked sei whale has so far shown a similar link for this species (Brown 1971), but Best and Gambell (1968) have argued that by analogy with fin whales and on purely geographical grounds it is reasonable to expect such a connection.

The evidence of the decline in the fin whale stock off Durban corresponds well with the fluctuations calculated for the Antarctic Area III population of this species (Chapman 1971). The Antarctic fin whales became reduced in numbers after the 1957/58 season, because of the increased catching effort applied to them in that area, but since 1963/64 the stock size has been more or less stable. The Area III stock of sei whales was at its maximum size in 1960, at the beginning of the period of more intensive sei whale catching in the Antarctic, and at the start of the 1965/66 season was still close to this maximum (Anon. 1968). Thereafter the stock size was reduced, so that in this species also the pattern of population change in the Antarctic agrees closely with the trends observed off Durban.

TABLE 3-1. Estimated stock sizes in Antarctic Area III and the catch per unit of effort (c.p.u.e.) in catcher-ton-days (c.t.d.) off Durban of fin and sei whales.

Fin whales			
Area III (Chapman 1971)	Stock size in thousands	Durban (Author's data)	Catch (per 10^5 c.t.d.)
1957/58 stock	66.5	1954-57 average	182
1962/63-1969/70 average	27.9	1963-70 average	40
Surviving stock	42%	Surviving stock	22%

Sei whales			
Area III (Anon. 1968)	Stock size in thousands	Durban (Author's data)	Catch (per 10^5 c.t.d.)
Stock up to 1964/65	23.0	1959-1965 average	68
1967/68	11.4	1967	14
Surviving stock	50%	Surviving stock	21%

The disturbing feature about these comparisons is that the degree
of depletion of both the fin and sei whale stocks off Durban is very
much greater than those suggested in the Antarctic, as indicated by
the figures in Table 3-1. In the Antarctic the fin and sei whale stocks
are estimated to have declined to 42 and 50% of their former levels,
respectively; off Durban the densities are just over one-fifth of their
previous values, so that neither species is now of major commercial
significance in the fishery.

REFERENCES

Anonymous. 1968. Report on the effects on whale stocks of pelagic
    operations in the Antarctic during the 1966/67 season and on the
    present status of those stocks. (Prepared by Fish Stock Evalua-
    tion Branch, FAO.) Rpt. IWC, 18:23-47.
Bannister, J. L., and Gambell, R. 1965. The succession and abundance
    of fin, sei and other whales off Durban. Norsk Hvalf.-Tid.,
    54:45-60.
Best, P. B., and Gambell, R. 1968. The abundance of sei whales off
    South Africa. Norsk Hvalf.-Tid., 57:168-174.
Brown, S. G. 1964. Whale marks recovered in the Antarctic whaling
    season 1962/63 and in South Africa 1963. Norsk Hvalf.-Tid.,
    53:277-280.
_____ 1971. Whale marking—progress report. Rpt. IWC, 21:51-55.
Chapman, D. G. 1971. Analysis of 1969/70 catch and effort data for
    Antarctic baleen whale stocks. Rpt. IWC, 21:67-75.
Gambell, R. 1968. Seasonal cycles and reproduction in sei whales of
    the southern hemisphere. Discovery Rpt., 35:31-134.
Laws, R. M. 1961. Reproduction, growth and age of southern fin
    whales. Discovery Rpt., 31:327-486.

B. SPERM WHALES OFF DURBAN

Ray Gambell

(The following is the slightly modified summary of a forthcoming Discovery Report of the same title.[1])

     Sperm whales have formed an important part of the catch taken in
the whaling operations off the coast of Natal since the start of the
industry in 1908. Since 1954 the catching activities from Durban have
included the use of whale-spotting aircraft, and both catch and sightings records have been used to follow changes in sperm whale densities.

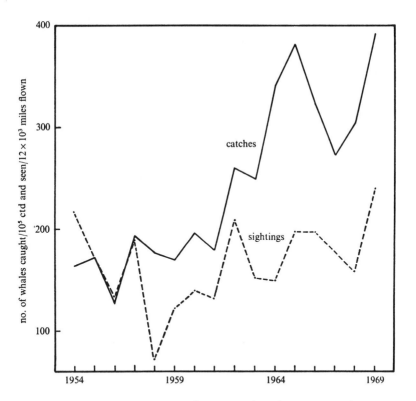

FIGURE 3-3. Annual catches and aircraft sightings per unit of effort
of sperm whales off Durban, 1954-1969. (c.t.d.: catcher-ton-days)

---

[1]It has come forth in the meantime: 1972. Discovery Rpt. 35:199-358.

The catch data per unit of effort, corrected by a tonnage factor
for the increasing efficiency of the vessels, are somewhat affected
by the selective hunting of baleen whales in preference to sperm up to
1959, and by certain changes in catching policy in some later years;
as shown in Fig. 3-3, they suggest a fairly steady increase in sperm
whale density. The aircraft sightings records, also plotted in Fig.
3-3, are unbiased by species selection and are thought to give a true
picture of the density of the whales in the sea; they indicate no
general trend of change in abundance. Considered in terms of three
size groups of whales (Fig. 3-4), the increased catches are attributed

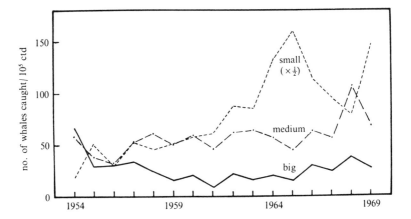

FIGURE 3-4. Annual catches per unit of effort of three size groups of
sperm whales off Durban, 1954-1969. Small: shorter than 11.6 m; medium,
11.9 to 13.7 m; big: longer than 13.7 m. (c.t.d.: catcher-ton-days)

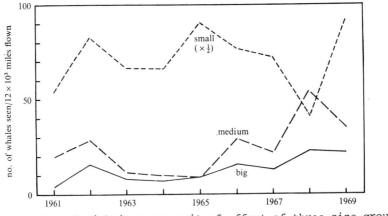

FIGURE 3-5. Annual sightings per unit of effort of three size groups
of sperm whales off Durban, 1961-1969. Sizes as in Fig. 3-4.

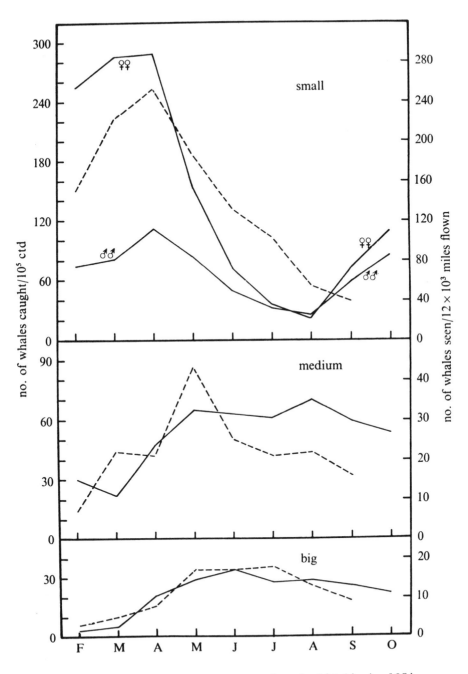

FIGURE 3-6. Average monthly catches (solid line), 1954
to 1969, and sightings (broken line), 1961 to 1969, per
unit of effort of various size groups of sperm whales
off Durban. Sizes and abbreviations as in Fig. 3-4.

to the capture of greater numbers of small animals (less than 11.6 m
or 38 ft in length); the catch per unit of effort of medium (11.9-13.7
m or 39-45 ft) and big whales (over 13.7 m or 45 ft) and the sightings
densities of all three size groups indicated in Fig. 3-5, show little
annual variation. The seasonal catch and sightings data correspond
closely in showing (Fig. 3-6) that small sperm whales (females and
small males) are most abundant off Durban in April, the numbers falling
to the lowest level in August or September. The medium and big sperm
whales, all males, have patterns of density fluctuations similar to
one another but opposite to that of the small whales. They are poorly
represented at the start of the whaling season in February and March,
but the densities increase to winter maxima from May to July or August
and then fall again towards the close of the whaling season in September or October.

There is little evidence of an overall movement of sperm whales
down the coast (heading 121° to 300°) from the sightings records of
single-size groups, as shown in Fig. 3-7. Whales in all three size
groups have a strong average movement up the coast (heading 301° through
360° to 120°) in the early part of the year. This movement slackens
and there is a less definite trend of direction indicated in June for
the small whales, in May and June for the medium, and in July for the
big whales. Thereafter the movement up the coast increases again.
Whales in mixed-size groups have a more marked movement down the coast
than the single-size groups, but follow broadly similar patterns of
fluctuations.

The histograms presented in Fig. 3-8 illustrate the fact that
big sperm whales are found as single individuals in more than half
the sightings of that size alone (geometric mean 2.0 per sighting) and
with small, and medium and small whales (geometric mean 1.5 and 1.6
big bulls per group respectively). When mixed with medium-sized
whales, less than one-third of the groups have a single big bull,
the geometric average being 2.6 big whales per group. Medium-sized
whales (Fig. 3-8) generally swim in groups of 4 to 6 when no other
size whales, or when big or big and small sperm whales, are present.
If mixed with small whales only, the medium-sized animals usually

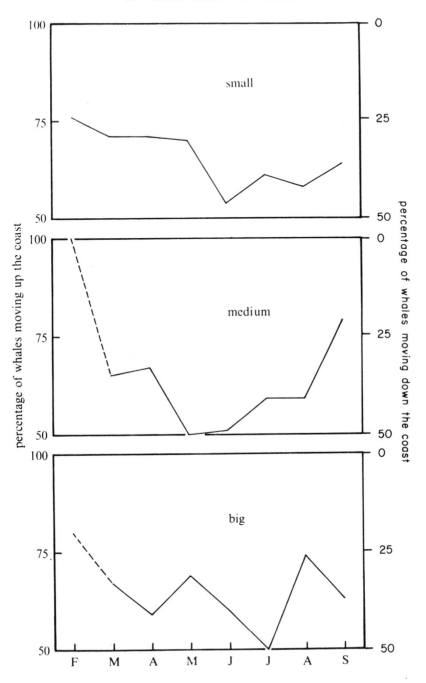

FIGURE 3-7. Percentage of three single-size groups of sperm whales seen heading up (301° through 360 to 120°) and down (121° to 300°) the coast off Durban, 1961 to 1969. Sizes as in Fig. 3-4.

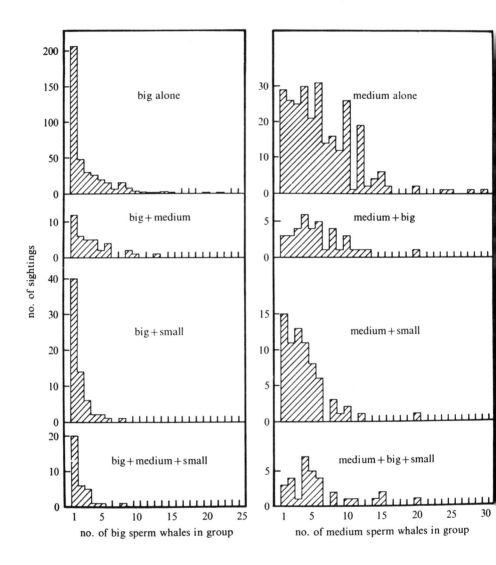

FIGURE 3-8. Frequency distribution of numbers of big and medium-sized sperm whales in aircraft sightings off Durban. Sizes as in Fig. 3-4.

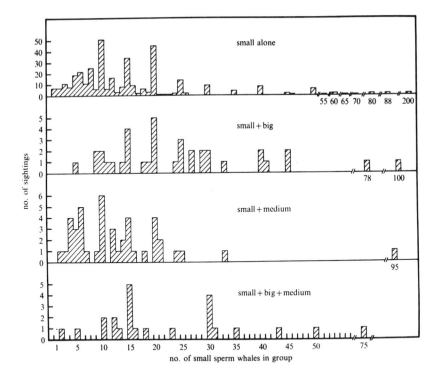

FIGURE 3-9. Frequency distribution of numbers of small sperm whales in aircraft sightings off Durban. Sizes as in Fig. 3-4.

number 1 to 3. Small whales alone, shown in Fig. 3-9, commonly form groups of 10 to 20. When mixed with big whales, there are usually 20 or more small animals, and 15 to 30 if medium-sized whales are present in addition. About 10 small whales are normally found if medium-sized sperm are also in the group. There is little segregation of the sexes in the groups of small whales off Durban, all-male groups representing 13% and all-female groups 26% of those sampled. It is concluded that there are on the average 10 mature females per big bull in the mixed groups, and the sexual classes of females are not segregated from one another.

Detailed study of the sizes and numbers of the ovarian follicles and corpora lutea and corpora albicantia in a large sample of females representing all the sexual classes, together with dated fetal records, point to a typical four-year reproductive cycle made up of $14\frac{3}{4}$ months gestation, 24 to 25 months lactation, and 8 to 9 months resting periods.

The average ovulation rate, allowing for the varying proportions of
the female population which ovulate and conceive at the main, post-
partum, mid-, and end-of-lactation estrous cycles, is calculated at
0.43 per year.

The average pregnancy rate is 19%. Nearly mature and older
females are less fertile than those with 3 to 7 ovarian corpora. The
catches show the effect of selection, particularly against whales in
the first year of lactation, since these animals represent only 9% of
the females taken instead of the expected 25%. Second-year lactation
whales comprise 23% of the catch, and resting females are much more
abundant than expected, at 48%, because of the reduced numbers of pregnant and first-year lactation whales.

Ages of sperm whales are determined from the number of dentinal
growth layers in maxillary and mandibular teeth, counted after bisec-
tion and acid-etching. The pulp cavity closes at an earlier age in
females than in males, but most teeth collected are considered suitable
for age-determination purposes. Evidence from the appearance of the
layers is not conclusive on their rate of formation, but the accumulation of 0.45 ovarian corpora per layer (Fig. 3-10) and data from long-

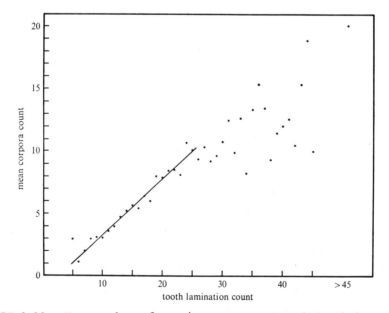

FIGURE 3-10. Mean number of ovarian corpora at each tooth-layer count,
and the fitted regression to the initial slope.

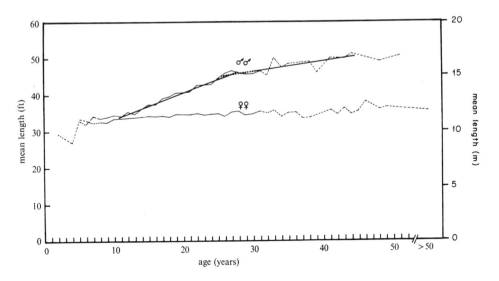

FIGURE 3-11. Mean body length at each age of male and female sperm whales caught off Durban (points representing fewer than ten observations are joined by broken lines).

term whale-mark recoveries point to an annual deposition.

Age-at-length curves, illustrated in Fig. 3-11, for the catch of females and the younger males are much affected by the minimum size limit, but the older males show a reduced rate of growth from 0.716 ± 0.008 to 0.278 ± 0.073 ft (0.218 ± 0.002 to 0.085 ± 0.022 m) per year in whales over 13.7 m (45 ft) long and 26 years old. Physical maturity is estimated at 15.2 m (50 ft) and age 40 years in males, and 11.0 m (36 ft) and 45 years for females. Sexual maturity in females occurs at a length of 8.8 m (29 ft) and age 7 years. The attainment of harem-master status by the bulls at a length of 13.7 m (45 ft) and an age of 26 years is associated with the reduced growth rate of the body and the continuing enlargement of the testes after this stage.

The age compositions of the catches show that the females are fully recruited to the fishery at about age 18 years, but the males enter more gradually from ages 14 to 23 years. Estimated instantaneous mortality is considered to be 0.08 for females and 0.20 for males, and natural mortality is considered to be 0.06 for both sexes.

External characters and morphometry are of little help in defining
stock limits, but a recovered whale mark shows a link between the males
in Antarctic Area III and those off Durban, which are considered part
of a local stock from recaptures in the same area a year after marking.
The limits of this stock are set at 20° and $70^{\circ}E$ longitude from a consideration of pelagic-catch density data. Estimates of exploitable
stock size from catches and fishing mortalities, mark recoveries, and
comparisons with the densities of the fin and sei whale stocks in the
area, give a range from 20,000 to 49,000 sperm whales, with equal numbers of males and females. The number of mature females is calculated
as 18,000 to 50,000, with a more probable narrower range from 18,000 to
30,000. This in turn leads to estimates for the sustainable yield of
females of 430 to 2,765 (430 to 1659 narrow range) and of males 500 to
2,435 (500 to 1,461 narrow range). The present catch of females off
Durban averages 877 whales per year and is within the sustainable yield
range, but the combined Durban and pelagic catches from the male stock
of about 2,000 per year are near the upper limit for a sustainable yield,
as is the fishing mortality calculated. The optimum yield in weight of
males would be obtained from animals of length 13.7 m (45 ft) and age
26 years.

CHAPTER 4

ON WHALE EXPLOITATION

IN THE EASTERN PART OF THE NORTH ATLANTIC OCEAN

Åge Jonsgård

Whaling in the eastern part of the North Atlantic Ocean includes
the catch of both large and small species of cetaceans. Because this
ocean is included in the regions in which factory-ship (pelagic) whaling for the large baleen whales is forbidden by the International
Whaling Commission, these species are taken only at shore stations.
Small cetaceans, however, are often worked up at sea, mainly on specially equipped fishing boats. The main raw materials utilized from
large whales are meat, blubber, and bones; from small cetaceans usually meat and blubber only. In general, the meat of both large and
small cetaceans is used for human consumption. In Norway, however,
only baleen whales are so utilized, the meat of toothed cetaceans being
used only for animal food.

Demand for whale products has often changed from one period of
time to another, resulting in decreasing or increasing whaling activity.
Increased whaling activity may have a depleting effect upon the stocks
of whales. In different areas of the eastern North Atlantic Ocean this
fact has been demonstrated several times during more than one hundred
years of modern whaling.

The following species of whales are mentioned in this chapter:

Fin whale	*Balaenoptera physalus* (Linné 1758)
Sei whale	*Balaenoptera borealis* Lesson 1828
Minke whale	*Balaenoptera acutorostrata* Lacépède 1804
Sperm whale	*Physeter catodon* Linné 1758
Bottlenose whale	*Hyperoodon ampullatus* (Forster 1770)
Pilot whale	*Globicephala melaena* (Traill 1809)
Killer whale	*Orcinus orca* (Linné 1758)

White whale   *Delphinapterus leucas* (Pallas 1776)
Narwhal   *Monodon monoceros* **Linné** 1758
White-sided dolphin   *Lagenorhynchus acutus* (Gray 1828)
Common porpoise   *Phocoena phocoena* (Linné 1758)

Large Whales

Three species of large whales are caught, namely fin, sei, and sperm whales (blue and humpback whales are protected). In recent years whaling has taken place from shore stations in Norway, the Faroe Islands, Iceland, Spain, Madeira, and the Azores. In Norway the whaling activity decreased considerably during postwar years, because the three west-coast stations ceased whaling operations from the middle of the 1950's onwards. The station in north Norway, however, has been operating continuously since whaling was readmitted in this area in 1948.

TABLE 4-1. Number of fin, sei, and sperm whales reported caught in different eastern North Atlantic waters in each of the seasons 1966 to 1970 (data from the Bureau of International Whaling Statistics).

| Whaling area | Species | Number of whales caught in— | | | | | Total whales caught |
		1966	1967	1968	1969	1970	
Norway	Fin	54	34	76	16	44	224
	Sei	1	0	0	1	0	2
	Sperm	36	22	1	111	51	221
Faroe Islands	Fin	4	0	6	0	0	10
	Sperm	1	0	6	0	0	7
Iceland	Fin	310	239	202	251	272	1,274
	Sei	41	48	3	69	44	205
	Sperm	86	119	75	103	61	444
Spain	Sperm	203	207	267	193	?	870
	Others[a]	80	80	92	122	?	374
Madeira	Sperm	113	85	78	83	44	403
Azores	Sperm	297	310	71	145	?	823

[a] Baleen whales, not specified.

Baleen whales have been caught by the stations in Norway (north Norway), the Faroe Islands, Iceland, and Spain, the catches in Iceland comprising as much as about 70% of the total catches. Sperm whales, however, constitute a considerable part of the catches in all the whaling areas, and only this species has been reported caught off Madeira and the Azores. See Table 4-1, which shows that altogether 1,508 fin, 207 sei (plus 374 baleen whales not specified), and 2,768 sperm whales have been taken in the eastern North Atlantic in the period 1966 to 1970.

Fin Whales

Up to the end of the 1950's it was generally assumed that the North Atlantic fin whales spent the summer or parts of the summer in high nort ern latitudes. Brown (1958), dealing with records of whales observed at sea by merchant ships and other vessels, stated that numbers of "rorqual had been observed in the summer in the Europe-North America zone between 30°N and 60°N (most of them between 40° and 50°). This shows that "ror-quals" (probably mainly fin whales) are found in almost all their area of distribution in the northeastern Atlantic in the summer too, although northward and southward migrations take place in the spring and autumn respectively. Brown suggested "that numbers of rorquals in the North Atlantic Ocean either miss the northern migration altogether, or get out of step with the main migration movements." This suggestion include the possibility that different populations of fin whales may be found in different North Atlantic waters.

Jonsgård (1958) compared data from fin whales caught off the west coast of Norway and off the Faroe Islands and stated that in both areas a decrease in the catch per unit of effort had taken place from 1951 onwards. He suggested a connection between fin whales inhabiting these waters. Jónsson (1964) compared the catches in the areas mentioned above with the Icelandic catches and found that the last mentioned had been rather stable. He suggested that fin whales caught off Iceland belonged to an independent population.

Further studies on fin whales in the eastern North Atlantic were made by Jonsgård (1966), who examined biological material collected at the shore stations in Norway and studied catch statistics from various whaling grounds. He found further indications that fin whales found off the Faroe Islands belonged to the same population (or popu-lations) as those found off western Norway. He also compared biological data on fin whales caught off western Norway and off north Norway and concluded that all indications pointed to the fact that different populations of fin whales inhabited these areas of whaling. Five years have passed since these conclusions were made, and experience from whaling and records of whales observed in the said areas during this period point to the fact that they were correct. Relatively few fin whales have been sighted off western Norway and off the Faroe Islands.

Jonsgård (1966) also compared catch data from Iceland with those from north Norway and found that although catches in both areas differed considerably from time to time, no definite change had taken place in the catch per unit of effort. Preliminary examination of catch data up to and including the season 1970 confirms this conclusion. Jonsgård indicated a possible connection between fin whales found in those two areas, since fin whales occurring off north Norway might very well undertake north and south migrations through the Denmark Strait. Markings of fin whales have been made to try and throw light upon this problem, but up to now in vain. Since 1965, Iceland and Norway have cooperated in marking fin whales in the Denmark Strait, but none of the marks has been recovered in animals caught off north Norway.

According to the whaling regulations, pelagic whaling for fin whales is not permitted in the North Atlantic Ocean. No doubt this important regulation more or less protects fin whales inhabiting waters which are too far off to be reached by the catchers of the shore stations. Jonsgård and Christensen (1968) observed surprisingly many fin whales off southeast Greenland in July and August 1968. On the 13th of July that year schools of fin whales were seen all around in position 63°27'N, 38°15'W, and it was roughly estimated that there were about 75 of them within 6 nautical miles of the vessel. Southeast Greenland waters are too far away for the Icelandic whaling operations, but 2 of the 14 fin whales marked in this area in 1968 have been caught by Icelandic catchers. Although this fact gives evidence for a connection between fin whales occurring off southeast Greenland and off western Iceland, the whales in question might belong

to a population (or populations) that migrates through the Denmark
Strait.

The possibility still exists that a separate population of fin
whales spends the summer off southeast Greenland. Mainly to look more
closely into this problem, 19 fin whales were caught in this area in
July 1970 under the provisions of Article VIII of the International
Convention for the Regulation of Whaling. A preliminary study of the
material collected shows that both recent immigrants and animals having
spent a period of time in cold waters were present; most of them had
fed on capelin. Electrophoretic studies are now being made on serum
and hemoglobin collected from those animals.

In conclusion it may be stated that different populations of fin
whales are found in the eastern North Atlantic. Fin whales occurring
off western Norway and off the Faroe Islands, which seem to belong to
the same population (or populations), have been considerably depleted
in postwar years, probably by overexploitation. The available data
do not indicate that fin whales are depleted in other areas of whaling.

Sei Whales

Little is known about sei whales in eastern North Atlantic waters.
At the end of last century, however, many sei whales were caught as
far north as off Finmark, and when the catch reached its peak in this
area in 1885, 720 animals were taken (Collett 1912). Sei whales are
seldom seen so far north nowadays, which is confirmed by the fact
that only three sei whales have been caught since 1948, when the shore
station in north Norway initiated whaling. Sei whales were rather
plentiful off western Norway during and after World War II, but in
the 1960's very few were observed. Although this species is known
for its irregular appearance, it cannot be denied that its disap-
pearance may be due to overexploitation. However, as a matter of
fact, relatively few sei whales have been caught in northeastern North
Atlantic waters since World War II. Altogether only 1,284 sei whales
have been reported caught in this area during the period 1946 to 1970
(Norway 220, Faroe Islands 224, Iceland 840), or on an average, 51
animals per season. If such diminutive catches have harmed the stocks

of sei whales, it can safely be concluded that the size of the sei
whale stocks just after World War II must have been very small. It
may be more likely that changes may have taken place with regard to
their route of migration. This explanation is in accordance with the
fact that sei whales were reported to be plentiful off the coast of
western Norway during World War II.

Sperm Whales

Whaling off the Azores and Madeira is based exclusively upon the
catch of sperm whales, and this species is also the main object of
Spanish whaling. Little is known about the sperm whale in these waters,
but the excellent biological studies carried out by Clarke (1956) at
the Azores should be mentioned. Data on catch statistics are incomplete, and almost nothing is known about the influence of the catch
upon the stocks.

In Norway and Iceland the catch of sperm whales is mainly dependent upon the number of fin whales inhabiting the whaling grounds
from time to time. Fin whales (and also sei whales) are preferred by
the whalers because they are more valuable. The relatively large
number of sperm whales, however, taken by the Norwegians in 1969 (111
animals and 3 catcher boats) and in 1970 (51 animals and 2 catcher
boats) is not due to an increasing number of sperm whales inhabiting
the whaling grounds. It can be fully explained by the increased
activity in whaling for sperm whales which took place during these
seasons because of the very high prices for sperm whale oil.

Small Whales

Whaling for small species of whales in the North Atlantic is
mainly carried out by three nations—Norway, Denmark, and Canada. The
Norwegian whaling is performed by specially equipped, licensed vessels, and the whaling area includes a great part of the northern North
Atlantic and adjacent Arctic waters. Four species are involved:
minke whales, which are the main object, and bottlenose, killer, and
pilot whales. Danish whaling includes whaling for pilot whales off
the Faroe Islands and the Eskimo whaling for different species of

small whales in Greenland waters. The Canadian whaling includes mainly
pilot whales taken off Newfoundland and white whales caught in Hudson
Bay. The numbers of the different species of small whales reported
caught by the different nations in each of the seasons 1966 to 1970 are
shown in Table 4-2.

TABLE 4-2. Number of different species of small whales reported caught
in different North Atlantic waters by different nations in each of the
seasons 1966 to 1970 (data from the Bureau of International Whaling
Statistics).

Nation	Species	Number of whales caught in—					Total whales caught
		1966	1967	1968	1969	1970	
Norway	Minke	2,153	2,175	2,733	2,391	2,307	11,759
	Bottlenose	340	264	384	485	535	2,008
	Killer	161	36	86	231	246	760
	Pilot	339	117	31	27	43	557
Denmark	Minke	19	21	32	231	– [a]	303
	Bottlenose	–	–	–	–	1	1
	Killer	2	–	–	–	–	2
	Pilot	1,702	1,982	1,749	1,395	388	7,216
	White	540	584	1,219	948	–	3,291
	Narwhal	–	–	–	220	–	220
	White-sided dolph.	–	–	–	–	59	59
	C. porpoises	–	–	–	1,328	–	1,328
	Unspecified	905	1,296	1,970	–	–	4,171
Canada	Minke	28	40	–	50	86	204
	Bottlenose	–	5	–	–	–	5
	Pilot	887	739	311	–	155	2,092
	White	482	390	582	178	72	1,704

[a] Information from Greenland for 1970 not yet received.

During the last three seasons the Norwegian whaling has moved
much farther to the west, now also including the waters off west Greenland and northeastern Canada, especially off Labrador. For this reason
small whales caught by other nations in western Atlantic waters will
be briefly mentioned in this paper. In recent years several papers
have been published on the biology of small whales in northern North
Atlantic waters, both from the western part (mainly by David E.
Sergeant) and from the eastern part (mainly by Åge Jonsgård). However, little is known about the effect of the catch upon the stocks

of small whales, except for minke whales in northeastern North Atlantic waters, which have been exploited every season since the end of
the 1920's. The following is a brief account of the exploitation and
its consequences for this species.

Minke Whales

Up to and including 1945, the Norwegian whaling for small whales
took place in Norwegian coastal waters, although the first whaling
attempts outside Norway (Faroe Islands and Shetland) were made just
before World War II. Just after the war a geographical extension to
Spitzbergen waters took place, and from about 1950 also further east
in the Barents Sea. About 1960 the catches began to decrease in the
Barents Sea, and a western extension took place mainly to the waters
off east and north Iceland. In the middle of the 1960's the first
vessel moved into the Denmark Strait, in 1968 into west Greenland
waters, and in 1969 as far west as Canadian waters, mainly off Labrador.

Although the biology of the minke whale is fairly well known and
statistical information including almost every whale taken is available,
it has been difficult to obtain exact data on the influence of whaling
upon stocks. This is mainly because no exact methods for age determination of minke whales are available. However, as will be briefly
shown, there are indications that in spite of thorough regulations a
decrease seems to have taken place in the stocks of minke whales inhabiting the Norwegian Sea and adjacent Arctic waters.

Fig. 4-1 shows the total catch of minke whales and the catch per

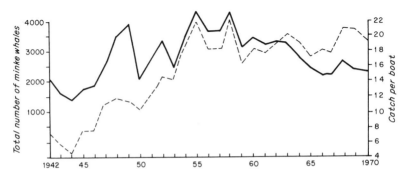

FIGURE 4-1. Total Norwegian catch of minke whales and the catch per
boat in each of the seasons 1942 to 1970. Total catch ——————;
catch per boat ------------.

boat in each of the seasons 1942 to 1970. The catches increased from
a little less than 2,000 animals at the end of World War II to 4,338
in 1958 and thereafter decreased to 2,307 in 1970. A similar trend
has taken place with regard to the catch per boat, except for the 1960's
when a slight increase occurred. This, however, may be due to the fact
that the number of licenses was gradually reduced from 183 in 1960 to
118 in 1970. This reduction has greatly increased the effort of the
remaining vessels, as mainly the smaller and less effective ones have
given up whaling. In 1960 the sum of the average size and average
engine power of the vessels was 150, increasing gradually during the
1960's to almost 300 in 1970. This increase, which is also due to some
new vessels being introduced and some older ones being enlarged, made
it possible to extend the whaling to areas farther west in the northern
North Atlantic.

Taking this development of whaling into consideration, it may be
argued that a greater increase in the catch per boat would have been
expected during the 1960's. Other factors not to be discussed on this
occasion come into the picture, and it is possible that the whaling
has been too intense in relation to the number of minke whales inhabiting the northern North Atlantic waters.

The catch statistics for minke whales caught in the northeastern
part of the North Atlantic show that calves are mainly found in a
restricted area from Möre in western Norway to Vesterålen in north
Norway, with a concentration area in Vestfjorden (the Lofoten area).
The calves, which are about half a year of age, stay in this area
almost all through the summer. Since there is no discrimination in
the catching of calves and older whales, changes in the number of
calves during a whaling period may indicate the effect of whaling on
the total northeastern North Atlantic stock (or stocks) of minke whales.

Fig. 4-2 shows the catches of minke whales and the catches of
calves (15 feet and less in length) in Vestfjorden in each of the seasons from 1942 to 1970. Until 1948 no decrease took place in the number
of calves caught, but from then on there was a gradual decline. This
may be due to overexploitation of minke whales inhabiting the northeastern Arctic waters, especially the Barents Sea. The large number

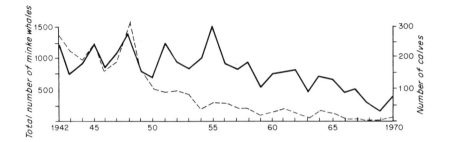

FIGURE 4-2. Total catch of minke whales and minke whale calves in
Vestfjorden (the Lofoten region) in each of the seasons 1942 to 1970.
Total catch ——————; calves ————————.

of minke whales caught in the Barents Sea after about 1950, which includes a surplus of mature females with fetuses, seems to have had a
reducing effect on the calves inhabiting Vestfjorden and thereby also
on the recruitment to the stocks of minke whales in those waters. In
1952 as many as 1,676 minke whales were taken in the Barents Sea, con-
stituting 49.8% of the total catch of minke whales that season.

In recent years Norwegian scientists have paid much attention to
the whaling for minke whales in northeastern North Atlantic waters, in
view of the fact that this species may need further protection. In
spite of the comprehensive regulations already introduced (see below),
an overexploitation seems to have taken place in this area. However,
the westward movement of the whaling activity in very recent years,
and also the reduced number of licenses issued, no doubt have spared
many minke whales in the northeastern North Atlantic. For this reason,
and also because the recruitment in this species is much better than
in all other species of baleen whales (minke whales give birth every
year), we have decided temporarily to watch the whaling carefully from
one season to another.

Other Species

As pointed out above, little is known of the effect of whaling
upon other species of small whales. During the 1960's Norwegian
whalers took more interest than previously in whaling for bottlenose
whales. Consequently, a biological research program was carried out
in 1967, when two biologists examined 52 animals on the bottlenose

whaling grounds off northeast Iceland. The extension of the Norwegian
whaling to Canadian waters in 1969, mainly based upon catching bottle-
nose, made further research necessary. Such research is to be carried
out by three Norwegian biologists in the 1971 season, probably in the
waters off Labrador.

The Norwegian Whaling Regulations for Small Whales

Several restrictions have since 1938 been enforced on the Norwegian whaling for small whales. From that year onwards the whaling
was made dependent upon a license issued by the Norwegian government,
and the licenses have been further restricted at various times since
then by increased qualification requirements. The geographical extension just after World War II with increased catches of minke whales
made it necessary to put further regulations into force. Since 1950
a closed season from the 1st to the 21st of July has been put into
force by the Norwegian authorities. In 1951 the International Whaling
Commission decided that from 1952 onwards, the season for the catching
of minke whales should be limited to six consecutive months per year.
In Norway the season opens on the 15th of March and closes on the 14th
of September, and the closed season in July has been retained. Further
restrictions were enforced by the Norwegian authorities in 1955, when
whaling for small whales north of latitude 70°N after the 30th of June
was prohibited. After 1961, this prohibition also covered the waters
south of 70°N and east of longitude 25°E as well.

REFERENCES

Brown, S. G. 1958. Whales observed in the Atlantic Ocean. Marine
    Observer, 28:142-145, 209-216.
Clarke, R. 1956. Sperm whales of the Azores. Discovery Rpt. 28:237-298.
Collett, R. 1912. Norges pattedyr. Kristiania, H. Aschehoug & Co.
    (W. Nygaard).
Jonsgård, Å. 1958. Taxation of fin whales (*Balaenoptera physalus* [L])
    at land stations on the Norwegian West Coast. Norsk Hvalf.-Tid.,
    47:433-439.
_____ 1966. Biology of the North Atlantic fin whale *Balaenoptera
    physalus* (L). Taxonomy, distribution, migration and food. Hval-
    råd. Skr., 49:1-62.
_____ and Christensen, I. 1968. A preliminary report on the "Harøybuen"
    cruise in 1968. Norsk Hvalf.-Tid., 57:174-175.
Jónsson, J. 1964. Whales and whaling in Icelandic waters. Aegir,
    22:1-13 (reprint).

CHAPTER 5

PRESENT STATUS OF

NORTHWEST ATLANTIC FIN AND OTHER WHALE STOCKS[*]

Edward Mitchell

With whaling for fin whales carried out for the first time on
the Nova Scotian coast in 1964, a program of research on the populations of large whales being fished was set in operation by the
Fisheries Research Board of Canada. The ultimate objectives of this
program are the estimation of population size and the assessment of
sustained yields under various exploitation regimes. First, population parameters, such as migration, reproduction, age, growth, and
population structure, have to be determined. These are being studied
in order to settle questions of population identity and discreteness,
so that estimates of population size and sustainable yields can be
made for each species. This paper is a summary of some data and an
account of research in progress relating to populations of whales in
the northwest Atlantic. The fin whale, *Balaenoptera physalus*, is
given special emphasis; sei, *B. borealis*, and sperm, *Physeter
catodon*, are given less emphasis commensurate with their relative
importance in the Nova Scotian and Newfoundland fisheries. Humpback,
Megaptera novaeangliae, blue, *Balaenoptera musculus*, bowhead, *Balaena
mysticetus*, little piked or minke, *Balaenoptera acutorostrata*, right,
Eubalaena glacialis, and bottle-nose, *Hyperoodon ampullatus*, whales
are mentioned briefly. The species are discussed in decreasing order
of commercial interest and importance.

---

[*] Paper presented in part, verbally, at government-industry conference
on whaling, Ottawa, 15 February 1971.

Balaenoptera physalus (Linné 1758)

Because the fin whale forms the basis for the present episode of the Canadian whale fishery, it is discussed in detail. Data and conclusions summarized below are tentative. Detailed studies are in progress on aspects of northwest Atlantic fin whale biology, including examination and interpretation of ovaries, testes, mammaries, vertebral epiphyses, ear plugs, baleen, and other samples. Final reports will be published separately on the male reproductive cycle, the female reproductive cycle, growth and age, tagging and migration, census and abundance, cumulative catch and availability, and measures of catch per unit effort.

Material

All of the fin whales examined to date have been obtained from commercial whaling operations in Nova Scotia and Newfoundland. These operations are presently carried out from three shore-based whaling stations (Fig. 5-1). The minimum legal length for fin whales in these waters (50 ft or 15.24 m) restricts the number of small and young whales available for sampling. The regulation is adhered to with varying degrees of selectivity by each of the stations, and the level of selection may well be affected by the abundance of whales on a seasonal basis. Blandford catches show the greatest number of fin whales under 50 ft (15.24 m) in length, Williamsport catches the smallest number. See Tables 5-1a to 5-1d.

Changes in the segment of the national quota allocated to each station on a yearly basis may result in a change in the level of selection. One example could be that a reduction in quota from one year to the next might lead to the average size of the catching being slightly greater because of enhanced selection for large and fat whales. Female fin whales accompanied by calves are likewise protected from capture, hence lactating females may be underrepresented in the catch.

Fin whales have been found to migrate parallel to the coast of Nova Scotia, from southwest to northeast through the season. During

FIGURE 5-1. Chart of a portion of the northwest Atlantic showing the
location of each of three whaling stations on the Canadian coast and
their catching fields. From north to south these stations are
Williamsport, Dildo (both in Newfoundland), and Blandford (Nova
Scotia). The dashed circles centered at each station represent
approximate 70-, 140-, and 210-mi catching ranges. Most stations
presently hunt between 50 and 150 mi out.

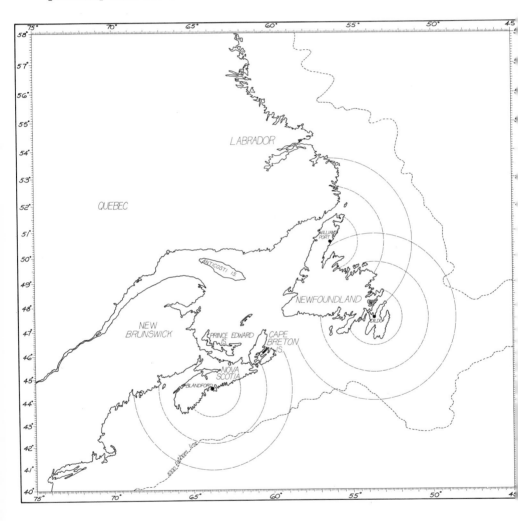

TABLES 5-1a, 5-1b, 5-1c, and 5-1d. The first three deal respectively
with fin whales at Blandford, Nova Scotia, in 1965 to 1971 (5-1a), at
South Dildo, Newfoundland, in 1966 to 1971 (5-1b), and at Williamsport,
Newfoundland, in 1967 to 1971 (5-1c). Table 5-1d summarizes the first
three. Parameters and abbreviations for all four are given in the
legend for Table 5-1a.

TABLE 5-1a. Length data in feet, and other parameters, on male, female, and total catch of fin whales at the Blandford, Nova Scotia,
whaling station during the most recent episode of fishing, 1965 to
1971. Whales were measured to the nearest six inches, rounded off to
the nearest foot. Totals represent fin whales examined by Fisheries
Research Board personnel, the number in the total catch being taken
from IWS (*International Whaling Statistics*). Lengths are given in
feet (only) for the following parameters in the total sample examined
by the FRBC:

n = number of fin whales examined;
$\bar{x}$ = arithmetic mean length in feet;
S.E.   = standard error of the mean;
S.D.   = standard deviation from the mean;
median = the length of the middle whale in the sample;
mode   = the length which occurs with the highest frequency
           in the sample;
range = the number of feet between the largest and the shortest
           whale in the sample;
%♀   = the percent of n comprising females in the sample.
Additional figures at the bottom of the table (% sample, n, $\bar{x}$, S.E.,
and S.D.) are given for all whales in the sample $\geqslant$ 50 ft in length
(i.e., the undersized whales were removed from the sample). Two or
three figures given for the mode indicate a bimodal or trimodal distribution. (Tables 5-1a to 5-1d have been updated to include 1971 data,
subsequent to preparation of the text of this paper.)

TABLE 5-1a.

	1965			1966			1967		
Canadian season	1/V-31/X			15/V-14/XI			1/VI-30/XI		
Station season	8/V-16/IX			7/VII-14/XI			12/VI-30/XI		
% time station operating	71.3			71.2			94.0		
Quota	No quota			No quota			300+25		
Number killed	108			264			326		
Number lost	0			0			1		
Number landed	108			264			325		
Number reported in IWC	108			263			Not reported		
% quota taken	No quota			No quota			100.3		
Number examined by FRBC	78			255			307		
% examined by FRBC	72.2			96.6			94.2		
In sample examined by FRBC:	Σ	♀	♂	Σ	♀	♂	Σ	♀	♂
n	78	43	35	255	134	121	307	166	141
$\bar{x}$	56.8	57.7	55.6	55.88	56.88	54.77	55.97	56.20	54.71
S.E.	0.5	0.9	0.5	0.313	0.485	0.358	0.264	0.332	0.332
S.D.	4.8	5.7	2.8	5.01	5.62	3.94	4.63	4.28	3.94
Median	57.5	60	56	56	58	55	57	58	56
Mode	60	60	56	58	62	58	58	61	58
Range	31	31	12	24	24	20	27	24	21
% ♀		55.12			52.54			54.07	
% sample		97.43			87.85			89.58	
n	76	42	34	224	116	108	275	149	126
$\bar{x}$	57.15	58.26	55.79	57.16	58.43	55.76	57.00	58.11	55.68
S.E.	0.463	0.710	0.453	0.255	0.386	0.269	0.221	0.320	0.253
S.D.	4.04	4.60	2.64	3.82	4.16	2.80	3.66	3.90	2.84

BLE 5-1a (cont'd.)

	1968	1969	1970	1971	All years
	1/VI-30/XI	1/VI-30/XI	1/VI-30/XI	15/IV-14/XI	-
	18/VI-18/XI	3/VI-29/XI	1/VI-20/XI	8/V-13/XI	-
	84.2	98.4	94.5	88.8	86.20
	262	224	150+20	40+110	-
	262	157	170	117	1,404
	0	0	1	0	2
	262	157	169	117	1,402
	262	157	Not available	Not available	-
	100.0	70.1	100.0	78.0	89.68
	261	154	169	117	1,341
	99.6	98.1	100.0	100.0	95.64

	♀	♂	Σ	♀	♂	Σ	♀	♂	Σ	♀	♂	Σ	♀	♂
»1	151	110	154	92	62	169	94	75	117	72	45	1,341	752	589
5	56.5	54.0	56.34	57.59	54.61	56.72	58.24	54.81	57.40	58.66	55.37	56.20	57.33	54.72
»3	0.5	0.4	0.407	0.544	0.535	0.364	0.501	0.438	0.416	0.553	0.490	0.132	0.190	0.158
.0	5.4	4.0	5.06	5.22	4.21	4.73	4.86	3.80	4.50	4.69	3.28	4.83	5.21	3.85
	58	55	57	59	55.5	57	60.5	56	58	59.5	55	57	59	55
	59	56	61	61	57	56	61	56	59	62	54	58	61	56
	22	18	23	21	18	22	22	15	23	23	14	33	33	22
	57.85			59.74			55.62			61.53			56.08	

	♀	♂	Σ	♀	♂	Σ	♀	♂	Σ	♀	♂	Σ	♀	♂
	85.45			88.32			91.72			96.59		89.63	90.29	88.79
3	131	92	136	83	53	155	88	67	113	70	43	1,202	679	523
91	58.00	55.41	57.58	58.59	56.01	57.54	58.95	55.70	57.72	58.98	55.67	57.22	58.41	55.68
63	0.366	0.308	0.339	0.466	0.292	0.322	0.442	0.360	0.397	0.520	0.466	0.112	0.160	0.125
2	4.18	2.95	3.96	4.26	40.1	4.01	4.15	2.94	4.23	4.35	3.05	3.91	4.18	2.86

TABLE 5-1b. Length data in feet, and other parameters, on male, female, and total catch of fin whales at the South Dildo, Newfoundland, land station, 1966 to 1971. Parameters and abbreviations as in Table 5-1a.

	1966			1967			1968		
Canadian season	15/V-14/XI			1/VI-30/XI			1/VI-30/XI		
Station season	17/V-22/X			9/VI-25/XI			1/VI-14/X		
% time station operating	91.3			92.9			91.3		
Quota	No quota			250-50			219		
Number killed	165			174			220		
Number lost	0			2			1		
Number landed	165			172			219		
Number reported in IWS	164			Combination			219		
% quota taken	No quota			69.6			100.45		
Number examined by FRBC	128			166			219		
% examined by FRBC	77.6			95.4			100.0		
In sample examined by FRBC:	Σ	♀	♂	Σ	♀	♂	Σ	♀	♂
n	128	83	45	166	106	60	219	98	121
$\bar{x}$	58.9	59.9	57.2	58.3	59.4	56.4	59.2	60.8	60.0
S.E.	0.4	0.6	0.4	0.4	0.5	0.5	0.3	0.4	0.3
S.D.	4.6	5.0	2.9	5.0	5.3	3.8	4.1	4.4	3.5
Median	59	60	57	58	60	57	59	62	59
Mode	59	59	56	60	65	60	59	62	57, 59
Range	21	21	14	28	28	16	22	22	17
% ♀		64.84			63.86			44.75	
% sample		98.44			95.79			97.27	
n	126	81	45	159	102	57	213	96	117
$\bar{x}$	59.12	60.19	57.2	58.88	59.96	56.87	59.55	61.04	58.31
S.E.	0.392	0.523	0.40	0.356	0.463	0.43	0.26	0.410	0.286
S.D.	4.40	4.70	2.9	4.49	4.68	3.27	3.81	4.02	3.09

TABLE 5-1b (cont'd.)

1969	1970	1971	All years
1/VI-30/XI	1/VI-30/XI	15/V-14/XI	-
1/VI-12/X	1/VI-29/XI	15/V-14/XI	-
90.2	99.5	100.0	94.20
188	225	160	-
188	181	117	1,045
0	2	1	6
188	179	116	1,039
188	Not available	Not available	-
100.0	80.44	73.12	84.72
123	179	116	931
65.4	100.0	100.0	89.60

Σ	♀	♂	Σ	♀	♂	Σ	♀	♂	Σ	♀	♂
123	59	64	179	77	102	116	39	77	931	462	469
57.6	58.6	56.8	59.0	60.8	57.7	59.2	62.53	58.07	58.77	60.08	57.55
0.4	0.7	0.5	0.3	0.5	0.4	0.32	0.839	0.318	0.147	0.232	0.158
4.9	5.7	3.8	4.4	4.8	3.5	3.43	5.24	2.79	4.48	4.98	3.42
59	57	58	59	61	59	59.5	61	58	59	61	56
59	52	59	59	63	59	60	60	58,59,	60	60	59
60	64	60						61			
23	23	16	23	20	18	20	15	13	29	29	18
47.97			43.02			33.62			49.62		
	96.75			98.89			100.00		97.74	97.84	97.66
119	57	62	177	77	100	116	39	77	910	452	458
57.94	59.00	56.89	59.91	60.74	57.88	59.2	62.53	58.07	59.03	60.37	57.71
0.426	0.716	0.439	0.320	0.544	0.329	0.32	0.839	0.318	0.139	0.218	0.150
4.64	5.41	3.46	4.26	4.77	3.29	3.43	5.24	2.79	4.20	4.63	3.21

TABLE 5-1c. Length data in feet, and other parameters, on male, female, and total catch
of fin whales at the Williamsport, Newfoundland, land station, 1967 to 1971. Parameters
and abbreviations as in Table 5-1a.

	1967			1968		
Canadian season	1/VI-30/XI			1/VI-30/XI		
Station season	7/VII-23/XI			25/VI-8/XI		
% time station operating	92.9			74.9		
Quota	250+25			219		
Number killed	272			219		
Number lost	1			0		
Number landed	271			219		
Number reported in IWS	Combination			219		
% quota taken	98.90			100.0		
Number examined by FRBC	261			219		
% examined by FRBC	96.0			100.0		
In sample examined by FRBC:	Σ	♀	♂	Σ	♀	♂
n	261	162	99	219	144	75
$\bar{x}$	58.24	59.40	56.11	60.28	61.65	57.65
S.E.	0.314	0.410	0.397	0.301	0.367	0.369
S.D.	5.08	5.22	3.95	4.46	4.41	3.19
Median	59	60	57	61	62.5	58
Mode	59	61	58	61 64	64	58
Range	27	27	19	21	21	15
% ♀		62.07			65.75	
% sample		97.32			99.55	
n	254	157	97	218	143	75
$\bar{x}$	58.53	59.92	56.28	60.33	61.24	57.65
S.E.	0.302	0.388	0.384	0.298	0.359	0.369
S.D.	4.82	4.87	3.78	4.40	4.29	3.19

TABLE 5-1c (cont'd.)

	1969	1970	1971	All years
	1/VI-30/XI	1/VI-30/XI	15/V-14/XI	-
	17/VI-17/X	22/VI-18/XI	29/VI-13/XI	-
	83.6	76.5	75.0	80.58
	188	225	200	1,107
	188	225	184	1,088
	0	0	0	1
	188	225	184	1,087
	188	Not available	Not available	-
	100.0	100.0	92.0	98.19
	188	225	184	1,077
	100.0	100.0	100.0	99.08

1969			1970			1971			All years		
Σ	♀	♂	Σ	♀	♂	Σ	♀	♂	Σ	♀	♂
188	121	67	225	115	110	184	110	74	1,077	652	425
60.36	61.52	59.25	59.44	60.86	58.38	59.28	60.45	57.55	59.50	60.53	57.56
0.228	0.358	0.297	0.236	0.388	0.284	0.318	0.439	0.364	0.135	0.176	0.161
3.82	3.95	2.43	3.54	4.16	2.97	4.32	4.61	3.13	4.44	4.50	3.33
60	62	58	60	61	59	60	61	58	60	61	58
61	61	58	60	64	60	61	65	58,59,60	61	61	58
20	20	11	20	19	15	21	19	16	27	27	19
64.36			51.32			59.78			60.54		
100.00			99.56			96.20			98.51	98.47	98.59
188	121	67	224	115	109	177	106	71	1,061	642	419
60.36	61.52	59.25	59.70	60.86	58.44	59.72	60.90	57.95	59.67	60.96	57.69
0.228	0.358	0.297	0.252	0.388	0.272	0.285	0.394	0.293	0.130	0.172	0.153
3.82	3.95	2.43	3.78	4.16	2.84	3.79	4.05	2.47	4.24	4.37	3.15

TABLE 5-1d. Length data in feet, and other parameters, on male, female, and total catch of fin whales at three Canadian land stations
(Blandford, Dildo, and Williamsport), 1965 to 1971. Parameters and
abbreviations as in Table 5-1a.

	1965 to 1971		
Canadian seasons	–		
Stations' seasons	–		
% time stations operating	86.99		
Total quota	–		
Total number killed	3,537		
Total number lost	5		
Total number landed	3,528		
Total number reported in IWS	–		
% total quota taken	90.86		
Total number examined by FRBC	3,349		
% examined by FRBC	94.93		
In sample examined by FRBC:	Σ	$\female$	$\male$
n	3,349	1,866	1,483
$\bar{x}$	57.96	59.20	56.40
S.E.	0.084	0.120	0.099
S.D.	4.86	5.22	3.84
Median	58	60	58
Mode	60	61	58
Range	37	37	24
% $\female$		55.72	
% sample	94.92	95.02	94.00
n	3,173	1,773	1,400
$\bar{x}$	58.56	59.83	56.90
S.E.	0.075	0.107	0.080
S.D.	4.24	4.51	3.22

this migration there may be some segregation of various categories
of fin whales. The date of the whaling season is then a factor in
whether the catch represents a fully representative sample of the
population. There is some evidence that the beginning and possibly
the end of the migration stream may not be adequately sampled by the
fishery.

Catches may also vary in composition latitudinally. Thus pregnant females carrying near-term fetuses migrating north past Nova
Scotia and reaching Newfoundland waters may give birth and be subsequently protected, and thus underrepresented in other catches.
(Calving grounds have not been identified for these stocks, however.)

Fisheries Research Board personnel have sampled the catch at
each of the three main whaling stations since 1965. In spite of
extensive sampling, only 91 fin whales which have both readable ear
plugs and the data from both ovaries represented in collections are
available from the Blandford catch for all years to 1969. This
represents 18% of the fin whales killed during that period of time.
Comparable figures for Dildo are 124 fin whales with readable ear
plugs and the data from both ovaries, representing 36% of the catch;
and for Williamsport 129 whales with readable ear plugs and the data
from both ovaries, representing 19% of the catch.

Identity of Populations

Geographic limits, ranges, and migrations. Differences in mean
length of fin whales sampled at each of three Canadian Atlantic stations indicate that there are latitudinal differences which might be
explained either by the presence of a cline in one population, or by
the presence of two or more discrete populations in the western North
Atlantic. For many parameters the corresponding figure for Dildo is
larger than that for Blandford, and the Williamsport figure is slightly
larger than that for Dildo. Use of data on reproduction, physical
maturity, and age will be mentioned below as it relates to the definition of two or more populations.

Six whale-tagging and census cruises have been carried out in
the summer months of 1966 and 1967, and the winter and spring months

of 1968, 1969, and 1971. The coverage of these expeditions ranges
throughout the western and central North Atlantic and has resulted in
extensive sighting of large and small cetaceans. One important conclusion resulting from these surveys is that fin whales are concentrated in the northwestern North Atlantic between the shore and the
1,000-fathom line, between latitudes $41^{\circ}20'$ and $57^{\circ}00'N$. This area
can be thought of as the summer feeding range of northwest Atlantic
fin whales.

Numbers of fin whales observed per 1,000 square nautical miles
have been calculated for regions of the North Atlantic Ocean. Inspection of these figures indicates that the density of fin whales
on the Canadian coast is approached only by the concentration of fin
whales in Denmark Strait, which is representative of a stock fished
by Iceland. Significant numbers of fin whales do not occur in the
tropical waters of the North Atlantic during northern summer months.
Thus there is strong evidence that fin whales of one or more populations concentrate on the Canadian continental shelf, and that this
concentration is spatially separated from other concentrations in
Denmark Strait in summer months.

A number of fin whales have been tagged in northwestern North
Atlantic waters (Fig. 5-2). Figures 5-3 and 5-4 show the position
of tags returned in Nova Scotian waters and Labrador-Newfoundland
waters respectively. These data are consistent in demonstrating
that a seasonal migration takes place on the continental shelf from
the region of Cape Cod in June and July up to and including waters
on the central Labrador coast to $57^{\circ}N$ latitude. Single whales evidently do not make this entire trip during the period June to November, but instead the data corroborate Kellogg's (1929) suggestion
that populations of fin whales on this coast are stratified and move
north and south so that grounds occupied by a southern population in
the summer are occupied by a northern population in the winter months.

As yet there is no conclusive evidence that this is the case,
but fin whales are known to migrate on the Nova Scotian shelf according to this schedule, and some sightings are accumulating for fin
whales in the Nova Scotian area during winter months. There are

FIGURE 5-2. Chart of a portion of eastern Canada and adjacent waters, showing position of marking and recapture for fin whales killed during the 1966 to 1969 seasons. Base is from American Geographical Society map 5a. Solid dots denote position of tagging, open circles position of recapture. Serial numbers of tags in squares indicate whales for which no detailed kill position is available, only position of catcher boat for day on which whale was taken.

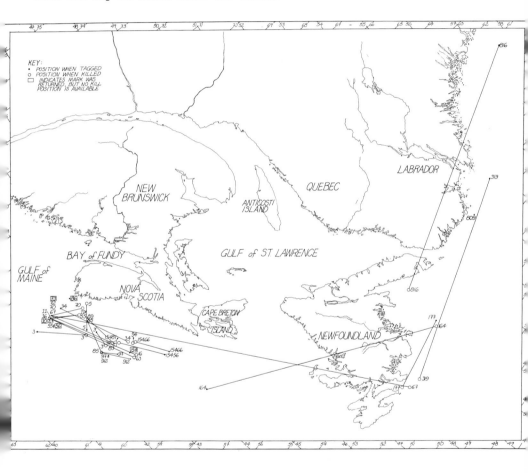

FIGURE 5-3. Detailed plot of points of tagging and of recapture for
all Nova Scotian tags for which both positons are available, 1966 to
1969. Solid lines represent the most reliable tracks, dotted lines
the least reliable. Each point is coded into a 17-unit calendar.
The calendar, arbitrarily taken as 1969, is as follows: *0* (a white
circle) = 12 July (any year) or earlier, *1* = 13-19 July, *2* = 20-26
July, *3* = 27-31 July and 1-2 August, *4* = 3-9 August, *5* = 30 July and
1-11 August, *6* = 12-23 August, *7* = 24-31 August and 1-4 September,
8 = 31 August and 1-6 September, *9* = 7-13 September, *10* = 14-20
September, *11* = 21-27 September, *12* = 28-30 September and 1-4 October,
13 = 5-11 October, *14* = 12-18 October, *15* = 19-25 October, *16* = 26
October and later. (Thus a circle which is exactly 3/4 black represents code 12, dates 28-30/IX and 1-4/X.) All points, whether representing tagging or kill, falling between 0 and 4 are included within
outline A, 4-8 within outline B, 8-12 within outline C, and 12-16
within outline D.

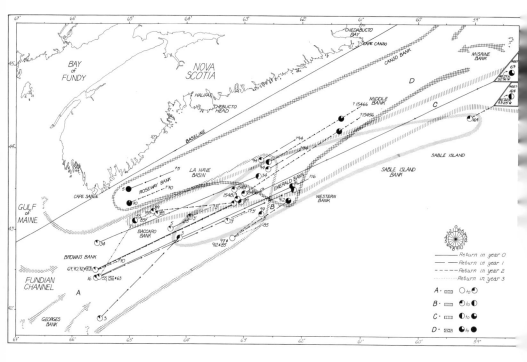

FIGURE 5-4. Detailed plot of points of tagging and of recapture for all Newfoundland and Labrador tags for which both positions are available 1966 to 1969. Connecting lines, date coding, and outlines are all as in Fig. 5-3. Tags 67 and 164 are included on both Fig. 5-3 and this figure.

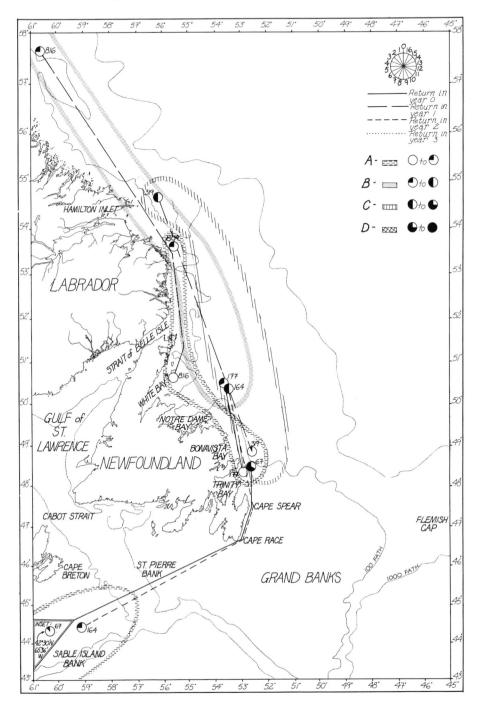

records of winter strandings of fin whales on the United States coast
to North Carolina (Brimley 1946; Caldwell and Golley 1965), a strand-
ing presumably of this species in May in Florida (Moore 1953), and
one record thought to be of a newborn fin whale near Galveston, Texas
(Gunter 1954). Thus there appears to be a migration along the eastern
North American continental shelf, with no evidence for onshore-offshore
migration of these fin whale populations to waters in the mid-North
Atlantic, or Denmark Strait area.

   Discreteness of populations. Numbers of fin whales have been
tagged in different 10° squares in the northwest North Atlantic Ocean.
Data examined from fin whales marked and tags returned by area in various parts of the northwest North Atlantic for the years 1960 through
1970 clearly indicate that there is some interconnection, on the order
of 10% of the stock off Nova Scotia, between the fin whale population
in the Nova Scotian area and that in the Labrador area during summer
months. No interchange has been demonstrated between the West Greenland area and the Labrador and Nova Scotian areas, nor between the
east Greenland-Iceland area versus the Labrador-Nova Scotia area. A
few fin whales have been tagged in the Gulf of St. Lawrence, and could
reasonably be expected to be returned in the Nova Scotian area early
in the season, but to date there have been no returns. These data
suggest that the fin whales in the Nova Scotian area and Labrador
area, notwithstanding problems of interconnection between them, are
clearly separable as a stock from the fin whale stock being fished in
Denmark Strait by Iceland.

   Throughout this report evidence is discussed for one versus two
or more discrete stocks of fin whales off the eastern Canadian coast,
but no conclusion can be reached at present on this question.

   Morphological differences. An external pigmentation (or color
pattern) scale is being devised in order to quantify small differences
found between individual whales, and possibly between populations, on
the Canadian east coast. Body measurements have also been taken from
a number of fin whales, and a discriminant function analysis will be
undertaken in order to describe further the possible differences between populations being sampled by the Blandford, Dildo, and Williamsport whaling stations.

Segregation within populations. The breeding and calving grounds for the fin whale populations fished off eastern Canada are unknown. Evidently all populations sampled by the three stations calve and breed at approximately the same time of the year, between December and April. It is possible that winter dispersion of the population is not random, but that individual whales generally migrate to the same region year after year, returning to the breeding areas where they were conceived and born. Such repetitive movements are indicated by summer, spring, and fall tag returns.

There is apparently some segregation during the migration schedule. Of 22 tags returned from whales marked in the Nova Scotian area, 2 were from whales that migrated to waters off Dildo, Newfoundland. Both of these animals were males. The indicated rate of movement in nautical miles per day graphed against the age of a whale in ear plug laminations indicates that older whales have moved along the migration path more quickly than younger whales of any sex, but the data are limited and this is not a conclusive finding. Partial segregation of some categories of fin whales in the migrating stream may be explained by varying times of departure from, or arrival on, the calving and breeding grounds.

Since there is a similarity in the time of birth and conception in both reputed stocks, it is important to know if the two stocks occupy the same or different breeding grounds in the period January through March. Additional whale tagging and census cruises in the vicinity of the eastern United States continental shelf and the Gulf of Mexico may settle this question.

The data from tagging show that, given the timing of the migration as now inferred, whales on the Labrador coast with tags in them are not available to the Nova Scotian fishery because they have not migrated to the catching field off Blandford before the end of the whaling season (late November). Conversely, whales tagged in the vicinity of Browns and Georges Bank may never be available to the Williamsport fishery, since they may not migrate that far north. They would be available only in part to the Dildo fishery, assuming that they were of the appropriate sex and age to make that lengthy migration, and that they arrived on the Dildo grounds during the whaling season.

Life History

Nutrition and growth. Most fin whales taken off Nova Scotia and off Newfoundland have stomachs full of krill or fish. Two species of krill are common (*Meganyctiphanes norvegica* and *Thysanoessa inermis*), and of fish, capelin (*Mallotus villosus*) form the main diet. During all months of the Canadian whaling season, fin whales are known to be feeding on the continental shelf.

Data on nutrition and growth will be presented in a subsequent study, along with details of feeding rates, feeding preferences, and other parameters.

Life History—Males

Reproduction. From histological examination of testes, it was possible to categorize sexual maturity in males in five stages, 0 (sexually immature), 1, 2, and 3 (stages of puberty), and 4 (sexually mature). The results of calculating the average age (ear plug laminations, one lamina defined as one dark or one light band, following Roe 1967) at each testis stage for each of the three whaling stations indicates that the average age at puberty is at 12.5 laminae at Blandford, 10.0 at Dildo, and 12.9 at Williamsport (Table 5-2).

TABLE 5-2. Length and age parameters of immature (0), puberal (1,2,3), and sexually mature (4) male fin whales at three Canadian Atlantic whaling stations.

Testis stage	Blandford	Dildo	Williamsport
	(1966+67+68+69)	(1966+67+68+69)	(1967+68+69)
	Average age (growth layers)		
0	6.9 (*n*=27)	4.8 (*n*=29)	5.9 (*n*=17)
1 + 2 + 3	12.5 (*n*=13)	10.0 (*n*=14)	12.9 (*n*=11)
4	34.4 (*n*=47)	24.2 (*n*=61)	20.4 (*n*=57)
	Average length (ft)		
	(1965 to 1968)	(1966 to 1968)	(1967 to 1968)
0	50.0	53.0	52.5
1 + 2 + 3	55.5	57.5	57.5
4	57.0	59.0	59.5

Sexually mature male fin whales from Nova Scotian waters might, according to one hypothesis, be migrating to the Dildo catching grounds, affecting the mean age of the landed catch (24.2 laminae, Table 5-2). Younger males from the Labrador grounds might remain in southern waters and be fished (affecting the average age of the puberal male catch at Dildo). Sexually mature males average 34.4 laminations at Blandford, while at Dildo the comparable number is 24.2 and at Williamsport 20.4. Since these numbers are derived from an open tail curve, they may not prove to be significant, but it is important that the catch at Blandford contains many more older males than at Dildo and at Williamsport. This may reflect earlier, heavy exploitation on the Williamsport grounds.

Point plots for the average length in feet of male fin whales at each of the three male sexual maturity classes show that at Blandford, sexually mature males are shorter by 2 ft (61 cm) than at Dildo and Williamsport. The fact that the average age of sexually mature males at Blandford is much greater than at Dildo and Williamsport indicates that they represent a different breeding group of whales, or that there is a cline in this feature. An alternative explanation would be that behavioral or density-dependent factors were influencing sexual maturity in many of these males, but if this were true it would be difficult to demonstrate at present.

The length at puberty is approximately 55.5 ft (16.9 m) at Blandford and 57.5 ft (17.5 m) at Dildo and Williamsport. These numbers were estimated from curves fitted by eye on scatter diagrams, but nevertheless are sufficient to demonstrate differences between Blandford on the one hand and Dildo and Williamsport on the other.

There are some data with which to examine the possibility of a male breeding season in fin whales. One good measure of a breeding season would be the apparent increase in the amount of sperm produced or some other reflection of testicular activity throughout the season at one or all of the stations. However, in order to obtain a base line for comparison, it must first be demonstrated that any apparent increase in sperm count or testicular activity is not due to an influx of different whales later in the season or to an increase in the

total number of mature whales. The percentage frequency of highly productive fin whale testes, based upon the parameter "amount of sperm
present" in the total catch of both mature and immature males, indicates that the same trend is obtained at each of the three whaling
stations, and that there is an apparent increase through the season
(from June to November) of the percentage frequency of highly productive fin whales in the samples. The trend is universal for all samples so far analyzed.

If immature males are excluded from the sample, or if only a
select sample of mature bulls is used, plots show the same increase
in apparent frequency of animals with high sperm count from June to
November.

The relation between length of males and weight of testes for
all whales sampled to 1968 shows the marked weight increase at maturity.

In summary, the data clearly show that there is a breeding season for male fin whales on the east coast of Canada. At the time of
the first samples, in June, there is a very low percentage of animals
with a high sperm count. This percentage increases throughout the
season. There are no data beyond November to indicate the month at
which the sperm count would peak. However, the data are in accord
with data from most other large whale populations, in which the breeding season is centered around the nonfeeding winter months.

Life History—Females

Reproduction. Taking the presence of a single corpus luteum or
corpus albicans of an early stage as the definition of the attainment
of sexual maturity in female whales, the average length at sexual maturity can be determined for the catch sampled at each whaling station.
(The number of corpora has been segregated into three classes: no
corpora, one corpus, and two or more corpora.) The average length
for all years in which data are available (Table 5-3), for a sample
that includes only those whales with a corpus luteum or corpus albicans
of stage *A* (whales that ovulated within one year before the time of
sampling), is 58.00 ft (17.6 m) at Blandford, 58.00 ft (17.6 m) at
Dildo, and 60.48 ft (18.4 m) at Williamsport.

TABLE 5-3. Reproductive parameters for female fin whales at three
Canadian land stations. Sample size refers to sample available for
percentages of whales in various reproductive states; sample size (*n*)
for mean length and mean number of ear plug laminations at first corpus
luteum or corpus albicans is less than 25, as indicated.

	(1967 to 1969) Blandford		(1967 to 1969) Dildo		(1967 to 1971) Williamsport	
	All	Mature only	All	Mature only	All	Mature only
Sample size	220		181		529	
1st corpus, $\bar{x}$ length	58.00 ft (*n*=12)		58.00 ft (*n*=11)		60.48 ft (*n*=23)	
1st corpus, $\bar{x}$ laminations	11.75 lam. (*n*=8)		11.70 lam. (*n*=9)		10.78 lam. (*n*=18)	
Pregnant	26.4%	43.3%	36.5%	57.9%	44.2%	58.5%
Lactating	10.5%	17.2%	6.1%	9.4%	14.7%	19.5%
Pregnant and lactating	3.6%	6.0%	0.0%	0.0%	1.3%	1.8%
Resting	20.5%	33.6%	20.4%	32.5%	17.2%	22.8%
Immature	39.1%		37.0%		23.9%	
Mature	60.9%		63.0%		76.1%	
Lost fetuses	50.0%		48.49%		67.95%	

The number of ear plug laminations has been plotted against the
number of corpora in both ovaries for the years 1966 to 1969. These
data are available for 91 female whales at Blandford, 124 at Dildo,
and 129 at Williamsport. A regression line has been fitted to the
resulting point diagram by the method of least squares. These data
indicate that there are 0.90 ovulations per two laminae at Blandford,
that is, there are 0.90 ovulations for a period of time represented
in ear plugs by two light bands (or two dark bands). Comparable
figures for Dildo and Williamsport are 0.82 ovulations and 0.68 ovulations.

The interpretation of laminae in absolute chronologic forms is
open to question, but the counts of laminae and their relation to
number of ovulations is a firm relationship within these data. Thus,

if the deposition of laminae is constant throughout the life of a
whale, and if the deposition of laminae proceeds at the same rate in
all the populations sampled by each station, then the ovulation rate
is different in the catch at each station. If a total of two growth
layers (two dark plus two light laminae) is taken as two years, then
Blandford whales ovulate approximately once every two years, and
Williamsport whales ovulate approximately once every three years.
The sample of Dildo whales likely represents some combination of these
two samples, an inference based on the tag return evidence, data from
age and length of whales, and other information. The conclusion can
be drawn that catches of female fin whales at Dildo represent a mixture of female fin whales sampled by both Blandford and Williamsport
whaling stations. The differing rates likely reflect earlier and
heavier exploitation of the stocks off northeast Newfoundland and
Labrador.

The frequency of the number of corpora by age class for the
three whaling stations (the percentage frequency of corpora present
in both ovaries compared to laminations) are similar. The Blandford
corpora-age frequency distribution comprises a greater number of animals having a high count of corpora of an age over 25 laminations than
do the animals at Dildo and Williamsport stations. There are relatively more old whales in the Blandford sample.

The data on fetal growth and timing are based on 53 fetuses from
Blandford, 64 from Dildo, and 67 from Williamsport. Table 5-3 gives
reproductive parameters for female fin whales at all stations. It
shows that at Blandford, 50.0% of fetuses were lost at sea (that is,
50.0% of the pregnant females were landed without fetuses, the fetuses
having been lost when the whales were slit open at sea to cool). Only
48.49% of fetuses were lost at Dildo, but 67.95% were lost at Williamsport.

Examination of the length of all fetuses of both sexes available
from the three stations, 1965 through 1970, as plotted against days
of the season for all years, shows a clear trend. Small fetuses (less
than approximately 200 cm) occur at all stations in June and July, and
fetuses less than 200 cm are generally not found in October and November

at any station. The data available show that the breeding and calving
season is well defined and is similar for all female fin whales sampled by Blandford, Dildo, and Williamsport stations. Attention is
thus focused on the need to determine winter breeding and calving
grounds for western North Atlantic fin whales. Published data on
strandings indicate that adult fin whales have stranded on the United
States coast as far south as Florida. A newborn fin whale washed
ashore in the Gulf of Mexico near Galveston, Texas, in February 1951,
according to Gunter (1954).

If the Labrador and Nova Scotian stocks of fin whales were discrete, but breed and calve in the same area, possibly the Gulf of
Mexico or the tropical or temperate United States Atlantic coast,
then isolating mechanisms other than reproductive timing must serve
to maintain population discreteness during the winter months. The
best data from tag returns support the hypothesis of stratification
of these stocks, implying separate breeding and calving grounds.

History of Exploitation

In Newfoundland waters, between 1903 and 1905 many shore stations
(Table 5-4) took 1,495 fin whales, an average of 498 per year. Using
Lucas' (1908) estimates for 1906 and 1907, and taking 50% of the total
catch as fin whales, 1,695 fins were taken between 1903 and 1907 for
an average of 339 fins per year. (This adjustment makes more nearly
equivalent the comparisons of catches during five years of the early
period and seven years of the later period.) In the period 1945 to
1951, from stations on the northeast coast of Newfoundland only, approximately 3,250 fin whales were taken for an average of 464 per
year (Table 5-5). The early episode resulted in a fall in catches,
which along with high licensing fees forced nine stations to go out
of business by 1907. Lucas (1908), Andrews (1916), and Sergeant
(1953), among others, have commented on this collapse. Thus it is
clear that the sustainable yield in these waters (all shores around
Newfoundland) is on the order of 400 fin whales per year or less.

TABLE 5-4. Whaling companies and time and area of operation in
Newfoundland and Labrador waters between 1898 and 1970. (After
Tønnessen 1967, p. 109, with modifications and updating.)

Place	Initial company or owner	Operating period	Not-operating period
Snook's Arm	Cabot Whaling Co.	1898-1914, 1918	1906
Balaena	Cabot Whaling Co.	1899-1914	
Chaleur Bay	Nfld. Steam Whaling Co.	1901-1906	
Rose-au-Rue	Nfld. Steam Whaling Co.	1902-1946	1908
Aquaforte	Anders Ellefsen	1902-1907	
Cape Broyle	Cape Broyle Whaling & Trading Co.	1903-1912	1909
Little St. Lawrence	St. Lawrence Whaling Co.	1903-1906	
Trinity Bay	Atlantic Manufacturing Co.	1904-1914	
Safe Harbour	Colonial Manufacturing Co.	1904-1906	
Riverhead (St. Mary's)	M. Cashin	1904-1906	
Beaverton	Henry J. Carle	1904-1915	
Dublin Bay (or Cove	Mic Mac Whaling Co.	1904-1914	
Cape Charles	Cape Broyle Whaling & Trading Co.	1904-1912	1911
Groais Island	Colin Campbell	1904-?	
L'Anse-au-Loup	Mic Mac Whaling Co.	1904-1905	
Harbour Grace	Nfld. Whaling & Trading Co.	1905-1910, 1956-1957	1907
Hawke Bay	Nfld. Whaling & Trading Co.	1905-1913	
Hawke Harbour	Daniel A. Ryan	1905-1915, 1947-1950, 1957-1958	
York Harbour (?Lark Hbr.)	Colin Campbell	1905	
Cape Broyle	Nfld. Whaling Co. (Chr. Hannevig)	1918	
Grady	British-Norwegian Whaling Co.	1927-1928	
Williamsport	Olsen Whaling & Sealing Co.	1947-1951	
Williamsport	Fisheries Products Ltd. (Atlantic Whaling Co.)	1967-1970	
South Dildo	Arctic Fisheries Products Co.	1947-1970	

TABLE 5-5. Catch statistics for Newfoundland and Labrador waters, 1898 to 1970. Data mainly from IWS, with modifications enumerated in footnotes. (M = minke whales, Rt. = right whales.)

Source	Year	Blue	Fin	Humpback	Sei	Sperm	Others	Total	Oil production (bbl of 170 kg)	Shore stations	Floating factories	Catchers
1,3	1898							47	162			
1,3	1899							95	1,266			
1,3	1900							111	2,580			
1,3	1901							258	3,798			
1,3	1902							472	7,650			
2	1903	225	345	287	-	-	1	858				
2	1904	264	690	281	39	1	-	1,275				
2	1905	265	460	161	2	6	-	894				18
2	1906	96	272	58	2	1	-	429				
2,3	1907						481	481			Sobraon 47 whales Labr.	
2	1908	36	345	24	-	1	-	406				
2	1909	48	385	79	2	4	-	518				
2	1910	33	295	56	-	-	-	384	8,580	5	-	5
2	1911	34	249	26	-	7	19	335	8,340	4	-	4
2,3	1912	60	202	22	-	5	-	289	8,000	10	-	10
2,3	1913	12	165	8	1	9	27	222	5,400	8	-	8
2,3	1914	5	142	13	-	1	-	161	3,100	7	-	7
2	1915	-	115	5	-	-	19	139	3,000	3	-	3
	1916	-	-	-	-	-	-	-	-	-	-	-
	1917	-	-	-	-	-	-	-	-	-	-	-
3	1918	-	-	-	-	-	101	101	2,500	1	-	2
3	1919	-	-	-	-	-	-	-	1,464	-	-	1
	1920	-	-	-	-	-	-	-	-	-	-	-
	1921	-	-	-	-	-	-	-	-	-	-	-
	1922	-	-	-	-	-	-	-	-	-	-	-
3	1923	-	66	3	-	1	-	70	1,600	1	-	1
3	1924	12	144	16	-	8	-	180	5,500	2	-	2
3	1925	12	270	35	4	10	-	331	8,400	2	-	3

TABLE 5-5 (cont'd.)

3	1926	10	329	18	3	-	-	360	11,600	2	-	3
3	1927	15	243	88	9	8	-	363	14,514	3	-	5
3	1928	58	358	21	23	48	-	508	20,580	3	-	7
3	1929	23	334	11	3	11	-	382	15,770	2	-	3
3	1930	23	282	7	1	8	-	321	13,100	3	-	5
	1931	-	-	-	-	-	-	-	-	-	-	-
	1932	-	-	-	-	-	-	-	-	-	-	-
	1933	-	-	-	-	-	-	-	-	-	-	-
	1934	-	-	-	-	-	-	-	-	-	-	-
3	1935	4	156	9	13	16	-	198	7,165	2	-	3
3	1936	20	146	10	2	14	-	192	7,186	2	-	3
3	1937	8	439	9	7	19	-	483	19,075	2	-	5
3	1938	-	-	-	-	-	1(Rt.)	-	-	1	-	1
3	1939	7	118	4	2	13	-	144	5,950	1	-	2
3	1940	1	64	7	-	6	-	78	2,950	1	-	1
3	1941	2	65	3	2	-	-	72	1,855	1	-	1
3	1942	4	62	1	4	-	-	71	1,855	1	-	1
3	1943	1	141	6	-	4	98	152	5,564	2	-	2
3	1944	5	231	10	1	17	-	264	8,963	2	-	4
3	1945	11	346	9	5	22	-	393	12,730	2	-	6
3	1946	11	502	5	-	11	-	529	19,561	2	-	6
3	1947	14	413	6	4	18	-	455	18,600	2	-	7
3	1948	57	669	15	4	14	49	808	30,478	3	-	9
3	1949	30	425	11	23	53	-	542	25,507	2	-	6
3	1950	15	409	16	16	29	-	485	20,508	2	-	7
3	1951	24	483	29	39	12	-	587	22,750	2	-	8
3	1952	-	1	1	-	-	-	2	70	-	-	-
3	1953	-	1	-	-	-	-	1	39	-	-	-
3	1954	-	-	-	-	-	-	-	-	-	-	-
3	1955	-	2	-	-	-	13	15	77	-	-	-

TABLE 5-5 (cont'd.)

Source	Year	Species of whale caught							Oil production (bbl of 170 kg)	Expeditions		
		Blue	Fin	Humpback	Sei	Sperm	Others	Total		Shore stations	Floating factories	Catchers
3	1956	–	7	–	2	13	–	22	867	1	–	1
3	1957	–	23	–	5	14	–	42	1,760	1	–	1
3	1958	–	55	4	–	7	–	66	1,580	1	–	1
3	1959	–	14	–	5	1	–	20	745	1	–	1
3	1960	–	1	–	–	1	–	2	27	–	–	–
3	1961	–	–	–	1	–	–	1	–	–	–	–
3	1962	–	–	–	–	–	–	–	–	–	–	–
3	1963	–	–	–	–	1	–	1	–	–	–	–
3	1964	–	1	–	1	–	–	2	–	–	–	–
3	1965	–	6	1	2	–	–	9	–	–	–	–
3	1966	–	164	–	–	2	–	166	4,327	1	–	1
3	1967	–	436	–	7	–	–	443	12,948	2	–	2
3	1968	–	438	–	4	–	–	442	15,252	2	–	3
3,4	1969	–	376	5	3	5	50(M)	439	–	2	–	3
3,4	1970	–	406	14	1	2	86(M)	509	–	2	–	3

(1) Millais 1907.

(2) *Journal of the House of Assembly of Newfoundland.*

(3) *International Whaling Statistics.*

(4) 5 humpbacks: 1 from Dildo (stranded), 4 from Williamsport under special permit.

This emphasizes the catch between 1945 and 1951, in which an average of 464 fin whales per year were taken on the northeast coast, in the same waters presently being fished by the Williamsport and Dildo land stations. The termination of whaling operations in 1951 has been explained as an economic decision, the price of oil having dropped. Nevertheless, there were clear signs that effort was shifting from fin whales to blue whales, sei whales, and humpback whales consistently from 1945 to 1951 (inclusive, see Table 5-5), and there are other indications of overexploitation as well. Sergeant (MS, 1966) concluded that decreasing length of the fin whale catch, the shift to sei (and other species), and the increased number of undersized fin whales (not a matter of better inspection, which remained at a consistent level throughout this fishery), all indicated overexploitation.

Also confirmatory is the fact that sei whales were not valuable in the late 1940's and early 1950's; the increasing sei whale catch therefore represents a depletion of fin whales.

Thus it can be concluded that 464 fin whales per year is above the sustainable yield for this northeast region of the Newfoundland coast. It must be emphasized that a sizable component (approximately 10% if the Nova Scotian and Newfoundland stocks were initially of equal size) of this catch by Rose-au-Rue (to 1946), Hawke Harbor, and Williamsport whaling stations may have been taken from the stock migrating northward from the Nova Scotian area. Thus the sustainable yield may be less than 464 whales minus 10% (464 - 46), or less than 418 fin whales.

Only three whaling stations have operated in the area of Nova Scotia and along the Nova Scotian and Quebec shores of the Gulf of St. Lawrence (Table 5-6). Catches in the Gulf of St. Lawrence (Table 5-7) were small and consistent between 1911 and 1915.

TABLE 5-6. Whaling companies and time and area of operation in Nova Scotia and along the shores of the Gulf of St. Lawrence (exclusive of Newfoundland coast) between 1898 and 1970. (Data from Tønnessen 1967, p. 109; and FRBC records.)

Place	Company or owner	Period
Seven Islands	Norwegian-Canadian Whaling Co.	1911-1915
Blandford	Karlsen Shipping Co.	1964-1970
Lower Saulnierville	Comeau Sea Foods, Ltd.	1965

TABLE 5-7. Catch statistics for Gulf of St. Lawrence area, 1910 to 1917, mainly from IWS. (Bl+F = Blue and fin whale statistics, combined.)

	Species of whales caught								Expeditions		
Year	Blue	Fin	Humpback	Sei	Sperm	Others	Total whales	Oil production (bbl)	Shore stations	Floating factories	Catchers
1910	–	–	–	–	–	–	–	–	–	–	–
1911	–	–	–	–	–	55 Bl+F	55	2,000	1	–	2
1912	–	–	–	–	–	ca.85 Bl+F	ca.85	3,333	1	–	2
1913	–	–	–	–	–	90 Bl+F	90	3,500	1	–	2
1914	–	–	–	–	–	78 Unspec.	78	3,390	1	–	2
1915	28	56	–	–	–	–	84	3,422	1	–	2
1916	–	–	–	–	–	–	–	–	–	–	–
1917	–	–	–	–	–	–	–	–	–	–	–

TABLE 5-8. Catch statistics for Nova Scotia, 1964 to 1970, mainly from IWS. (Abbreviations as in Table 5-5.)

	Species of whales caught								Expeditions		
Year	Blue	Fin	Humpback	Sei	Sperm	Others	Total whales	Oil production (bbl)	Shore stations	Floating factories	Catchers
1964	–	56	–	–	4	19(M)	79	957	1	–	2
1965	–	135	–	–	–	12(M)	147	2,667	2	–	4
1966	1	263	–	8	–	–	272	8,068	1	–	3
1967	–	309	–	55	2	15(M)	381	11,051	1	–	3
1968	1	262	–	100	–	–	363	8,824	1	–	3
1969	1	157	2	149	–	–	309	7,033	1	–	2
1970	–	170	1	93	25	–	289	7,168	1	–	2

The catch in Nova Scotian waters between 1964 and 1970, inclusive, comprises a total of 1,349 fin whales with an average of 193 per year (Table 5-8). There is some evidence that this catch is too high, and that the sustainable yield from this population is less than 193.

Population Structure

Sex ratio. At the Blandford whaling station the sex ratio (percentage of females in the catch examined by Fisheries Research Board personnel, Table 5-1) was 55 in 1965, 53 in 1966, 54 in 1967, 58 in 1968, 60 in 1969, and 56 in 1970. The average percent of females in this entire sample was 56. Comparable percentages for the Dildo whaling station, years 1966 through 1970 inclusive, are 65, 64, 45, 48, and 43, with an average of 52% of females in the examined catch. Similar percentages for the Williamsport station, 1967 to 1970, are 62, 66, 64, and 51, with an average of 61% of females in the observed catch.

These data indicate that the percentage of females in the catch at Blandford has remained relatively stable with minor fluctuations. The percentage of females in the Dildo catch fell from a high of 65% in 1966 to a low of 43% in 1970, and it should be emphasized that this is the only whaling station presently catching more male fin whales than females. A similar drop occurred in the Williamsport catch between 1969 and 1970, from 65 to 51%.

Thus selection of whales has changed somewhat, or the composition of that portion of the population being fished has changed in the last two or three years off northeastern Newfoundland. An examination of the mean length of whales taken is instructive in this regard.

Size composition. Length frequency histograms have been plotted by sex for the three whaling stations for all years, and for the total sample for all years by station. Examination of the female length frequency histograms for all years shows that the distribution is most symmetrical for the Williamsport station, has a larger component of smaller whales at Dildo, and an even larger component of small whales at Blandford. The three histograms are gradational, one to the other. The same figures for males at the three stations show a similar grada-

tion with a less pronounced difference between stations in the percentage of smaller male fin whales. The peaks for Williamsport and
Dildo are similar, while that for Blandford is at a lower length.

There is a systematic bias in the data from the Blandford fishery,
which can be emphasized as follows: 135 fin whales have been captured
that are less than 50 ft (15.24 m) in length, comprising 11% of the
total catch between 1965 and 1970 inclusive. Comparable figures for
Dildo and for Williamsport are: 21 whales less than 50 ft or 2.6%
of the Dildo catch, 1966 through 1970; and 9 whales less than 50 ft,
which is 1% of the total catch at Williamsport, 1967 through 1970
inclusive. Thus in considering mean length of the sampled whales,
allowance must be made for the fact that 10% of the Blandford sample
is undersized and not represented to the same extent in the northern
Newfoundland samples.

The data for mean length of whales at each of the three stations,
1967 through 1970 inclusive, have been corrected. The 11% of undersized whales (below legal minimum length) at Blandford has been sub-
tracted from the catch. The mean length of this corrected sample is
57.17 ft (17.4 m). Similar corrections were performed for the Dildo
and Williamsport catches, with 2.6% and 1% of the catch being sub-
tracted (those whales under 50 ft). The resulting mean lengths are
58.96 ft (17.97 m) and 59.64 ft (18.18 m), respectively.

A test on these three means (corrected for undersized whales)
was performed using the Student-"t" test at the 0.001 level of significance. The differences between the means at Blandford and Dildo
are significant. The same applies for comparisons between Dildo and
Williamsport, and Williamsport and Blandford.

The differences between mean lengths at each of the three stations imply that, not considering gunner selection, there are either
three stocks being fished; a number of isolated schools available to
each station; or that within one latitudinally dispersed population
there may be a cline in such characters as mean length.

The mean length at Blandford (Table 5-1a) has remained relatively constant throughout this fishery. However, the mean length
at Blandford in 1970 increased slightly over the previous four years,

presumably due either to increased selectivity for larger whales or to a change in availability of larger whales.

The mean length at Dildo (Table 5-1b) for the entire catch between 1966 and 1970, and that at Williamsport for the period 1967 to 1970, likewise appear to have remained stable.

The increased range of length and the decrease of 1 ft (30 cm) in the mean length at Williamsport in 1970 may be interpreted in one of two ways. Either there was decreased selectivity during catching because of an increased quota, or this is the beginning of a trend that might reflect overexploitation.

The comparison of a catch corrected to omit undersized whales at each of three stations has been carried out for males and females separately on a station-to-station basis. The values of a Student-"t" test on males and females separately at Dildo and Williamsport, using the same level of significance (0.001) as for the total sample (all whales and both sexes) show that there are less significant differences between these two stations for the female sample than for the total sample, and that there is even less significance for males than for females. The levels of significance for females lie between 0.05 and 0.025, those for males between 0.05 and 0.10. The level for the females is marginally significant. That for the males is usually considered as reflecting no significant difference between the mean lengths. The difference between means in the total sample of both sexes from both stations is highly significant. This is due to the large sample size of 1,678 fin whales.

Similar tests on females between Dildo and Blandford show significance in differences at the 0.001 level, and at the same level the males are also demonstrated to be significantly different in mean length.

Age composition. Ear plugs have been collected from 1,352 fin whales at three whaling stations between 1966 and 1969. Of these, 370 are from the Blandford fishery, and 128 or 35% proved readable. The Dildo fishery yielded 524, of which 164 or 31% were readable. A total of 461 ear plugs was collected from the Williamsport fishery, of which 181 or 40% were readable. This makes for a total of 473 readable ear plugs, or 35% of the total collected.

Good data are available, especially from the young fin whale component at Blandford, since a high percentage of whales (11% of the catch, 1965 to 1970) are under 50 ft (15.24 m) in length. Age curves for Blandford probably show a better approximation of the population than do age curves for the other stations, because Blandford has a heavier representation of the younger age classes. These results will be published in a separate study.

Physiological condition. Some biological parameters for females are presented in Table 5-3. The percentage of pregnant females within the sample of all females in the catch varies greatly: 26.4% at Blandford, 36.5% at Dildo, and 44.2% at Williamsport. However, these percentages are more comparable if taken as the percentage of mature females (only) that are pregnant. These are 43.3% at Blandford, 57.9% at Dildo, and 58.5% at Williamsport.

The percent of mature females judged to be lactating when landed is 17.2% at Blandford, 9.4% at Dildo, and 19.5% at Williamsport (Table 5-3). These data probably indicate that selection is different at each station, with the Dildo station being the most selective and the Williamsport station being the least selective and taking the greatest percentage of lactating females.

The data of Table 5-3 show a decrease in the percentage of mature females (from Blandford north to Williamsport) judged to be in a resting state: 33.6% at Blandford, 32.5% at Dildo, and 22.8% (of mature females) at Williamsport.

Because of minimum legal length requirements, the immature percentage of the catch will always be smaller than in the entire population. The percentage of immature females in the catch at Blandford is 39.1, that at Dildo 37.0, and at Williamsport 23.9. The Blandford figure may be related to the fact that 11% of the catch at Blandford for that period is below minimum legal length.

The differences between the Dildo and Williamsport figures may be of more significance, since both stations fish only 1 to 3% below legal minimum length. The greater percent of mature females caught at Williamsport (Table 5-3) may be related to the larger average

length of the whales caught at Williamsport (Table 5-1). This is part
of the evidence for segregation within a single stock of whales.

Population Density

  Fishing effort and catch per unit effort. Mr. K. R. Allen is
making a detailed study of the catch per unit effort for each of the
Canadian Atlantic whaling stations and is presenting his results in
separate studies (Allen 1970, 1971).

  Changes in abundance. Sighting data from catcher-vessels for
eastern Canadian waters are available for the vessels: *Fumi* (Williamsport); *West Whale 8* and *R. D. Evans* (Dildo); *Chester* and *Thorarinn*
(Blandford). Reported sightings of fin whales for the *Chester* were
884 for 1968, 714 for 1969, and 465 for 1970. The *Thorarinn* reported
1,746 sightings of fin whales in 1970. The *West Whale 8* reported
sightings of an increased absolute number of fin whales between 1969
and 1970. The *Fumi* reported sightings of 1,532 fin whales in 1969 and
1,477 in 1970. Thus there is an adequate data base. These sightings
have been related to sighting effort in terms of hunting days or hours
of hunting and chasing.

  For the Williamsport area, data are available from the *Fumi* for
1968 through 1970. For each hour of hunting and chasing in 1968,
1.16 fins were sighted, 1.52 were sighted in 1969, and 1.49 were
sighted in 1970. Considered on a working day basis, 10.0 fin whales
were sighted per working day in 1968, 12.9 in 1969, and 10.9 in 1970.

  For Dildo, the data from the *R. D. Evans* are incomplete at present
but show 7.0 fins per working day in 1968 and 6.4 in 1969. The *West
Whale 8*, on a working day basis, shows 7.9 fin whales sighted per
working day in 1968, 6.3 in 1969, and 6.1 in 1970. In 1966, the *Kyo
Maru 17* worked out of Dildo and sighted 4.5 fins per working day.

  For Blandford, sightings from the *Chester* are available 1966
through 1970. These show 7.8 sightings of fin whales per day in 1966,
6.3 in 1967, 7.3 in 1968, 6.2 in 1969, and 6.8 in 1970. There is a
trend here of a drop in sightings between 1966 and 1969 and an increase
from 1969 to 1970, based on a steady drop in the absolute numbers of
encounters with whales, from a high of 884 to a low of 465. On this

evidence, availability of fin whales to the Blandford fishery presumably increased in 1970.

The *Thorarinn* has worked only one year at Blandford, but its
efficiency is apparently much higher than that of other eastern
Canadian catchers, since it reported sighting 13.6 fin whales per day
in that season.

Changes in the density of fin whales sighted per 1,000 square
miles of ocean searched may or may not prove to be significant after
statistical tests have been applied to the data. The data from each
of the tagging and census cruises have been broken down by area and
are presently being broken down by month and by direction of travel
of ship relative to migration stream of whales.

Mr. K. R. Allen is presently studying the changes in abundance
indicated by sightings and catch per unit effort, hence these data
will not be commented on further here.

Changes in distribution. Catches have been plotted for each of
the three stations for most recent years. There has been a consistent
trend in the position of catch that is not related to seasonal movements of fin whales, or to fluctuations on a year-to-year basis such
as could be expected in a population of whales on rich feeding grounds.

Data from the Blandford fishery indicate that in early years
whales frequenting inshore waters were heavily fished, but for up to
four years, tag returns show that the same groups of whales migrated
through the catching field, northward early in the season and southward by November. There is no evidence that these whales learned to
avoid the catching area or moved farther offshore as a result of hunting pressure.

At Blandford catch plots show that the catch positions have become successively more dispersed between the years 1966 and 1970.
This has not been caused by any change in the behavior of migrating
fin whales, but may be a result of lowered density of fin whales in
inshore waters, forcing catchers to search larger areas in order to
capture fin whales of 50 ft (15.24 m) in length or longer.

The catch plots for Dildo and Williamsport land stations show
that the catching grounds have moved progressively farther away from

the stations.   These data will be discussed in detail in a subsequent
report.

Mortality

   Instantaneous total mortality coefficients. Estimates of total
mortality should be made from the age composition of catches at each
station, weighted by the total effort expended in obtaining these
catches.   These data will be reported in a separate study.

   The total mortality coefficient (Z) can be calculated by weighting whale sightings per 100 catcher-boat hunting hours in each of
several years to find the survival rate (S).   The sighting data require some reconsideration and weighting in order to make comparisons
on a season-to-season basis for the same vessel on the same grounds.
Discussion of this method is also deferred to a later study.

   Total mortality can be derived from length-frequency distribution of males and females in catches by applying an age-length key
and making corrections for changes in catch per unit effort.   Again,
these data will be reported in a separate study.

   Natural mortality and fishing mortality coefficients. The total
mortality coefficient (Z) comprises natural mortality (M) and fishing
mortality (F).   One approach to the estimation of natural mortality
in fin whales is the determination of age composition in a random
sample of a virgin population.   The Blandford fishery, starting in
1964, sampled whales that presumably had never been fished before.
This approach might be used here, if the identity or discreteness
of the Nova Scotia stock could be resolved.

   No ear plugs were collected from the 1964 fin whale fishery at
Blandford, and there are only a very few available from the 1965
fishery.   Only 24 readable plugs are available from the 1966 catch,
and 140 from the 1967 fishery.   The 1966 and 1967 catch could be
taken in order to derive an age-length key which could then be applied to the previous two years (191 fin whales) to determine the
age composition of the virgin stock.   This comparison will be reported on in a future study.

Recruitment

Stock-recruitment relationship. Mr. K. R. Allen is presently
working up catch-effort data in relation to catch, and he will present
his results in another study.

Estimates of recruitment from pregnancy data. A more direct
measure of recruitment can be made from pregnancy data and juvenile
mortality than is the case with stock-recruitment relationships.
There is a seasonal breeding cycle, as demonstrated by evidence from
testes and from a fetal growth curve. The gestation period in fin
whales is approximately 11.5 months. Thus the proportion of mature
females carrying fetuses during the northward migration gives a
measure of the birth rate. The percentage of pregnant females in the
catch of mature females should equal the number of lactating females
in the population sampled. (In fact, a high percentage of lactating
females have been captured at some stations but these are omitted
from consideration here.) Pregnant females and resting females are
equally available for catching (although one possible error here is
that pregnant females may be selected for since they are fatter).
The percentage occurrence of resting females in the mature female
catch can be observed by inspection of landed whales. Thus, drawing
on the data of Table 5-3, the following relationship defines birth
rate from the composition of the mature female stock (following
Chittleborough 1965, p. 117):

$$\text{birth rate} = \frac{\text{no. pregnant}}{\text{no. pregnant} + \text{no. lactating} + \text{no. resting}} .$$

This relationship, calculated from the data of Table 5-9, is
equivalent to an average annual birth rate of 0.389:

Blandford	0.358
Dildo	0.389
Williamsport	0.418

If the average birth rate is 0.389 and the sex ratio of fetuses is 0.5,
then 100 mature females will give birth to 19.45 female calves annually.
The number of these calves that will ultimately be recruited to the
fishable stock at the age of three ear plug laminae (at a length of

TABLE 5-9. Estimation of relative numbers of pregnant, lactating, and
resting females in mature female fin whale population off eastern
Canada. Taking data from Table 5-3, the percentage of pregnant females and resting females in the sample is available. Taking this
percentage of pregnant and resting females as an absolute number in
order to arrive at a ratio between the two, and assuming that the
number of pregnant females in the population equals the number of
lactating females at any one time, the figures below can be taken to
represent the relative numbers of the three reproductive states in
the population.

Station	No. pregnant	No. lactating	No. resting
Blandford	43	43	34
Dildo	58	58	33
Williamsport	59	59	23

50 ft or 15.24 m) is

$$19.45e^{-3M}$$

where M is average natural mortality in the first three age categories
and $e = 2.7183$ (a mathematical constant, base of natural logarithms).

The above differences in birth rates probably represent the result of differing histories of exploitation of the three fishable
stocks. It is interesting that the birth rate is higher at Williamsport, but that the calculated ovulation rate is much lower than that
at Blandford. The ovulation rate is expressed in laminations, and it
is possible that the rate of deposition of laminae is different in the
stock available to Williamsport compared with that available to Blandford.

Population Size

Independent population estimates have been made on the basis of
tag-recapture experiments and direct censuses. The first tag return
estimate was based upon the simple ratio: number of whales tagged/
number of tags recovered = total number in population/number captured.
Of 76 fin whales tagged on the 1966 cruise of the *William S.*, 3 were
undersized and therefore not in the population being hunted in year 0,
so the number effectively marked (n_m) was 73. Up to 10 November 1966,

4 tags had been recovered by the Nova Scotian (Blandford) and Newfoundland (Dildo) whaling stations (n_r). To this same date, the two stations had taken a total of 372 fin whales (n_c). Then,

$$N = \frac{n_m \times n_c}{n_r} = \frac{73 \times 372}{4} ,$$

or N = 6,790 fin whales over 50 ft (15.24 m) in length.

Of the 61 fin whales tagged in western North Atlantic waters in 1967 (Mitchell 1970), 55 were tagged west of Cape Farewell. Only 50 of these were of legal size (50 ft, 15.24 m long or longer).

Of the 76 fin whales tagged in 1966, 4 were recovered from a final catch of 427. The 1967 catch of 748 yielded only 1 tag (of year 0) shot in 1967, but 6 tags (of year 1) shot in 1966. Utilizing data for both years:

$$N = \frac{(50 - 6 + 73 - 4)\ 748}{1 + 6} ,$$

or N = 11,984 fin whales over 50 ft (15.24 m) in length.

This calculation is corrected for 6 fins tagged after 13 August 1967 that were moving south and not available to the fishery in the 1967 season, and for the 4 returns of 1966. (It is the revised estimate of 10,566 cited in Allen 1970.) For a number of reasons, the calculation is high and must be considered the maximum possible estimate of the total fin whale population off eastern Canada.

Eleven tags were returned in the 1968 season. At this time, 22 tags (comprising 16%) had been returned from 137 tagged fin whales. An estimate of population size based on all of these returns was 11,610 fin whales over 50 ft (15.24 m), the legal limit (data mainly from returns in years 1 and 2).

The first strip census was based upon detailed logs of the 1966 tagging cruise of the *William S.* The track of the ship was plotted on appropriate charts, and the number of whales sighted on a given day was recorded. In order to calculate the area searched, the elapsed time run in daylight over a known course (T_e) was divided into the time on watch (T_o). The resulting percentage of time actually on watch ($T_\%$), multiplied by the distance traveled (D_t), gives

the distance in nautical miles over which a whale watch was kept (D_s).
The average visibility, knowing that a whale could be sighted at a
maximum of 4 mi away on the best of days, was doubled (V) and multiplied by the distance searched (D_s), giving the area searched (A_s).
That is:

$$\frac{T_o}{T_e} = T_\%$$

$$D_t \cdot T_\% = D_s$$

$$\text{and } D_s \cdot V = A_s.$$

Since fin whales were not seen far off the continental shelf,
south of Cape Cod, or north of about 57°N on the Labrador coast, calculations have been limited to the area of the continental shelf between
57°N and Cape Cod. Calculated from chart USHO 0955, the area of the
continental shelf in this region was determined to be 386,900 sq mi.
Addition of the daily totals of fin whales seen per unit area searched
reveals 238 fins (N_f) seen in 13,903 sq mi (A_s). (After sighting,
every whale was approached closely for identification and tagging.)
Then the simple ratio of area searched divided by total area equals
number of fins seen divided by total number of fin whales:

$$\frac{A_s}{A_{total}} = \frac{N_f}{X} \text{ , where } X = \text{total number of fin whales,}$$

or $\dfrac{13,903}{386,900} = \dfrac{238}{X}$, $X = 6,620$ fin whales of all sizes.

Since then, the area of the continental shelf has been recalculated
and is taken as 420,910 sq mi, resulting in a revised estimate of
7,205 fin whales. Fin whales comprised approximately 58% of all
whales sighted.

The 1967 cruise of the *Polarstar* searched approximately 18,221
sq mi on the continental shelf between 57°N and Cape Cod, sighting
137 fin whales. The resulting estimate was 3,162 fin whales, which
comprised approximately 57% of all whales sighted.

These and subsequent data have been broken down by region, by
direction of travel, and by time. Details will be presented in a
future report. Present estimates, based on data from the *William S.*

(July to October 1966) and the *Polarstar* (July to October 1967; May to June 1969) cruises, are approximately 340 fin whales in the Gulf of St. Lawrence and approximately 2,800 in the remainder of the Nova Scotian area.

Stock Assessment

In 1966, two independent estimates of fin whale population size (one for legal-sized fins only, the other for all fins) were made. The ratio of effective reproduction to stock size (r) minus the ratio of natural deaths to stock size (M) in an exploited stock can be taken as sustainable catch. As an approximation, figures on sustained yield from Antarctic fin whale data were used (Chapman, Allen, and Holt 1964):

$$r - M \qquad = 0.12,$$
$$0.12 \times 6,790 = 814 \text{ fins},$$
$$0.12 \times 6,620 = 794 \text{ fins}.$$

Both analyses indicated that about 800 fin whales might be killed per year off the entire Canadian east coast during the initial cropping of the whale stock without seriously damaging it. This conclusion was also in accord with Sergeant's (MS, 1966) statement that sustained yield in the northern Newfoundland and Labrador area was approximately 400 fin whales per annum.

In 1967, the estimate from tag-recapture experiments was 11,984, that from strip census data 3,162 fin whales.

The mean of the above estimates of the fin whale population on the continental shelf between Cape Cod and 57°N is approximately 7,200 whales. If we take the value of $r - M$ as 0.08 (Allen 1970), the sustainable yield would be about 560 fin whales for the entire northwest Atlantic population of fin whales at this stock size.

There has been no significant fishery for fin whales in the Nova Scotian region in the past (Tables 5-6 to 5-8), but the northern Newfoundland and Labrador area has been fished episodically since the 1890's. Catches (Table 5-5) averaging over 250 fin whales per year have been landed over long periods, with two periods averaging nearly 500 per year. Picking two time series (1903 to 1907 and 1945 to 1951), it can be shown (see History of Exploitation, above) that the sustainable yield is of the order of 418 fin whales per annum or less if the stock was initially of about the same size as the Nova Scotian stock.

The population of fin whales in the Nova Scotian area may or may not be distinct from the population off northern Newfoundland and Labrador. When the sustainable yield is considered from both of these populations, the total is close to that based upon census and tag-recapture data.

Allen (1971) used catch and effort data from the Blandford fishery for the period 1966 to 1969 to assess the available stock of fin whales. Taking $M = 0.04$, and $r = 0.05$ of the parent stock, Allen calculated the initial stock (Table 5-10) and concluded with an extrapolated estimate of the stock at the beginning of the 1970 season of 484 fin whales. He further concluded on the basis of changes in abundance (Table 5-11) that the effect of recent catching at the

TABLE 5-10. Stock assessment of fin whales available to the Blandford, Nova Scotia, land station during the summer season, 1 June to 30 November in years 1966 through 1969 (from Allen 1971, p. 65).

Year	Catch	Catch/day	Initial stock
1966	263	3.26	1,248
1967	309	2.23	1,035
1968	262	1.93	759
1969	144	1.33	569

TABLE 5-11. Changes in abundance of fin whales from catch per unit of effort data at eastern Canadian land stations, 1968 to 1969 (from Allen 1971, p. 65).

Station	1968	1969	Ratio 1969:1968
Williamsport (catch/hr)	0.29	0.27	0.92
Dildo (catch/day)	1.51	1.39	0.92
Blandford (catch/hr)	0.32	0.20	0.62
Blandford (catch/day)	1.93	1.33	0.69

Blandford land station is quite different from that at the two northern Newfoundland stations, and that this is evidence for two distinct stocks of whales. Allen estimated the stock available off northern Newfoundland by assuming that the 8% reduction in catch per unit effort at Williamsport and Dildo (Table 5-11) was the direct result of catching. Then he calculated S_{68}, the mean stock in 1968, from

$$\frac{S_{68} - \dfrac{481 + 311}{2}}{S_{68}} = 0.92,$$

and concluded that S_{68} was approximately 5,000. Recruitment and natural mortality were ignored in this calculation.

Canadian national quotas for fin whales have been steadily reduced in the light of additional data on the stock size of fin whales (Tables 5-1a to 5-1d).

Balaenoptera borealis Lesson 1828

Three hundred ninety-six landed sei whales have been examined by FRBC personnel at Blandford in the years 1967 through 1970. The mean length of the total catch remained stable (44.7, 43.9, 44.5, and 44.5 ft; or 13.6, 13.4, 13.6, and 13.6 m respectively), as did the mean length of the males and females when considered separately.

The percentage composition of the catch (in terms of female whales) has dropped from 41 to 37 between 1967 and 1970 (e.g., 41%, 41%, 39%, and 37%). This trend may be due to any number of reasons, such as normal fluctuations in the availability of sei whales to the fishery, and changing emphasis in the fishing of sei whales at Blandford at different months within the season.

A plot of length of male sei whales against occurrence throughout the season shows that for the years 1967 through 1970 there has been a relatively consistent availability of sei whales in June and mid-July, and again in mid-September and early to middle October.

The length of female sei whales plotted against time of occurrence throughout the season at Blandford reveals a generally similar pattern to that for males, with some minor differences. The seasonal occurrence is approximately the same, June-July, and September-October.

In both males and females this pattern in the years 1968 to 1970 does not correspond exactly with that in 1967, a year in which some sei whales were taken in mid-season. This may be related to changes in effort.

One of the most striking differences in the data is the complete absence of females in the 1970 catch before 25 August. There was sei whale hunting effort in June and July of 1970, as demonstrated by the catch of a minimum of seven sei whales which were examined by the Fisheries Research Board. Perhaps this absence of females in the early 1970 catch demonstrates nothing more than that the availability of sei whales in these waters is partially controlled by hydrographic and other conditions.

Sei whale ear plugs are difficult to read, and those of the population sampled from Blandford are no exception. Ear plugs have been collected from the Blandford fishery, and attempts have been made to read them.

Although a sample of only nine fetuses is available from all sei whales examined at Blandford, the length of these fetuses plotted against time of occurrence throughout the season indicates a well-defined breeding season for the population being fished. Calving occurs in winter months.

Additional reproductive information is available for 52 females from the fishery between 1966 and 1969, which demonstrates that sei whales have been captured that do not show corpora lutea or corpora albicantia at lengths ranging from 35 ft (10.7 m) to 48 ft (14.6 m). One animal 38 ft long (11.6 m) showed a total of four corpora in both ovaries. Sei whales have been recorded in this fishery with only one corpus at lengths between 42 ft (12.8 m) and 48 ft (14.6 m). The greatest number of whales with one or more corpora (14 whales, or 27% of the sample) occurs at 47 ft (14.3 m). These whales have ovaries showing between 1 and 13 corpora. The greatest number of corpora observed in the ovaries of one whale is 17.

Geographic limits, migration, and ranges are not certainly known for the population being fished. Sei whales have been tagged in waters from the Labrador Sea to northern Venezuela (Table 5-12), but to the

TABLE 5-12. Some cetaceans tagged on Fisheries Research Board cruises in the North Atlantic to April 1971.

Cruise	Fin	Sei (and Bryde's)	Blue	Hump-back	Right	Minke	Sperm	Pilot	Killer	Bottle-nose	Total
William S. (1966)	76	14	2	8	2	2	62	1	2	0	169
Polarstar (1967)	61	1	1	31	0	1	19	0	0	0	114
Polarstar (December 1968 through January 1969)	1	0	0	0	0	0	0	0	0	0	1
Polarstar (February 1969)	0	6	0	33	0	2	9	0	0	0	50
Polarstar (May–June 1969)	34	5	3	4	0	3	1	6	0	2	58
West Whale 8 (March 1971)	7	2	0	0	0	0	10	0	0	0	19
Total	179	28	6	76	2	8	101	7	2	2	411

end of the 1970 season, no tags have been returned in over 396 whales
examined.

Sei whales have been fished off the Labrador coast (Table 5-5)
in late season, August through November (Sergeant, MS, 1966). Presumably sei whales migrate northward, in part through the Blandford
catching field, in June through mid-July and appear on the southern
coast of Newfoundland in August and September (Millais 1906, quoted
by Kellogg 1929) on the way northward. Andrews (1916) suggested that
occurrences on the southern Newfoundland coast were sporadic. Observations from the May-June 1969 research cruise of the *Polarstar* confirm numbers of sei whales in the Labrador Sea and Davis Strait.
Availability of sei whales to Canadian shore stations may be epi-
sodic, and catch records (Tables 5-5 and 5-8) reflect not only this,
but also the relative desirability of the species when and if available. (Fin whales have always been favored over sei on this coast.)

The species is under study, but since there is presently no evidence of depletion of the stock, and because the history of sei whale
fishing is sporadic due to its ancillary position relative to the fin
whale fishery, no recommendation has yet been made for a national
quota.

Physeter catodon Linné 1758

Only males were taken by Canadian shore stations in the period
1945 to 1951 (Table 5-5), and these were from the entire catcher
range mainly in late season (early August to early November). Earlier
records indicate relatively few were taken by American sperm whalers
on the Grand Bank (Townsend 1935) in the nineteenth century.

Because of fluctuations in the price of sperm whale oil and the
problems attendant on mixing mysticete and sperm whale oils in a small
plant, the recent Canadian catch has been small (Table 5-8). As in
past fisheries on the Newfoundland and Labrador coasts (Table 5-5),
the catch is composed exclusively of males. Between 1964 and 1970,
three shore stations captured 39 males, averaging 47.7 ft (14.5 m) in
length. Twenty-five of these, averaging 48.8 ft (14.9 m) were taken

by the Blandford station in late August to mid-October 1970, after the quota of fin whales was caught. The largest whale landed was 59 ft (18.0 m) in length.

One hundred one sperm whales have been tagged in waters from Labrador and West Greenland to the eastern United States and the Caribbean (Table 5-12). To date, none has been returned in the Canadian fishery.

Bull sperm whales migrate along the continental slope and come within catcher range of all Canadian shore stations. Pods have been observed feeding on the slope as far north as Cape Chidley on the Labrador coast (on the 1967 cruise of the *Polarstar*).

There is no national quota set for sperm whales in Canada. The species is presently under study.

Megaptera novaeangliae (Borowski 1781)

The humpback is a migratory whale that frequents both inshore waters and open ocean waters. An early estimate of the numbers of humpbacks in the northwest Atlantic based on cumulative catch data is approximately 1,500 whales (Sergeant, MS, 1966; see Table 5-5). Dispersal is not well documented and requires study before populations can be further identified and assessed. Allen (1970) suggested that there has been an increase in population size from the depleted level resulting from the earlier fishery.

Seventy-six humpbacks have been tagged in the western North Atlantic from the coast of South America, through the Caribbean and waters of eastern United States and Canada to Bermuda and West Greenland (Table 5-12). Only one tag has been returned to date. This tag was implanted 15 August 1967 on the West Greenland coast, and recovered approximately 25 August 1968, approximately 120 nautical miles due north of the position of tagging.

The general migration routes of humpback whales have been previously mapped (Kellogg 1929; Mackintosh 1965; Tomilin 1957). The tracing of particular stocks and the determination of movements of specific herds in the western North Atlantic is not yet possible, mainly due to lack of tag returns and shipboard sightings. The

Fisheries Research Board program has resulted in a number of strip
censuses and other shipboard observations on humpback whales, in addition to tagging in the main areas of concentration.

The significance of tags returned from various catching fields
is of prime importance in the study of humpbacks begun in 1969. The
most information per tag returned can be obtained from whales on the
northward migration, the least on the southward. This is due to the
latitudinal position of the catching fields relative to the areas in
which whales have been tagged. The land stations capable of returning
tags with the most importance from lengthy northward migrations are,
in order: West Greenland, Williamsport, Dildo, and Blandford. For
lengthy southward migrations, only Blandford can return any tag with
three likely points of occurrence, such as a whale tagged off West
Greenland and/or off central Labrador.

One high-priority question is, Do west Atlantic humpback whales
stay on the Canadian coast or cross over to West Greenland? Another
important question relates to the migration rate of humpbacks. The
movements and interrelationships of Caribbean, Bermudan, eastern
United States, Nova Scotian shelf, and Grand Banks humpbacks should
be resolved at least in part before any proposals are put forth regarding population size and sustainable yields from stocks available
to eastern Canadian whaling stations.

A total of 21 humpback whales was captured in 1969 and 1970 from
Blandford, Williamsport, and Dildo land stations. Thirteen (62%) of
these were females. Of the females, 7 (54%) were pregnant, 3 (23%)
were lactating, 1 (8%) was immature, and 2 (15%) were judged to be
resting. The permit under which these whales were taken called for
40 humpback whales over 45 ft (13.7 m) in length. The sample of 21
actually collected ranged from 32 to 47 ft (9.8 to 14.3 m) in length.

Fetuses were collected from the 7 pregnant females. The length
of each fetus was plotted by time of occurrence against days of the
season. These few data points indicate a slightly later reproductive
cycle than is the case with sei and fin whales in the same regions.
Thus birth would be in mid to late winter. This is corroborated by
sightings of newborn calves off Puerto Rico on the February 1969

cruise of the *Polarstar*. Williamson (1961) observed a suckling calf on the Grand Bank on 8 March 1961.

The humpback whale is presently receiving total protection in the North Atlantic Ocean.

Balaenoptera musculus (Linné 1758)

Blue whales were fished extensively in the North Atlantic Ocean in both the nineteenth and twentieth centuries. The northward migration on the western side was fished in the twentieth century from within the Gulf of St. Lawrence (Table 5-7), along the Newfoundland and Labrador coasts (Table 5-5), and pelagically in Davis Strait (Jonsgård 1955a).

Jonsgård stated that the pelagic whaling between 1922 and 1934 was excessive, ultimately affecting catches at Canadian shore stations, and concluded that the same stock was involved and that it was distinct from the northeast Atlantic stock. He further emphasized that a relatively small pelagic catch of 360 blue whales over 12 seasons had a great effect on shore station catches off Newfoundland, Labrador, and Iceland, and that accordingly very few blue whales inhabited these waters.

Sergeant (MS, 1966) picked a time series in the early period of Newfoundland whaling for a cumulative catch estimate of the available stock. He estimated that approximately 1,500 blue whales were killed between 1898 and 1915. Allen (1970) assumed that abundance of fin whale stocks did not change appreciably between 1903 and 1951, and used the ratio of blue whale to fin whale catches as an index of blue whale stock size. He concluded that initial stock size was slightly over 1,100 blue whales and that the maximum sustainable yield was approximately 100 whales.

Since it is likely that all of these whales were on migration to or from Davis Strait, this estimate probably approximates the entire virgin northwest Atlantic stock. Subsequent episodes of fishing (Tables 5-5 and 5-7) slowed recovery of the stock. It may have been depleted more than ever by fisheries through 1951 (Jonsgård 1955a).

The apparent abundance of blue whales as reflected in catches in the
northwest Atlantic is not evidence of an increase in stock size, but
a manifestation of a transitory concentration. Jonsgård concluded
that this situation has subjected the stocks to increased depletion.

As judged from cumulative sightings recorded by whale catchers,
and from estimates based upon strip censuses, the number of blue
whales in the northwest Atlantic is relatively small. At the present
time, the stock can probably be numbered in the very low hundreds, at
most. The species has apparently not recovered as rapidly as has the
humpback whale.

Three blue whales taken by error in the Canadian fishery were
examined by Fisheries Research Board personnel between 1966 and 1969.
No tags were recovered. (Six blue whales were tagged between 1966
and 1969, Table 5-12.) The reduced state of the stock is such that
no scientific sampling has been or will be contemplated that involves
the killing of blue whales for examination in these waters. However,
stranded blue whales and those killed in error will still be examined
for tags and biologically sampled. Only with continued full protection is there hope for recovery of the northwest Atlantic population.

Balaena mysticetus Linné 1758

The former range of the bowhead whale is reflected in the distribution of known whaling grounds throughout the Arctic and northwest
Atlantic (Lubbock 1937; Townsend 1935). The fishery in Davis Strait
and waters to the north was first attempted after 1718 by Dutch whalers,
who alone sent 748 ships (Jenkins 1921). The stocks were greatly depleted by the middle of the nineteenth century, and the fishery ceased
completely when the last whaling ship left Dundee in 1913 (Lubbock
1937).

The species was hunted nearly to extinction. The chief fishery
in Davis Strait had been south of Disco Island, West Greenland
(Jenkins 1921). In 1966, on the cruise of the *William S.* (Mitchell
1968), I discussed sightings of whales with experienced West Greenland
whalers. They had never seen a bowhead whale, but knew all other

species of large and small Cetacea occurring in Davis Strait. The
stock, near the center of this fishery, must have been nearly completely extirpated, but appears to be recovering slightly. There are
reports of this species recently on the West Greenland coast, where
it is said to occur seasonally off Egedesminde and Disco Island (Anon.
1965). The hunting of bowhead whales on this coast (an 8-m long
young whale was killed, Anon. 1965) has led to a polemic in the
Grønlandsposten (Anon. 1968; Øynes 1968).

Bowheads are now completely protected from commercial exploitation. Eskimos are allowed to fish them for food. Approximately 10
are landed annually in the western Arctic (Mansfield 1971) by Eskimos
hunting from skin boats with bomb-lances, and less commonly with
shoulder guns. Johnson, Fiscus, Ostenson, and Barbour (1966) state
that approximately 25% of the whales struck are secured, and estimated
the total annual kill in Alaskan waters by Eskimos at 30 to 40 animals, mainly immature individuals. This annual kill is larger than
the world stock in the estimate presented by McVay (1971). Extra-
limital strays are known (Nishiwaki 1970). On the basis of the observed sustained kill and sightings by hunters and others of up to
177 whales per season in a migration stream (Mansfield 1971), the
western Arctic stock appears to be recovering well and must be
counted in the high hundreds or low thousands.

The geography of the Bering and Chukchi Seas is such that a
great percentage of the western Arctic bowhead population must pass
within sight of active whaling centers such as Point Hope and Point
Barrow, Alaska. The migration is more diffuse in the eastern Arctic,
where there is little active whaling. A bowhead was killed about
1946 in Cumberland Sound (Anderson 1947), and a young animal was taken
in northern Foxe Basin in 1963 (Sergeant 1968). Several were observed
off Southampton Island in 1926 and 1927 (Sutton and Hamilton 1932).
Mansfield (1971) has summarized sightings from land, shipboard, and
aircraft of bowheads in the Canadian Arctic that have accrued during
the last 15 years. He found sightings concentrated in areas of former whaling activity, suggested that the species might winter in
Hudson Bay (see also Sergeant 1968), and stated that pods of from 5

to 20 whales have been commonly observed. Up to 50 bowheads were
counted at one locality in a single season. Population size of the
eastern Arctic stock may be smaller at present due to over 250 years
of overexploitation, or the stock may be recovering as is the western
Arctic stock, but this is difficult to assess because of more widespread distribution in geographically different and remote areas.

Eubalaena glacialis (Borowski 1781)

The North Atlantic right whale formed the basis of an extensive
fishery that resulted in the near extinction of the species by 1880
(Jenkins 1921). In the northwest Atlantic the stock was fished as
early as 1631, but the right whale fishery of the United States coast
had virtually collapsed by 1750 (Allen 1908). Effort then shifted to
Newfoundland waters (Starbuck 1878), while the species was still recorded in small numbers from the United States coast (Allen 1916).
Subsequently, one or two were taken every few years from the eastern
coast of the United States until the early 1900's (Allen 1908).

The species has been considered rare, and has been fully protected for many years in these waters. Although possibly never very
common off Florida (see True 1904; Allen 1908), recent sightings
(Moore 1953; Moore and Clark 1963; Layne 1965) indicate that it is
reoccupying the southern end of its range. Layne suggested that an
actual increase in numbers is indicated by the data. Should this be
the case, it becomes more important to monitor the small humpback
whaling effort from Bequia and neighboring islands in the Grenadines
for the possible occurrence of right whales in the catch (see Coffin
1970, and references in Caldwell, Caldwell, Rathjen, and Sullivan
1971; and Cubillas 1971 for Cuba).

Sightings (Neave and Wright 1968, 1969; and see Schevill 1968),
strandings (Sergeant, Mansfield, and Beck 1970) and an illegal capture (Sergeant, MS, 1966) in Canadian waters indicate a similar,
moderate recovery at the north end of the migration range. The species is regularly seen by whalers out of the Blandford, Nova Scotia,
land station, and two were tagged on the 1966 cruise of the *William S.*

(Mitchell 1968, 1970). Estimates based upon sightings and strip census data indicate a population numbered only in the tens of individuals
in the region off eastern Canada in summer months.

Pelagic whaling in the North Atlantic, such as that announced by
Tønnessen (1970, p. 290) on the part of Norwegian small factory-catcher
vessels could pose a threat to this and other scarce species unless
adequate inspection safeguards are provided for offshore whaling.

Eschrichtius robustus (Lilljeborg 1861)

The California grey whale was nearly exterminated by whalers in
the North Pacific, but has recovered and presently numbers approximately 11,000 (Rice and Wolman 1971). Fraser (1970) has apparently
contributed to the question of the identity of Dudley's (1725) scrag
whale from the northwest Atlantic by demonstrating that Eschrichtius
robustus very probably survived to historic times in the Atlantic and
was known to Icelandic naturalists. The northeast Atlantic and Icelandic stock of grey whales was completely exterminated by commercial
exploitation, according to one possibility outlined by Fraser (1970).
The northwest Atlantic stock may have suffered the same fate before
it was clearly distinguished from other species of whales.

Smaller Cetacea

The discussion above has dealt with the larger species found in
the northwest Atlantic, in approximate order of their present commercial importance. Below some remarks are offered upon smaller species now being fished in the same region, also in descending order of
importance. Canadian and Norwegian whalers have recently fished or
are fishing the combination of minke, bottle-nose, killer, and pilot
whales from small catching vessels. From a beginning on the coast
of Møre in the late 1920's (Jonsgård 1968), the Norwegian small-whale
fishery (Østby 1953; Anon. 1963; Jonsgård 1955b) has expanded across
the North Atlantic to Icelandic and West Greenland waters (Jonsgård
and Christensen 1968). At least three Norwegian ships were active in
the northwest Atlantic along the Canadian coast in the 1969 season,

and the numbers are apparently increasing. The Canadian fishery was
prosecuted for a short time from the Blandford land station in Nova
Scotia, beginning in 1962, but has been most active from the Dildo
land station in the region of Trinity Bay, Newfoundland.

Balaenoptera acutorostrata Lacépède 1804

The present Canadian fishery for minke whales began in 1947 in
Conception, Trinity, and Bonavista Bays on the northeast coast of
Newfoundland. One to three Norwegian-type catching vessels mounting
50-mm cannons have been used (Sergeant 1963). In this operation minke
whales were towed to shore for processing at the Dildo land station,
giving good opportunity for examination and sampling. Other techniques have been used in the Nova Scotian effort, which began in 1962
and which was supplanted beginning in 1964 by a large-scale fishery
for fin whales (Table 5-8).

Sergeant's study (1963) was based upon catch statistics cover-
ing approximately 360 animals, and biological samples from approximately 43 whales caught mainly in June and July. He concluded that
northwest Atlantic whales matured sexually at 24 ft (7.3 m) (females)
or at 22-23 ft (6.7 to 7 m) (males); that mating occurred over several
winter months, with birth in early winter; and that one calf per annum
is produced and lactation is short. Sergeant also judged that the
earliest weaning females migrated north earlier than those calving and
weaning calves in the late winter.

Furthermore, Sergeant compiled literature and new records in
describing the distribution of minke whales from Ungava Bay and Hudson
Strait through Davis Strait to Florida (including the Gulf of Mexico
side). Apparently the northwest Atlantic stock of minke whales summer
along the coast between Ungava Bay and Cape Cod, and winter in waters
south to Florida and offshore. Sergeant concluded from length fre-
quencies that the Newfoundland catch consisted mainly of mature animals, and he reaffirmed Schwartz's (1962) view that immature whales
are distributed more southerly in summer months.

Between 1955 and 1969, approximately 530 minke whales were captured in Newfoundland waters (data from *International Whaling Statistics*)

This average take of 35 whales per annum has had no discernible effect on the stock.

Sergeant (1963) gave no estimate of the size of the minke whale population. Estimation of population size from ship census data involves difficulties not apparent with other mysticete species. The minke whale tends to approach ships, especially stationary vessels, while most other mysticete species tend to avoid or ignore ships at comparable ranges. This "ship-loving" tendency introduces an important bias into the resulting data that is related to the size, type, and acoustic characteristics of the survey vessel, and to other features.

Since 1966, attempts have been made to examine minke whales landed at shore stations in eastern Canada. Eight have been tagged in widely different regions of the western North Atlantic (Table 5-12). To date, no tags have been recovered in the Canadian fishery.

Hyperoodon ampullatus (Forster 1770)

The bottle-nose whale was fished off Nova Scotia from 1962 to 1967. In those six years 87 bottle-nose were taken, mainly in the area of Sable Island and the edge of the Grand Bank. Bottle-nose whales were not tagged until the May-June 1969 cruise of the *Polarstar* (Table 5-12), on which 2 of these whales were also captured and dissected.

Concentrations of bottle-nose whales occur mainly in early summer months near Sable Island, and along the edge of the shelf around Newfoundland and along Labrador. Sergeant and Fisher (1957) mention sightings and captures in Newfoundland waters. Lindsay (1911) described captures from schools off Cape Chidley and Frobisher Bay by one of the Dundee vessels sailing north to hunt right whales after the Newfoundland seal fishery in 1884.

There is a long history of exploitation of this species, and a widely scattered literature on its natural history and early catch statistics. The present state of the stock in the northwest Atlantic is entirely unknown, but numbers appear to be sufficient to support one or two small and localized fisheries.

Globicephala melaena (Traill 1809)

Pilot whales (or "potheads") have been hunted from small vessels
in inshore waters more intensively in eastern Canadian waters (Sergeant
1962) than in waters off the Faeroe Islands (Williamson 1949, 1970;
Degerbøl 1940; Joensen 1962). Natural factors, such as the correlation
between squid catches and high densities of pilot whales (Sergeant and
Fisher 1957), and intensified hunting resulted in a peak catch of
nearly 10,000 whales in 1956. Catches since then have fallen steadily
to a low of approximately 100 annually. This collapse was explained
by Sergeant (1962) as a result of changed hydrographic factors, but it
appears to have resulted instead from overexploitation of a local stock.
Entire schools were driven ashore, giving Sergeant opportunity to study
all age and length classes of both sexes.

Tables in *International Whaling Statistics* and Sergeant and Fisher
(1957) show that a total of 47,078 pilot whales were killed in Newfoundland waters between 1951 and 1961. Between 1953 and 1957 alone,
30,090 of these were taken. Catches dropped in 1962 and 1963, climbed
to almost 3,000 in 1964, but have declined steadily since that time
to the low hundreds. Thus, ignoring recruitment and mortality, the
localized stock fished by the Newfoundland whalers can be roughly
estimated at about 50,000 animals, and it presumably has not recovered
very well from the initial overexploitation.

It is a moot point whether other stocks from Davis Strait, Denmark Strait, and elsewhere will migrate into the rich feeding areas
off Newfoundland before the local stock recovers. In spite of the
decline in numbers, sightings are numerous in the region of Davis
Strait, the Grand Bank, the Nova Scotian shelf, the Laurentian Channel, the Gulf of Maine, and elsewhere.

Orcinus orca (Linné 1758)

The killer whale has not been fished commercially off eastern
Canada, other than small catches ancillary to the fishery for minke
and bottle-nose whales. The *International Whaling Statistics* and
Sergeant and Fisher (1957) list only 7 killer whales taken in Nova

Scotian and Newfoundland waters between 1947 and 1968.

Sightings are common in these waters. Two killers were tagged in 1966 off Cape Farewell on the *William S.* cruise, and another one was captured and dissected on the Canadian coast on the May-June 1969 *Polarstar* cruise. The species is commonly seen on the whaling grounds and according to Sergeant and Fisher (1957) movements may be in part associated with migrations of mysticetes and seals. Jonsgård and Lyshoel (1970) found the distribution and migration of the northeast Atlantic population to be dependent on herring distribution and movement. Migrating trends and population size are unknown for the northwest Atlantic stock.

Delphinapterus leucas (Pallas 1776)

The population of white whales in the St. Lawrence estuary is probably distinct from that in Hudson Bay and is now essentially unexploited (Sergeant and Brodie 1969). Vladykov (1944) studied this stock in an earlier fishery. Catches in some years were in the hundreds, with 1,800 in a good season. An estimate might be in the very low thousands, probably less than 5,000 whales in this population.

Acknowledgments

Detailed acknowledgments will be given in a number of papers being prepared on the population dynamics of North Atlantic whales, but special thanks are in order for Mr. V. M. Kozicki, Mrs. B. Osborne, and Miss B. Mason for technical assistance with this summary manuscript.

REFERENCES

Allen, G. M. 1916. The whalebone whales of New England. Mem. Boston Soc. Nat. Hist., 8:105-322.
Allen, J. A. 1908. The North Atlantic right whale and its near allies. Bull. Amer. Mus. Nat. Hist., 24:277-329.
Allen, K. R. 1970. A note on baleen whale stocks of the north west Atlantic. Rpt. IWC, 20:112-113.

_____ 1971. A preliminary assessment of fin whale stocks off the
     Canadian Atlantic coast. Rpt. IWC, 21:64-66.
Anderson, G. 1947. A whale is killed. Beaver, Outfit 277:18-21
     (March).
Andrews, R. C. 1916. Whale hunting with gun and camera. New York,
     Appleton and Co.
Anonymous. 1963. The small whale fisheries 1962. Norsk Hvalf.-Tid.,
     52:57-70.
_____ 1965. Grønlandshvalen. Inuit, Godthaab, no. 2:4-5 (September).
_____ 1968. Grønlandshval paa bestilling. Grønlandsposten,
     Godthaab, no. 10, p. 3.
Brimley, C. S. 1946. The mammals of North Carolina. Installment
     No. 18. Carolina Tips, Carolina Supply Co., 9:6-7.
Caldwell, D. K., Caldwell, M. C., Rathjen, W. F., and Sullivan, J. R.
     1971. Cetaceans from the Lesser Antillean island of St. Vincent.
     Fish. Bull., 69:303-312.
_____ and Golley, F. B. 1965. Marine mammals from the coast of
     Georgia to Cape Hatteras. J. Elisha Mitchell Sci. Soc., 81:24-32.
Chapman, D. G., Allen, K. R., and Holt, S. J. 1964. Special Com-
     mittee of Three Scientists, Final Report. Rpt. IWC, 14:39-92.
Chittleborough, R. G. 1965. Dynamics of two populations of the hump-
     back whale, *Megaptera novaeangliae* (Borowski). Australian J. Mar.
     Freshwater Res., 16:33-128.
Coffin, P. 1970. The beautiful barkeep of Bequia. Look magazine,
     15 December 1970, pp. 56-61.
Cubillas, V. 1971. Una captura insólita . . . en Cuba. Mar y Pesca,
     No. 65 (febrero):32-37.
Degerbøl, M. 1940. Mammalia. København, pt. 2, vol. 3, The zoology
     of the Faroes.
Dudley, P. 1725. An essay upon the natural history of whales with
     a particular account of the ambergris found in the *Sperma Ceti*
     whale. Phil. Trans. Roy. Soc. London, 33 (387):256-269.
Fraser, F. C. 1970. An early 17th century record of the Californian
     grey whale in Icelandic waters. Invest. on Cetacea (Pilleri),
     2:13-20.
Gunter, G. 1954. Mammals of the Gulf of Mexico. U.S. Fish. Bull.,
     55:543-551.
International Whaling Statistics. Oslo, Grøndahl and Son. Nos. 1
     (1930) to 66 (1970).
Jenkins, J. T. 1921. A history of the whale fisheries from the
     Basque fisheries of the tenth century to the hunting of the
     finner whale at the present date. London, Witherby.
Joensen, J. S. 1962. Grindadráp i Føroyum 1940-1962. [Pilot whales
     (*Globicephalus melaena* Traill) killed in Faroe (1940-1962).]
     Fróðskaparrit (Annal. societ. scient. Faeroensis), 11:34-44.
Johnson, M. L., Fiscus, C. H., Ostenson, B. T., and Barbour, M. L.
     1966. Marine mammals. *In* N. J. Wilimovsky and J. N. Wolfe,
     eds., Environment of the Cape Thompson region, Alaska. Oak
     Ridge, Tennessee, U.S. Atomic Energy Comm. Pp. 877-924.
Jonsgård, Å. 1955a. The stocks of blue whales (*Balaenoptera musculus*)
     in the Northern Atlantic Ocean and adjacent Arctic waters. Norsk
     Hvalf.-Tid., 44:505-519.

_____ 1955b. Development of the modern Norwegian small whale industry. Norsk Hvalf.-Tid., 44:698-718.

_____ 1968. A review of Norwegian biological research on whales in
the northern North Atlantic Ocean after the Second World War.
Norsk Hvalf.-Tid., 57:164-167.

_____ and Christensen, I. 1968. A preliminary report on the
"Harøybuen" cruise in 1968. Norsk Hvalf.-Tid., 57:174-175.

_____ and Lyshoel, P. B. 1970. A contribution to the knowledge of
the biology of the killer whale *Orcinus orca* (L.). Nytt Magasin
for Zoologi, 18:41-48.

Kellogg, R. 1929. What is known of the migrations of some of the
whalebone whales. Ann. Rept. Smithsonian Inst., 1928 (publ.
2981), 467-494.

Layne, J. N. 1965. Observations on marine mammals in Florida waters.
Bull. Florida State Mus., 9:131-181.

Lindsay, D. M. 1911. A voyage to the Arctic in the whaler *Aurora*.
Boston, Dana Estes and Co.

Lubbock, B. 1937. The Arctic whalers. Glasgow, Brown, Son and
Ferguson, Ltd.

Lucas, F. A. 1908. The passing of the whale. New York Zool. Soc.
Bull., suppl. no. 30:445-448.

Mackintosh, N. A. 1965. The stocks of whales. London, Fishing News
(Books) Ltd.

McVay, S. 1971. Can leviathan endure so wide a chase? Nat. Hist.,
70:36-41, 68-72.

Mansfield, A. W. 1971. Occurrence of the bowhead or Greenland right
whale (*Balaena mysticetus*) in Canadian Arctic waters. J. Fish.
Res. Bd. Canada, 28:1873-1875.

Millais, J. G. 1906. The mammals of Great Britain and Ireland.
London, Longman's, Green and Co. Vol. 3.

_____ 1907. Newfoundland and its untrodden ways. New York,
Longman's, Green and Co.

Mitchell, E. 1968. North Atlantic whale research. Fish. Council
Canada, Ann. Rev. 1968:45, 47-48.

_____ 1970. Request for information on tagged whales in the North
Atlantic. J. Mammal., 51:378-381.

Moore, J. C. 1953. Distribution of marine mammals to Florida
waters. Amer. Midland Nat., 49:117-158.

_____ and Clark, E. 1963. Discovery of right whales in the Gulf
of Mexico. Science, 141:269.

Neave, D. J., and Wright, B. S. 1968. Seasonal migrations of the
harbor porpoise (*Phocoena phocoena*) and other Cetacea in the
Bay of Fundy. J. Mammal., 49:259-264.

_____ _____ 1969. Observations of *Phocoena phocoena* in the Bay of
Fundy. J. Mammal., 50:653-654.

Nishiwaki, M. 1970. A Greenland right whale caught at Osaka Bay.
Sci. Repts. Whales Res. Inst., 22:45-62.

Østby, H. 1953. The Norwegian small whale hunting. Norsk Hvalf.-
Tid., 42:698-718.

Øynes, P. 1968. "Grønlandshval paa bestilling." Grønlandsposten,
Godthaab, 1968, no. 11:11.

Rice, D. W., and Wolman, A. A.   1971.   The life history and ecology
    of the gray whale (*Eschrichtius robustus*).   Amer. Soc. Mammal.,
    spec. pub. no. 3.
Roe, H. S. J.   1967.   Seasonal formation of laminae in the ear plug
    of the fin whale.   Discovery Rpt. 35:1-30.
Schevill, W. E.   1968.   Sight records of *Phocoena phocoena* and of
    cetaceans in general.   J. Mammal., 49:794-796.
Schwartz, F. J.   1962.   Summer occurrence of an immature little piked
    whale, *Balaenoptera acutorostrata*, in Chesapeake Bay, Maryland.
    Chesapeake Science, 3:206-209.
Sergeant, D. E.   1953.   Whaling in Newfoundland and Labrador waters.
    Norsk Hvalf.-Tid., 42:687-695.
_____   1962.   The biology of the pilot or pothead whale *Globicephala
    melaena* (Traill) in Newfoundland waters.   Bull. Fish. Res. Bd.
    Canada, 132.
_____   1963.   Minke whales, *Balaenoptera acutorostrata* Lacépède, of
    the western North Atlantic.   J. Fish. Res. Bd. Canada, 20:1489-1504.
_____   MS, 1966.   Populations of large whale species in the western
    North Atlantic with special reference to the fin whale.   Fish.
    Res. Bd. Canada, circ. no. 9.
_____   1968.   Whales.   *In* Science, History and Hudson Bay.   Ottawa,
    Dept. of Energy, Mines and Resources.   Chap. 7, pt. 5, pp. 388-396.
_____   and Brodie, P. F.   1969.   Body size in white whales *Delphinapterus
    leucas*.   J. Fish. Res. Bd. Canada, 26:2561-2580.
_____   and Fisher, H. D.   1957.   The smaller Cetacea of eastern
    Canadian waters.   J. Fish. Res. Bd. Canada, 14:83-115.
_____   Mansfield, A. W., and Beck, B.   1970.   Inshore records of
    Cetacea for eastern Canada, 1949-68.   J. Fish. Res. Bd. Canada,
    27:1903-1915.
Starbuck, A.   1878.   History of the American whale fishery from its
    earliest inception to the year 1876.   Rept. U.S. Commissioner
    Fish and Fisheries, pt. 4, 1875-1876.
Sutton, G. M., and Hamilton, W. J., Jr.   1932.   The mammals of South-
    ampton Island.   Carnegie Mus. Mem., 12:1-111.
Tomilin, A. G.   1957.   Zveri SSSR i prilezhashchikh stran [Mammals
    of the USSR and adjacent countries].   Moskva, Izdatel'stvo
    Akademi Nauk SSSR.   Vol. 9, Kitoobraznye [Cetacea].
Tønnessen, Joh. N.   1967.   Den moderne hvalfangsts historie.
    Sandefjord, Norges Hvalfangstforbund.   Vol. 2.
_____   1970.   Norwegian Antarctic whaling, 1905-68:   an historical
    appraisal.   Polar Record, 15:283-290.
Townsend, C. H.   1935.   The distribution of certain whales as shown
    by logbook records of American whaleships.   Zoologica (N.Y.),
    19:1-50.
True, F. W.   1904.   The whalebone whales of the western North Atlantic.
    Smithsonian Contrib. Knowledge, 33:1-332.
Vladykov, V. D.   1944.   Chasse, biologie et valeur économique du
    marsouin blanc ou béluga (*Delphinapterus leucas*) du fleuve et
    du golfe Saint-Laurent.   (Études sur les mammifères aquatiques,
    3.)   Québec, Dept. Pêcheries.

Williamson, G. R. 1961. Winter sighting of a humpback suckling its
    calf on the Grand Bank of Newfoundland. Norsk Hvalf.-Tid.,
    50:335-336, 339-341.
Williamson, K. 1949. Notes on the caaing whale. Scottish Naturalist,
    61:68-72.
_____ 1970. The Atlantic Islands, A study of the Faeroe life and
    scene. Ed. 1, 1948. London, Routledge and Kegan Paul.

CHAPTER 6

WHALES AND WHALE RESEARCH IN THE EASTERN NORTH PACIFIC

Dale W. Rice

In an earlier review of Pacific coast whaling and whale research
(Rice 1963), I wrote that "the southern hemisphere stocks of baleen
whales have continued to decline at an accelerating rate. . . . Fleets
of factory ships and catcher boats represent a capital investment of
millions of dollars. If they can no longer operate at a profit in the
Antarctic, some of them may turn to the North Pacific."

From 1954 through 1961, the number of floating factories operating
in the North Pacific remained at three, and the number of whales killed
—including those taken by 24 to 30 shore stations—varied between
8,000 and 13,000. By 1963 the number of floating factories had increased to seven. The number of whales killed increased each year,
reaching a peak of 24,150 in 1968. By 1967, the North Pacific had sur-
passed the Antarctic as the world's major whaling ground, and during
the two seasons 1969 and 1970 the North Pacific catches have been
almost double those of the Antarctic.

The most recent phase of whaling in the United States began in
1956 when the Del Monte Fishing Company established a shore station
on San Francisco Bay at Point San Pablo, in Richmond, California. In
1958 a second station was established on the point by the Golden Gate
Fishing Company. The latter closed in 1966. The catch at these stations (Table 6-1) has consisted mainly of fin, sei, humpback, and
sperm whales, along with a few blue, giant bottlenose, and killer
whales. The highest annual catch was 338. A company in Astoria,
Oregon, took 13 whales between 1961 and 1965. The United States
catch has constituted only 0.4 to 3.2% of the total North Pacific
catch.

TABLE 6-1. Number of whales taken by California shore stations, 1956-1970.

Season	Shore stations	Catcher-boats	Number of whales						Total
			Blue	Fin	Sei	Hump-back	Sperm	Giant bottlenose	
1956	1	2		3		133	9		145
1957	1	3		22	1	199	14	1	237
1958	2	4	26	109	2	115	8	1	261
1959	2	5	5	108	37	140	17	2	309
1960	2	5	1	138	47	67	16	2	271
1961	2	5	2	118	51	62	101	4	338
1962	2	5	2	123	22	39	60		247[a]
1963	2	5	6	17	96	55	77	1	253[a]
1964	2	5	2	148	13	27	63	1	254
1965	2	5	4	114	22	4	97	2	243
1966	2	5		42	62		69	1	175[a]
1967	1	3		44	3		101		150[b]
1968	1	3		38	14		84		136
1969	1	3		31	10		68		109
1970	1	3		5	4		64		73

[a] Includes one killer whale.
[b] Includes two killer whales.

Review of Research

History and Methods

In 1958 a research program on the larger species of whales was undertaken by the Marine Mammal Biological Laboratory of the former Bureau of Commercial Fisheries (since October 1970, the National Marine Fisheries Service). The objectives of this program (Rice 1963) were to determine the basic features of the life history and ecology of each species, to define the geographical distribution and migrations of each stock, and to ascertain the size and sustainable yields of each stock. Research methods have included (1) examination of the whales

brought into the shore stations; (2) marking of whales from chartered
catcher boats; (3) observations of living whales from vessels and from
shore; and (4) analysis of catch statistics.

Beginning in 1959, a biologist has been on duty at the whaling
stations throughout most of each season. In 12 years 2,191 whales have
been examined (Table 6-2). This number includes 73% of the whales taken
during the regular whaling season, plus 316 gray whales taken in winter
under special scientific permits.

TABLE 6-2. Number of whales examined at California shore stations,
1959-1970.

Species	Males	Females	Total
Blue	7	13	20
Fin	325	357	682
Sei	101	183	284
Humpback	129	105	234
Gray	166	150	316
Sperm	392	245	637
Giant bottlenose	13	0	13
Killer	4	1	5

At the whaling stations, data routinely recorded for each whale
include body length, sex, date and locality of capture, blubber thickness, condition and depth of mammary gland, diameter of uterine cornua,
sex and length of fetus, weight of testes, species and quantity of food
in stomach, species and numbers of ectoparasites and endoparasites,
and degree of fusion of vertebral epiphyses. Material collected for
further analysis in the laboratory includes the ovaries, the ear plugs
of baleen whales and the teeth of sperm whales (for age determination),
histological specimens of the testes, uterus, and mammary glands, samples of the stomach contents and parasites, and pathological specimens.
A detailed description of procedures has been published by Rice and
Wolman (1971).

A whale marking program was instituted in 1962 (Table 6-3). We
have done most of our marking during the winter in the area between

TABLE 6-3. Number of whales effectively marked, and number of marked whales recovered, 1962-1970.

Species	Number marked	Number recovered	Remarks
Blue	76	0	Protected since 1966
Fin	56	8(14%)	
Sei	10	2(20%)	
Bryde's	19	0	Unexploited stock
Humpback	44	0	Protected since 1966
Gray	5	0	Protected
Sperm	176	3(2%)	
Total	386	13	

San Francisco (38°N latitude) and the Islas Revillagigedo (18°N latitude), so that we can determine which breeding stocks contribute to the populations that are exploited on the various summer grounds.

All observations of whales—as well as other marine mammals—made during the marking cruises have been recorded in detail: species, date and time, position, number in group, and any other pertinent information such as direction of travel, behavior, feeding, and presence of calves. The approximately 5,000 sightings of about 100,000 individuals provide considerable data on the distribution and relative abundance of 26 species of cetaceans and 9 species of pinnipeds. Since the taking of blue and humpback whales was banned, the catcher-boat captains have been logging all sightings of these species during whaling operations; these records, if maintained through the years, will provide information on population trends in these species. Counts of migrating gray whales are regularly made from shore, and migrating bowhead whales have been counted from the ice edge. Aircraft have also been used to a limited extent to observe the latter two species.

Statistics of the whaling industry are the most detailed available for any fishery. The species, sex, body length, and date and position of capture are on record for virtually all of the whales taken in postwar years. Detailed effort data are also on record. Except

for catches off Baja California, the prewar statistics for modern-style
whaling in the eastern North Pacific are not as complete, but records
are available of the total number of whales of each species taken each
year by most stations. The major gap in the record is in the Californi
statistics for the years 1930 to 1936, which give annual totals, but
do not record the species. For the old-style American fishery for
sperm and right whales, from 1784 to 1925, the area of operation and
the production of sperm oil, whale oil, and whalebone for each voyage
have been recorded (Starbuck 1878; Hegarty 1959). Townsend (1935)
examined the logbooks from 1,665 old-style whaling voyages and prepared a classic series of charts on which were plotted, by month, the
positions where 53,877 sperm, right, humpback, and bowhead whales were
killed. Maury (1851) had earlier prepared a similar but less detailed
chart for sperm and right whales. These long series of detailed statistical data, when analyzed in the light of our biological knowledge
of whales, provide an unusually detailed picture of past and current
trends in whale populations.

Since whales wander widely over the high seas, and each stock may
be exploited in several parts of its range and by more than one nation,
research objectives cannot be fully met by a study confined to one portion of the range of a stock. In 1962, just prior to the rapid expansion of the North Pacific whale fishery, at the urging of the late
Remington Kellogg, then United States Commissioner to the International
Whaling Commission (IWC), the Scientific Committee of the IWC appointed
a North Pacific Working Group consisting of biologists and mathematicia
from the four nations then engaged in whaling in the North Pacific—
Canada, Japan, the Soviet Union, and the United States. This group was
asked to coordinate national research efforts, to exchange data, and
to determine the sustainable yields of each stock of whales.

Data on catch, effort, and length-frequency distribution for the
postwar years were exchanged in 1964, and data for subsequent seasons
have been exchanged annually. Catch data include the number of whales
of each species and sex taken during each month in each 10-degree
square of latitude and longitude. Effort data include the horsepower
and tonnage of each catcher-boat and the number of days that it hunted

in each 10-degree square each month. Length-frequency data for each species and sex are reported for each 20-degree zone of longitude. Data on the number of whales effectively marked each year in each 10-degree square, and full particulars on all recoveries of marked whales, are exchanged. (A whale is considered "effectively" marked if the mark is entirely imbedded in it.) Certain other tabulated biological data have been made available by each country. These data have included age/length keys based on counts of growth layers in ear plugs or teeth, and on counts of corpora albicantia in the ovaries.

This exchange of data has enabled mathematicians from each country to make independent estimates of population sizes and sustainable yields of each stock of whales.

Accomplishments

The rorquals (blue, fin, sei, and humpback whales) received priority in our research efforts in the early 1960's because they are more vulnerable to overexploitation than are sperm whales. We have in preparation a comprehensive report on the life history and ecology of the rorquals in the eastern North Pacific. The basic features of the biology of the rorquals—especially fin and humpback whales—throughout their ranges are fairly well known, and population assessment methods are well established. The North Pacific Working Group's assessments convinced the IWC of the need for extending complete protection to blue and humpback whales, beginning in 1966. The group's estimates of the sustainable yields of fin and sei whales led to the imposition of catch limits on these two species for the first time in 1969.

Sperm whales, the most important species in the North Pacific fishery, are more abundant and far more widely distributed than the baleen whales, but they present more complex management problems. Because of their polygyny, marked sexual dimorphism in size, and partial geographical and social segregation of the sexes, sperm whale catches consist of about 75% males, almost all adult; only a few of the older females are killed. Until recently, knowledge of the reproductive cycle and social behavior of the sperm whale was sketchy. The Scientific Committee of the IWC recommended that special scientific

permits be issued for taking sperm whales under the legal length limit,
so that samples of females and younger males could be studied. Since
1966 the United States has taken 189 sperm whales under special scientific permits. A preliminary report on our sperm whale studies was
submitted to the special meeting on sperm whale biology convened by
the IWC Scientific Committee in March 1970, and a full report will soon
be published. Intensive sperm whale research has also been conducted
in many other parts of the world during the past 20 years. Our understanding of the biology of this species is now reaching the point where
we should soon be able to prescribe rational management regulations.

Studies on protected species have also been undertaken. A mono-
graph on the life history and ecology of the gray whale was recently
published (Rice and Wolman 1971). Soviet biologists have also studied
the reproductive cycle and parasites of gray whales taken by natives
in the Bering and Chukchi Seas (for example, Zimushko 1969). I spent
April and May of 1961 and 1962 in Eskimo whaling camps near Point Hope
and Barrow, Alaska, to study bowhead whales. Because of problems of
logistics and communications, and the small number of bowheads taken,
further expenditure of our limited time and money did not seem justi-
fied. The bowhead remains the least-studied species of large cetacean.

Status of Species

The following accounts include all nine of the species of baleen
whales that occur in the eastern North Pacific, plus the two largest
species of toothed whales. I have briefly summarized what is known
of the distribution and movement of each species, and its population
size and trends. Most species of baleen whales appear to have a tendency to migrate more in coastal than in mid-ocean waters, so the concept of more or less discrete eastern and western North Pacific breeding stocks is reasonable pending further results from marking programs,
blood-type studies, etc. Sperm whales range widely over the oceans,
and separate stock units cannot be discerned until more marked animals
are recovered.

Blue Whale (Balaenoptera musculus)

Distribution. One of the world's last remaining sizable stocks

of blue whales congregates from February to early July each year along
the west coast of Baja California. The presence of blue whales in
these waters was known to nineteenth-century whalers, who rarely tried
to capture them because they were too swift to pursue and kill with
open boats and hand harpoons or bomb-lances. One of the few who tried
and succeeded was Captain Scammon (1874), who wrote:

> Several days trial were made in the brig *Boston*, in 1858,
> off Cerros [= Cedros] Island, to capture these animals.
> It was in the month of July, and the sea, as far as the
> eye could discern, was marked by their huge forms and
> towering spouts. . .
>
> On a second voyage of the *Page*, six of these immense
> creatures were taken by the bomb-gun and lance, off the
> port of San Quentin, Lower California, where the moderate
> depth of water was favorable for their pursuit. Large
> numbers of them were found on this ground, where they
> had been attracted by the swarms of sardines and prawns. . .

The whales in this area were first exploited by whalers using
modern methods in the winter of 1913 and 1914, when three catcher-boats
operating from the floating factory *Capella I* killed 83 of them (Tønnessen 1967). Ingebritsen (1929) wrote:

> In 1913, I carried on whaling operations from Magdalena
> Bay, Lower California. At the end of October the blue
> whale came from the north and proceeded southward along
> the shore. Then in April, May, and June it came north-
> ward again.

Each winter and spring season from 1924/25 to 1928/29, and again
in early 1935, one or two floating factories operated mostly from Bahía
Magdalena, but also from other points along the west coast of Mexico.
Between Cabo San Lucas and Isla Cedros, Baja California, these expeditions took 47 to 239 blue whales each season, and a total of 989. Since
then blue whales have not been exploited off Baja California.

During our whale marking cruises we found blue whales along the
entire west coast of Baja California, from 23°33'N latitude to 32°05'N

latitude. We also saw two east of Isla Cerralvo on the west side of
the mouth of the Gulf of California on 8 February 1966, and on moving
north midway between Cabo San Lucas and the Islas Tres Marías on 16
February 1967. All of the blue whales that we observed were less than
80 km from shore, some only 3 km. Many were in shallow water between
50 m and 200 m deep.

Small numbers of blue whales were taken off Baja California in
October, but catches were practically nil from November through January. The catch statistics and our observations show that the major
influx of blue whales begins in late February, the greatest numbers
are present in April, and that departure is nearly complete by early
July.

The whereabouts of these blue whales during the remainder of the
year, and their relation to populations on whaling grounds farther
north, is problematical. During the 1965 whale marking cruises, we
effectively marked 49 blue whales—26 in February and March, and 23
in June. None of these marks was recovered during the 1965 whaling
season. These negative results are inconclusive. Killing of blue
whales has been banned in subsequent seasons.

Although the blue whales are gone from Baja California waters by
late June or early July, they usually do not show up off central California until late September. If they head north upon leaving Baja
California, they must pass central California far offshore. In May
1963 Berzin and Rovnin (1966) observed blue whales migrating north at
41° to 42°N latitude, 130°W longitude, off the coast of northern California. Catches off Vancouver Island show two peaks of abundance, in
June and September.

In more northerly waters, pelagic whaling has revealed three
major summer concentration areas for blue whales: (1) the eastern
Gulf of Alaska, from 130° to 140°W longitude; (2) the area south of
the eastern Aleutians, from 160° to 180°W longitude; and (3) the area
from the far western Aleutians to Kamchatka, 170° to 160°E longitude.
Catches in the eastern Aleutian area show a peak in June, whereas
those in the Gulf of Alaska show a peak in July. This difference
tends to support Berzin and Rovnin's (1966) hypothesis that some of

the northbound migrants turn west at about 50°N latitude and proceed
directly to the eastern Aleutians, while others continue north along
the coast into the Gulf of Alaska. Out of 15 Japanese and Soviet
recoveries of blue whales marked on the summer grounds, five demonstrated movement between the three major areas noted above; one whale
even moved from the eastern Okhotsk Sea to the Gulf of Alaska (Ivashin
and Rovnin 1967; Anon. 1967). The remaining 10 whales were recovered
in the area where they were marked.

From the above facts I postulate that the majority of this stock
of blue whales leaves Baja California waters in May and heads north,
passes central California far offshore, and arrives on the whaling
grounds off Vancouver Island in June. From there at least some must
proceed to the eastern Aleutians or into the Gulf of Alaska. They
leave the latter areas in August and pass Vancouver Island in September, central California in late September and October, and Baja California in October. There are no data on their movements from November
through January, but they must be either farther offshore or farther
south. Our observation of a northbound animal north of the Islas Tres
Marías in February gives a little support to the second alternative.

Population. The catch of blue whales off Baja California from
1924/25 to 1928/29 averaged 188 per year. The catch per unit of effort (gross number of catcher's day's work) showed no downward trend.
During the same five-year period, blue whale catches in California,
British Columbia, and Alaska averaged 101 per year. Annual catches
fluctuated but showed no downward trend; in fact, an upward trend is
suggested, even though effort remained about constant. During this
period the population of blue whales in the eastern North Pacific as
a whole apparently sustained an average kill of 289 animals per year.
To do this would require a total population of about 6,000 animals,
since the recruitment can hardly have been greater than 0.05. Aside
from this five-year period in the late 1920's no single year's catch
has ever exceeded 271, except for a catch of 440 in 1963. In 1964
Doi, Nemoto, and Ohsumi (1967) calculated that the summer blue whale
populations in the three main pelagic whaling areas had dropped to
about 1,420 from a postwar initial stock of about 2,430. This estimate

did not include populations east of 140°W longitude.

The above data suggest that blue whales were never very abundant
in the eastern North Pacific, and their population size has not decreased
very markedly.

Off Baja California I have seen blue whales among—and possibly
feeding on—shoals of pelagic red crabs, *Pleuroncodes planipes*, which
are often abundant in the inshore waters. Although there is not a very
marked seasonal variation in the abundance of these crabs, the presence
of the blue whales does coincide with the period of greatest crab abundance (Longhurst 1967).

Since this seasonal aggregation of blue whales takes place close
to the overpopulated megalopolis of southern California, it offers an
opportunity for commercial whale-watching cruises. Cruises to observe
gray whales and seal rookeries are already a rapidly growing business
in this area. It is probably the only place in the world where the
average citizen has a fair chance of seeing a live blue whale with a
reasonable expenditure of time and money.

Fin Whale (Balaenoptera physalus)

Distribution. The winter grounds of fin whales in the southeastern
North Pacific, as elsewhere in the world, remain very poorly known. We
have found fin whales from 35°30'N latitude off the Big Sur coast of
California south to 22°50'N latitude west of Cabo San Lucas, Baja California. The only area where they were encountered repeatedly in any
numbers was west of the Channel Islands off southern California. Although we saw none far offshore, many must spend the winter out at
sea, because the number observed in the immediate offshore waters is
insufficient to account for the entire eastern North Pacific population. On 20 May 1966 Kenneth C. Balcomb (personal communication)
observed 8 to 12 fin whales in the mid-Pacific at 17°54'N latitude,
158°48'W longitude, and Berzin and Rovnin (1966) reported some at 37°N
latitude, 138°W longitude in February 1964. In the summer fin whales
range in the immediate offshore waters around the North Pacific, from
Chukchi Sea southward along the Asian side to Japan, and along the
American side to southern California and sometimes central Baja California.

Eight fin whales that we marked in the winter (November to January) off southern California were recovered in the summer (May to July) off central California (one), Oregon (four), British Columbia (one), and in the Gulf of Alaska (two).

Off central California the pattern of seasonal changes in the abundance of fin whales varies from year to year, but a general pattern is evident. Numbers reach a peak in late May or early June, after which they fall off. There is usually a second influx later in the summer, which is more prolonged and more variable in its timing than the early peak.

Population. During their first two years of operation (1956 and 1957), the California catcher-boats did not go far enough offshore to find many fin whales, because humpback whales were so abundant closer inshore. By 1958 the increasing scarcity of the latter species forced the whalers to turn their attention to fin whales. Although the catch fluctuated considerably from year to year, there was an overall downward trend in catch per gross catcher's day's work from about 0.14 in 1958 to 1960 to about 0.06 in 1968 to 1970, a reduction of about 57% (Table 6-4). This figure agrees with Ohsumi, Shimadzu, and Doi's (1971) estimate that the entire eastern North Pacific stock of fin whales decreased by 55% during this period, from about 20,000 to 9,000 recruited animals. This stock is well below the level of maximum sustainable yield.

Sei Whale (Balaenoptera borealis)

Distribution. The winter distribution of sei whales is even less known than that of fin whales. We found sei whales widely but sparsely scattered from $35°30'N$ latitude off Point Piedras Blancas, California, south to $18°30'N$, 600 km offshore in the vicinity of the Islas Revillagigedo. Nowhere did we find them regularly or in any numbers. Perhaps the majority spend the winter in far offshore waters.

The summer distribution of sei whales is similar to that of fin whales, except that they rarely go north of the Aleutian Islands. Off our coast they range south to the area west of the California Channel Islands, and we saw one at $27°13'N$, off Baja California.

TABLE 6-4. Baleen whale catch (C) and catch per unit of effort (C/E) at California shore stations, 1956-1970.

Year	Gross number of catcher's day's work	Blue		Fin		Sei		Humpback	
		C	C/E	C	C/E	C	C/E	C	C/E
1956	317		0	3	0.010		0	133	0.420
1957	505		0	22	0.044	1	0.002	199	0.394
1958	703	26	0.037	109	0.155	2	0.003	115	0.164
1959	920	5	0.005	108	0.116	37	0.040	140	0.152
1960	915	1	0.001	138	0.151	47	0.051	67	0.073
1961	915	2	0.002	118	0.129	51	0.056	62	0.068
1962	915	2	0.002	123	0.134	22	0.024	39	0.043
1963	915	6	0.007	17	0.019	96	0.105	55	0.060
1964	915	2	0.002	148	0.164	13	0.014	27	0.030
1965	871	4	0.005	114	0.131	22	0.025	4	0.005
1966	797			42	0.053	62	0.078		
1967	549			44	0.080	3	0.005		
1968	421			38	0.090	14	0.033		
1969	532			31	0.058	10	0.019		
1970	368			5	0.014	4	0.011		

Off central California sei whales are usually present only during the late summer and early autumn. In some years a few arrive in late May or June, but in most years they do not show up until early July.

One sei whale that we marked off southern California in November 1962 was killed off Vancouver Island in August 1966. Another, marked in the same general area in June 1965, was killed about 1200 km off the Washington coast in July 1969.

Population. As with fin whales, the California whalers made no real effort to take sei whales until 1949, after the humpbacks had been depleted. Since that year the sei whale catches have fluctuated more than the fin whale catches, but there has been a similar overall downward trend in catch per unit of effort from 0.05 in 1959 to 1961 to 0.02 in 1968 to 1970, a reduction of 60% (Table 6-4). Ohsumi, Shimadzu, and Doi (1971) estimated that the eastern North Pacific sei

whale stock decreased from roughly 40,000 to 28,000 recruited animals
during this period, but that the stock was still at or above the level
of maximum sustainable yield.

Seven percent of the sei whales taken off central California have
been infected with a unique disease that results in the progressive
shedding of the baleen plates and their replacement by an abnormal
papilloma-like growth. In one case all baleen plates were missing.
The epidemiology and etiology of this disease are unknown, but the
collecting of two afflicted whales from one pod suggests that it is
contagious. Histological studies revealed the presence of tiny granu-
lar structures that resemble bedsonia (*Chlamydia* sp.). Attempts by
the Naval Biological Laboratory in Oakland to isolate the causative
agent were unsuccessful. The loss of such a highly specialized feeding apparatus as the baleen plates would appear to make it impossible
for a whale to feed. Yet none of the diseased animals appeared emaci-
ated and most had anchovies (*Engraulis mordax*), sauries (*Cololabis
saira*), or jack mackerel (*Trachurus symmetricus*) in their stomachs;
none had euphausids or copepods. Healthy sei whales feed on crustaceans as well as fish. The ultimate effects of this disease are unknown. Because of its high incidence and severity, it might cause
significant mortality. Strangely enough, this disease has never been
found in sei whales elsewhere in the world, nor in any other species
of whale, with the possible exception of an Antarctic fin whale (Tomilin and Smyshlyaev 1968).

Sei whales are also much more prone to heavy infestations of
parasitic helminths than are other baleen whales. The liver fluke
Lecithodesmus spinosus causes hardening of the tip of the liver. The
stomach worm *Anisakis simplex*, although usually abundant, is not normally pathogenic. In one case, however, many worms had invaded the
liver, which was undergoing pathological degeneration. Perhaps both
these helminths sometimes kill the host.

Bryde's Whale (Balaenoptera edeni)

Distribution. On the west coast of Baja California, Bryde's
whales inhabit the inshore waters from 26°12'N latitude south to Cabo

San Lucas. They also range all across the southern end of the Gulf of California north, at least as far as we went, to $20°40'N$ latitude on the western side and Mazatlán, Sinaloa, on the eastern side. Robert L. Brownell (personal communication) found a stranded specimen in the northern gulf. We saw a few as far south as the Islas Tres Marías.

Bryde's whales appear to be year-round residents in these waters, as we encountered them on two cruises in June and September as well as during the winter and early spring cruises.

Population. Until recent years, Bryde's whales were not distinguished from sei whales in the International Whaling Statistics. The floating factories that operated on the west coast of Mexico in 1913/14, 1924/25 to 1928/29, and 1935 reported taking 121 "sei" whales between Bahía San Juanico, Baja California, and the Islas Tres Marías, Nayarit. Because sei whales are scarce in this area, whereas Bryde's whales are common, I believe that most if not all of the animals reported as "sei" whales were probably Bryde's whales. Indeed, the companies' daily catch records for 1925/26 list some of them as "Brydehval" or as "sei (Brydehval)," and all 34 whales taken that season that were reported as sei whales in the International Whaling Statistics were reported as Bryde's whales by Kellogg (1931) and Radcliffe (1933).

This stock of whales has not been exploited since 1935 and may be assumed to be at the carrying capacity of the area. Insufficient data are available to make a quantitative assessment.

Minke Whale (Balaenoptera acutorostrata)

Distribution. During the winter we have found minke whales widely spread all the way from central California south to the Islas Revillagigedo. They are most abundant in the vicinity of the Channel Islands off southern California.

In the summer they occur from at least as far south as $26°45'N$ latitude off Baja California, north to the Chukchi Sea. They are fairly common off central California, but are much more abundant in Alaskan waters. A sizable population inhabits the inside waters of Puget Sound.

has never been attempted in the eastern North Pacific as it has off
Norway and Japan. The eastern North Pacific population of minke whales
is probably at the carrying capacity, but there are insufficient data
for a quantitative assessment of its size or sustainable yield. In
recent years pelagic expeditions in the Antarctic and one shore station
in South Africa have taken significant numbers of minke whales because
of a scarcity of larger species of baleen whales.

Humpback Whale (Megaptera novaeangliae)

Distribution. During the winter months, most humpback whales
congregate in warm waters close to continental coastlines or oceanic
islands. Their distribution at this season in the eastern North Pacific is well known from the logbook records of the nineteenth-century
American whaleships, the catch records maintained by the floating
factories that operated off Mexico in the 1920's and 1930's, and our
observations during the past decade. These winter grounds include
three somewhat discrete areas:

(1) The west coast of Baja California, chiefly from Isla Cedros
south to Cabo San Lucas, and around the cape at least as far north as
Isla San José. A few may also be found at this season farther north
along the west coast as far as Ensenada, and rarely to southern California.

(2) The mainland coast of west-central Mexico, from southern
Sinaloa to Jalisco, especially in the vicinity of the Islas Tres Marías
and Isla Isabela, Nayarit, and Bahía Banderas, Jalisco.

(3) The far offshore Islas Revillagigedo, including Isla San
Benedicto, Isla Socorro, and Isla Clarión.

Humpback whales also winter around the main Hawaiian Islands. I
observed some off Oahu in February 1966. According to Kenneth S.
Norris (personal communication), they are regularly seen around the
main islands.

In the western North Pacific, humpback whales are known to winter
around the Mariana Islands, around the Bonin Islands, and from southern Honshu, Kyushu, and South Korea southwest through the Ryukyu Islands to Taiwan (Omura 1950; Nishiwaki 1959).

Humpbacks occur all summer off central California. From there, their summer range extends around the entire North Pacific in the immediate offshore waters as far as Japan, and north through the Bering Sea into the Chukchi Sea.

Migration of humpback whales between summer grounds in the eastern Aleutians and the winter grounds in the Ryukyu and Bonin Islands has been demonstrated by eight Japanese mark recoveries (Anon. 1967). Although none of the 28 humpbacks that we marked off California and Mexico from June 1963 to June 1965 had been recovered by the end of the 1965 whaling season (after which the hunting of humpbacks was banned), it is probable that some of the humpbacks that winter in this area migrate far enough north to mingle with the western Pacific stock on the summer grounds.

Population. Because of their coastal habits, humpback whales are particularly vulnerable to exploitation by shore stations. Off Baja California, the number of humpbacks killed per gross catcher's day's work dropped steadily from 0.41 in 1924/25 to 0.03 in 1928/29, a 93% reduction. Off central California, the number dropped from 0.42 in 1956 to 0.005 in 1965, a reduction of almost 99% (Table 6-4). Decreases were also apparent in the catches at the Alaskan shore stations that operated in prewar years, and at the British Columbia stations in both pre- and postwar years. By the early 1960's the only area remaining in the North Pacific where large numbers of humpbacks congregated in the summer was around the eastern Aleutians and south of the Alaska Peninsula, from 150° to 170°W longitude. Large pelagic catches in that area in 1962 and 1963 reduced the population to an estimated 2,100 (Doi, Nemoto, and Ohsumi 1967). An additional 588 humpbacks were killed in 1964 and 1965. The remaining population probably represents the bulk of both the eastern and western North Pacific breeding stocks.

We made a survey over the entire eastern North Pacific winter grounds between 26 January and 15 March 1965, a time of year when the majority of the animals should be there. Our two vessels spent a total of 68 days cruising; we encountered only 33 groups of humpbacks, totaling 102 individuals: 10 on the west coast of Baja California,

65 along the coast of southern Sinaloa, Nayarit, and Jalisco, and 27
around the Islas Revillagigedo. More recent but less extensive cruises
revealed similar numbers. It is difficult to extrapolate these counts
to obtain an estimate of the total population on the winter grounds.
Since humpbacks concentrate in coastal waters during the winter, I
believe that we saw a fairly large proportion of the population. If
so, the entire eastern North Pacific stock now numbers only a few
hundred individuals.

Gray Whale (Eschrichtius robustus)

Distribution. The migrations of the eastern Pacific stock of
gray whales between their Arctic summer grounds and their Mexican
winter grounds is well known and has been summarized most recently by
Rice and Wolman (1971).

Population. In the winter of 1969/70 we estimated that the population size was around 11,000 and had remained stable since 1967/68
(Rice and Wolman 1971). The 1970/71 census yielded an estimate almost
identical with those of the preceding three years.

Black Right Whale (Eubalaena glacialis)

Distribution. The "Kodiak Ground," which encompassed the entire
Gulf of Alaska from Vancouver Island to the eastern Aleutians, was
renowned in the nineteenth century as one of the best areas for hunting right whales during the summer. At that season a few could be
found in the southern Bering Sea and all across the North Pacific
above 50°N latitude.

Their winter grounds have been somewhat of a mystery (Scammon
1874). In other parts of the world the females with calves resort to
coastal bays in the winter, but none has ever been found doing so in
the eastern North Pacific. Only a few right whales have been found
during the winter and spring months off the west coast—some as far
south as Punta Abreojos, Baja California.

Population. The nineteenth-century American whalers almost succeeded in completely exterminating the right whale in the eastern
North Pacific. How close they came is apparent from the fact that
from 1905, when modern whaling methods were introduced on the west

coast, to 1937, when right whales were given legal protection, only
24 were killed by the whaling stations in Alaska and British Columbia.
Omura, Ohsumi, Nemoto, Nasu, and Kasuya (1969) have summarized recent
sightings on the summer grounds, and Pike and MacAskie (1969) reported
several sightings off British Columbia. Rice and Fiscus (1968) reviewed the status of right whales in the southeastern North Pacific.
The lack of any additional sightings since then further confirms our
opinion that this stock numbers only a few individuals and has not
noticeably increased in the past 35 years.

Bowhead Whale (Balaena mysticetus)

Distribution. Bowheads spend the winter in the loose, southern
edge of the pack ice, which usually extends across the central Bering
Sea from Kuskokwim Bay, Alaska, west-southwest to the northern shores
of the Kamchatka Peninsula, USSR.

As soon as the ice north of the Bering Strait begins to break up
in the spring, the whales migrate northward through the open leads.
Many follow the shore lead (between the fast ice and the pack ice).
The first whales pass Point Hope, Alaska, in early April, and Point
Barrow in late April. In the summer they are distributed in the shallow waters of the northern Bering Sea, the Chukchi Sea, and the Beau-
fort Sea east to about Banks Island. During the autumn freeze-up,
they retreat south of the Bering Strait.

Population. American whaling ships first went through Bering
Strait into the Arctic Ocean in 1848. From 1868 on, the Arctic Ocean
was the principal resort of the North Pacific fleet. In 1870 the
fleet reached a peak of 53 vessels that took about 487 bowhead whales
(estimated on the basis of whalebone production). Starting in 1884,
many shore whaling stations were established in northwestern Alaska
by white men who employed Eskimos. Many writers have stated, without
documentation, that the bowhead whale population was greatly reduced
during this period. The statistics show that during the peak of the
fishery, from 1868 to 1884, the catch per vessel fluctuated but showed
no downward trend. An estimated average of 219 whales was killed
each year (excluding 1871 and 1876, when most of the fleet was lost

in the ice). If the population was stable, the fishery mortality could hardly have exceeded 5%, so the population may have been around 4,000 or 5,000. The fishery collapsed when the bottom fell out of the whalebone market after 1909.

The maritime Eskimos of Arctic Alaska have hunted bowhead whales for perhaps 50 centuries. Today the Eskimos living in the villages of Point Hope, Barrow, and Wainwright hunt whales each year during April and May. Temporary camps transported by dogsled are established at the edge of the fast ice. The whales are chased in easily transportable skin-covered umiaks, about 6.6 m long, propelled with paddles. The only major innovation since prehistoric times is the adoption of darting-guns and shoulder-guns which fire bomb-lances. Formerly, bone-headed harpoons were used. An annual average of about 10 bowheads are killed and recovered. For each whale recovered about 3 or 4 are struck and lost; some of these may die, so the total number of whales killed off Alaska is probably about 20 per year. Bowheads are rarely killed off Siberia (Zimushko 1969). Catch statistics for Barrow since 1928, compiled by Maher and Wilimovsky (1963), and for Point Hope since 1890, compiled by the late Don C. Foote (personal communication), give no indication that the population size has changed during this century. These facts imply a minimum population of about 400.

Each year between early April and early June, between 100 and 200 whales are observed migrating past Point Hope and Barrow. The highest rate of migration that I observed was at Point Barrow on 11 May 1962, when 25 whales passed during the 23-hour period from 0030 to 2330 hours. These counts are made only during periods when the whaling camps are occupied; whaling is suspended when the shore lead closes, and when it becomes more than 2 km wide. In the early part of the whaling season it may be too dark to see whales during the midnight hours, but often they can be heard blowing. Many more whales pass too far offshore to be seen from the fast ice. Pilots flying for the Arctic Research Laboratory at Barrow told me that they have seen bowheads in leads and polynyas as far as 80 km offshore in May. The number of whales observed from the whaling camps is thus only a small proportion of the total population.

Sperm Whale (Physeter catodon)

Distribution. During the winter, sperm whales are scattered
across the entire North Pacific below $40°$N latitude. From November
through April we have frequently encountered breeding schools over the
continental slope off the coast of California from $33°$ to $38°$N latitude. South of California, except for two large bulls at Isla Guada-
lupe, we encountered sperm whales only in the area south of Cabo San
Lucas and west of the Islas Tres Marías. Many were taken the year
round in the latter area in the nineteenth century, but we found only
a few, all males. The old records also indicate that the area around
the main Hawaiian Islands was a year-round concentration area.

During the summer, sperm whales may be found anywhere in the
North Pacific. The area of greatest population density extends from
the southwestern Bering Sea and northern Gulf of Alaska south to $50°$N
latitude, dipping to below $40°$N latitude on the American coast. Other
major summer grounds lie between $25°$ and $35°$N latitude from $180°$ longitude west to Japan, and thence northward along the entire Asiatic coast.

The summer range of the population that winters off California
is indicated by the recovery of three whales that were marked off
southern California in January. These included a male taken off
northern California in June, an animal of unknown sex off Washington
in June, and a female in the western Gulf of Alaska in April.

On the whaling grounds off central California sperm whales are
common from early April until the middle of June, reaching a peak in
mid-May. They are again common from the end of August to the middle
of November, reaching a peak in mid-September. Very few are present
in mid-summer. We found no sperm whales between San Francisco and
Bahía Magdalena during cruises in May, June, and September. The two
annual peaks of abundance suggest that the whales are moving north
through the whaling grounds off San Francisco in the spring and are
returning south in the autumn.

Japanese and Soviet mark recoveries reveal considerable longitudinal dispersal of sperm whales. The mating season extends from
late winter to late summer, so a female might mate almost anywhere
within her year-round range. Furthermore, the harem bulls apparently

do not remain long in a particular breeding school. These facts suggest that all North Pacific sperm whales comprise one widely interbreeding population.

Population. During their first five years of operation (1956 to 1961), the California whaling stations took sperm whales only when they could not find baleen whales. In 1962 there was a sharp rise in the price of sperm oil, and baleen whales were becoming less common, so the whalers expended more effort on sperm whales. Since the data are inadequate to determine the number of catcher's day's work spent hunting sperm whales, I have used simply the catch per vessel per season as a measure of the abundance of legal-sized male sperm whales (Table 6-5). The catch per boat averaged 12.6 from 1961 to 1965.

TABLE 6-5. Male sperm whale catch, effort, catch per unit of effort, and mean body length, at California shore stations, 1956-1970.

Season	Catch of male sperm whales[a]	Number of catcher-boats	Whales per catcher-boat	Mean body length[b] Feet	Meters
1956	9	2	4.5	43.2	13.18
1957	14	3	4.7	44.4	13.54
1958	8	4	2.0	45.7	13.94
1959	17	5	3.4	42.6	12.99
1960	14	5	2.8	44.7	13.63
1961	59	5	11.8	42.0	12.81
1962	46	5	9.2	41.4	12.63
1963	75	5	15.0	41.4	12.63
1964	54	5	10.8	39.6	12.08
1965	82	5	16.4	40.8	12.44
1966	37	5	7.4	40.5	12.35
1967	29	3	9.7	41.1	12.54
1968	23	3	7.7	38.7	11.80
1969	21	3	7.0	40.4	12.32
1970	20	3	6.6	40.1	12.23

[a] Excluding whales taken on special scientific permits, 1966 to 1970.

[b] Excluding whales less than 34.5 ft (10.52 m) long taken on special scientific permits, but including whales longer than 34.5 ft taken out of season on permits.

Thereafter it dropped and reached a low of 6.6 in 1970. The mean body
length of legal-sized males in the catch dropped from 44.0 ft (13.42 m)
during 1956 to 1960 to 39.7 ft (12.11 m) during 1968 to 1970. This
decrease in the availability of adult males off California agrees with
Ohsumi, Shimadzu, and Doi's (1971) calculation that the recruited stock
of male sperm whales in the entire northern North Pacific dropped from
134,000 in 1964 to 64,000 in 1970.

From 1956 through 1965, only 18% of the sperm whales killed off
California were females. From 1966 through 1970, when special scientific permits allowed the taking of animals shorter than the legal
minimum length, 53% of the catch was females. In the North Pacific
as a whole, females make up about 25% of the catch, and the number
taken has never been so high as the estimated sustainable yield (Ohsumi,
Shimadzu, and Doi 1971).

Giant Bottlenose Whale (Berardius bairdi)

Distribution. This species is endemic to the North Pacific. It
ranges from St. Matthew Island in the Bering Sea south in the eastern
Pacific to $32°30'N$ latitude off southern California, and in the western
Pacific to $34°00'N$ off Japan.

Seasonal movements are poorly understood. California catches
suggest two peaks of abundance, in July and October. Off British Columbia, the majority have been killed in August.

Thirteen of 15 (87%) taken off California were males. Likewise,
off British Columbia, 92% of those killed were males (Pike and MacAskie
1969), and in most areas of Japan males predominate in the catch (Omura,
Fujino, and Kimura 1956). Since females average larger than males, the
preponderance of males in the catches in certain areas suggests a partial geographical segregation of the sexes.

Population. Although bottlenose whales are regularly encountered
off central California, they are not common there. Because of their
relatively small size, whalers in the eastern North Pacific rarely
bother to kill them. Only 15 were taken off California from 1956 to
1970, and 29 off British Columbia from 1953 to 1967.

Summary and Conclusions

The cetacean fauna of the eastern North Pacific includes 11 species of "large" whales—here defined as species that attain a body length greater than 9 meters. Included are 9 of the world's 10 species of baleen whales (order Mysticeti) and the two largest species of toothed whales (order Odontoceti). Seven of these species are virtually worldwide; these include the sperm whale, the black right whale, and all the rorquals (family Balaenopteridae) except Bryde's whale, which is more or less circumtropical in coastal waters. Two species, the gray whale and the giant bottlenose whale, are endemic to the North Pacific. The bowhead whale is restricted to Arctic waters.

Since baleen whales are more abundant in the highly productive coastal waters, and most species migrate annually between higher latitude summer feeding grounds and lower latitude winter breeding grounds, the eastern North Pacific stocks of each species are probably more or less discrete from the western North Pacific stocks. More data are required to delimit stock units of toothed whales.

From the standpoint of their conservation, the species of large whales may be divided into three categories: (1) legally protected species whose population sizes are (or were) small; (2) species that are currently taken commercially; and (3) species that have remained virtually unexploited.

Five species of large whales are now afforded complete protection from commercial whaling by the 1946 International Convention for the Regulation of Whaling. The black right whale was formerly abundant in the eastern North Pacific, but was so heavily exploited in the nineteenth century that its population was reduced to perhaps no more than a few dozen individuals and has not noticeably increased since. The humpback whale, also formerly abundant, has been reduced in recent years to only a few hundred individuals. The blue whale population probably numbers no more than 2,000, but never was very great. The bowhead whale population does not appear to have been seriously reduced by nineteenth-century exploitation, but it numbers only a few thousand at most in its limited Arctic range. The gray whale population, much reduced in the nineteenth century, has greatly increased and is now

stable at about 11,000. A few bowheads and gray whales are killed by
Eskimos.

Three species are currently being exploited commercially in the
eastern North Pacific, under the regulation of the International Whaling Commission. The fin whale population has been reduced during postwar years by 55%, from about 20,000 to 9,000 recruited animals (that
is, animals of legally harvestable size), a number well below the optimum size. The sei whale population has been reduced during the same
period by 30%, from about 40,000 to 28,000 recruited animals, which is
about the number that will produce the maximum sustainable yield. The
sperm whale stocks in the eastern and western North Pacific probably
comprise one interbreeding population that may number several hundred
thousand; during the past decade the number of legal-sized males
(> 10.5 m long) has been reduced by 50%, but females have been under-
harvested.

Three species have been virtually unexploited in the eastern
North Pacific, and their population sizes are unknown. These species
are the Bryde's whale, the minke whale, and the giant bottlenose whale.

REFERENCES

Anonymous. 1967. Summarized result of the whale marking in the North
    Pacific [by] the Whales Research Institute, Tokyo. Rpt. IWC,
    17:116-119.
Berzin, A. A., and Rovnin, A. A. 1966. Raspredelenie i migratsii
    kitov v severovostochnoi chasti Tikhogo Okeana, v Beringovom i
    Chukotskom Moryakh. Iszv. Tikhookean. Nauchno-Issled. Inst.
    Ryb. Khoz., 58:179-208.
Doi, T., Nemoto, T., and Ohsumi, S. 1967. Memorandum on results of
    Japanese stock assessment of whales in the North Pacific. Rpt.
    IWC, 17:111-115.
Hegarty, R. B. 1959. Returns of whaling vessels sailing from Ameri-
    can ports. New Bedford, Mass., Old Dartmouth Historical Society.
Ingebritsen, A. 1929. Whales caught in the North Atlantic and other
    areas. Cons. Perm. Int. Explor. Mer., Rapp. Proc.-Verb. Réun.,
    55:1-26.
Ivashin, M. V., and Rovnin, A. A. 1967. Some results of the Soviet
    whale marking in the waters of the North Pacific. Norsk Hvalf.-
    Tid., 56:123-135.
Kellogg, R. 1931. Whaling statistics for the Pacific coast of North
    America. J. Mammal., 12:73-77.

Longhurst, A. R. 1967. The pelagic phase of *Pleuroncodes planipes* Stimpson (Crustacea, Galatheidae) in the California Current. Calif. Coop. Oceanic Fish. Invest. Rep., 11:142-154.

Maher, W. J., and Wilimovsky, N. J. 1963. Annual catch of bowhead whales by Eskimos at Pt. Barrow, Alaska, 1928-1960. J. Mammal., 44:16-20.

Maury, M. F. 1851. Whale chart. U.S. Navy Hydrographic Office, H.O. Miscel. No. 8514.

Nishiwaki, M. 1959. Humpback whales in Ryukyuan waters. Sci. Rep. Whales Res. Inst., 14:49-87.

Ohsumi, S., Shimadzu, Y., and Doi, T. 1971. The seventh memorandum on the results of Japanese stock assessment of whales in the North Pacific. Rpt. IWC, 21:76-89.

Omura, H. 1950. Whales in the adjacent waters of Japan. Sci. Rep. Whales Res. Inst., 4:27-113.

_____ Fujino, K., and Kimura, S. 1956. Beaked whale *Berardius bairdi* of Japan. . . Sci. Rep. Whales Res. Inst., 10:89-132.

_____ Ohsumi, S., Nemoto, T., Nasu, K., and Kasuya, T. 1969. Black right whales in the North Pacific. Sci. Rep. Whales Res. Inst., 21:1-78.

Pike, G. C., and MacAskie, I. B. 1969. Marine mammals of British Columbia. Fish. Res. Bd. Canada Bull., 171:1-54.

Radcliffe, L. 1933. Economics of the whaling industry with relationship to the convention for the regulation of whaling. 73rd Congr., 2nd sess., Senate Comm., Washington, D.C., U.S. Govt. Printing Office.

Rice, D. W. 1963. Pacific coast whaling and whale research. Trans. N. Amer. Wildlife Natur. Resource Conf., 28:327-335.

_____ and Fiscus, C. H. 1968. Right whales in the southeastern North Pacific. Norsk Hvalf.-Tid., 57:105-107.

_____ and Wolman, A. A. 1971. Life history and ecology of the gray whale. Amer. Soc. Mamm., Spec. Publ. 3.

Scammon, C. M. 1874. The marine mammals of the northwestern coast of North America. San Francisco, John H. Carmany & Co.

Starbuck, A. 1878. History of the American whale fishery. Rep. U.S. Comm. Fish and Fisheries for 1875-1876, pt. 4:1-779.

Tomilin, A. G., and Smyshlyaev, M. I. 1968. O nekotorykh faktorakh smertnosti kitov. Byll. Mosk. Obshch. Isp. Prirody Otd. Biologii, 73:5-12.

Tønnessen, J. N. 1967. Den moderne hvalfangsts historie. Sandefjord, Norges Hvalfangstforbund. Vol. 2.

Townsend, C. H. 1935. The distribution of certain whales as shown by logbook records of American whaleships. Zoologica, 19:1-50.

Zimushko, V. V. 1969. Nekotorye dannye po biologii serogo kita. *In* Morskie mlekopitayushchie, V. A. Arseniev, B. A. Zenkovich, and K. K. Chapskii, eds. USSR, Akad. Nauk.

CHAPTER 7

RESEARCH ON WHALE BIOLOGY OF JAPAN

WITH SPECIAL REFERENCE TO THE NORTH PACIFIC STOCKS

Hideo Omura and Seiji Ohsumi

Brief History of Whale Research

In Japan the Whales Research Institute was founded in 1946 as a
private organization, with field work in both the Antarctic and North
Pacific Oceans. The method of observations on whale carcasses on the
floating factories and at the land stations was generally in line with
that adopted by the Discovery Committee (Mackintosh and Wheeler 1929).

In the earlier days, however, much effort was expended on measurement of body proportions, examination of sexual condition of both sexes,
and weighing of the whale body. From these studies, combined with other
research, it was made clear that the baleen whale stocks in the northern and southern hemispheres are separated from each other, and that
the average body lengths at sexual maturity of both sexes of the blue,
fin, and sei whales in the northern hemisphere are shorter by a meter
or so than those in the southern hemisphere. The collection of data
concerning body weight from various species of whales has contributed
to the calculation of whale catches in terms of weight (Crisp 1962).

In 1955 the International Whale Marking Scheme was adopted by the
International Whaling Commission at its sixth annual meeting, and since
1955/56 season Japan has carried on whale marking in the Antarctic using
the Discovery mark under this scheme. As of the end of 1969/70 season
a total of 2,032 whales had been marked by Japanese expeditions.

Whale marking in the North Pacific was inaugurated in 1949, using
a mark somewhat shorter than the Discovery with a flat tip. By the
end of 1970 4,291 whales were marked effectively, of which 405 whales
were recaptured (9.5%).

The presence of conspicuous laminations in the core of the ear plug of fin whales was first reported by Purves (1955), though he could not ascertain whether lamina formation is a semiannual or an annual event. Since then our efforts have been largely directed towards the collection of ear plugs and counting of their laminations. In 1959 a new and very effective method for collection of ear plugs was improved in Japan (Omura 1963), which enabled us to collect plugs from nearly every whale taken. More than ten thousand ear plugs were collected annually in several years since the 1959/60 Antarctic season. All of the age data thus collected from the Antarctic were sent to the Special Committee of Three for stock assessment in the form of an age-length key.

The problem of whether lamination is a semiannual or an annual event had long been unsolved, though the rate of two laminae a year was accepted by various authors, and this rate was adopted in the stock assessment work done by the Special Committee of Three. Ohsumi (1962, 1964), however, showed that the rate is probably near one a year, based on the material obtained with long-term whale marks. The matter was finally settled by Roe (1967) from a study of formation of laminae throughout the year.

Age study of the sperm whale was also one of the main items of the research, and a vast number of mandibular and maxillary teeth were collected for population study (see, for example, Nishiwaki, Hibiya, and Ohsumi 1958); Ohsumi, Kasuya, and Nishiwaki (1963) estimated that the lamina formation is one a year. This estimation was confirmed by Best (1970).

Taxonomic studies were also carried out. The occurrence of Bryde's whale in the seas off Japan was proved and the validity of the species was established (Omura 1966). The pygmy blue whale in the Antarctic was identified as a subspecies of the ordinary blue whale (Ichihara 1966). An assessment of the stock of this subspecies was also made (Ichihara and Doi 1964). The black right whale was studied, obtaining the sample by special permit for scientific research (Omura 1958 and subsequent papers). The fin whale in the North Pacific was separated into four subpopulations by serological study, combined with the result

of whale marking, and, further, the fin whale in the East China Sea
was confirmed as an independent subpopulation from the above (Fujino
1960). In the smaller cetaceans, a new species of *Mesoplodon* (*M.
ginkgodens*) was discovered (Nishiwaki and Kamiya 1958).

Food of whales and oceanographic conditions of the whaling grounds
were also studied in detail (see, for example, Nemoto 1959; Nasu 1966).

This is a brief summary of research conducted by the Whales Research Institute; in 1966 the situation changed. In that year a group
for whale research was formed at the Far Seas Fisheries Research Laboratory, Fisheries Agency of the Japanese government, and since then stock
assessment and related work on whales has been undertaken by this group
in collaboration with the Tokai Regional Fisheries Research Laboratory.
On the other hand the Whales Research Institute, a private organization,
has reduced its activities.

Biological research on whales is now also carried out at some of
the universities, for example, at the Ocean Research Institute of the
University of Tokyo and at Nagasaki University.

Results of researches are mainly published in the Scientific Reports of the Whales Research Institute.

Stock Assessment and Its Biological Basis

The biological materials for stock assessment are obtained from
five categories of sources:

a. Observation and collection of samples from whale carcasses

Observation of whales on the factory ship or at the land stations
is made by whaling inspectors, including a biologist nominated as in-
spector, in cooperation with the crew of the expeditions. A series
of observations is carried out on the spot as routine work. In these
are included observations on species, sex, body length, position and
date of catch, stomach contents, mammary glands, fetus if any, thickness of blubber, and weight of testes. Ovaries, ear plugs, and teeth
are preserved for laboratory work. Samples of stomach contents are
also collected for examination.

b. Special permits for scientific research

Basic matters for population study are sometimes obtained from whales protected by the Convention, for example, from undersized whales or protected species. In the past, studies on the formation of laminae in the ear plugs, which needed samples also from the younger whales, composition studies of a school of sperm whales, osteological and other research on the right and pygmy blue whales, etc., were carried out under the special permits for scientific research.

c. Whale marking

Whale marking is important to obtain knowledge of whale movements, identification of stock units, age determination, and mortality. Japan has carried on whale marking since 1949 in the North Pacific and since the 1955/56 season in the Antarctic, as stated before.

d. Sighting of whales

One or more of the catcher-boats of each Japanese expedition have as a special mission solely to engage in whale sighting and observation of sea conditions, in much broader areas than the other catchers. Naturally the object is to serve the expedition, but the sighting data thus collected are useful for the estimation of population size, because this gives a source independent of the other methods. The estimation of the population of the protected species, such as blue, humpback, right, and gray whales, or of the unexploited species, for example the minke whale in the Antarctic, is largely dependent on sighting. In Japan the sighting data have been accumulated and analyzed since 1965. A theoretical method of population estimate from sighting data was basically reevaluated (Doi 1970) and has now been developed so as to reach the stage of practical use in estimating the absolute number of whales in the sea (Doi 1971c).

e. Catch statistics

The International Bureau of Whaling Statistics supplies the necessary statistical information for stock assessment. For the North Pacific stocks the necessary data are exchanged among the member countries of the North Pacific Group of the International Whaling Commission. But in the computation of the catch per unit of effort (c.p.u.e.) or the index of relative abundance, adjustments for tonnage (or horsepower) of catchers, weather, latitudinal difference or difference in length of days, and preference of species by catchers are needed.

Most of these problems still remain unsolved. The c.p.u.e. cannot always be regarded as giving the true index of abundance. Standardization of catching effort and effective overall catching intensity are urgently needed.

These are the five main sources for the stock assessment work of Japan. The materials obtained from these five sources are treated as shown in the arrow diagram, Fig. 7-1.

The object of the stock assessment is to obtain the values of the sustainable yield (s.y.) and the maximum sustainable yield (m.s.y.) for the rational management of whaling. In the estimation of these values it is very important to know the numbers of recruits corresponding to the size of the matured female stock, hence reproduction. In practice such a relation is difficult to obtain from the age composition of the catch, because of the lower reliability of c.p.u.e. used in the calculation (see Fig. 7-2).

In order to overcome such difficulties, an improved method of assessment of whale population was devised, setting a theoretical model (Doi, Ohsumi, and Nemoto 1967; Doi and Ohsumi 1969; Doi, Ohsumi, Nasu, and Shimadzu 1969; Doi 1971a; but see also Fig. 7-2). In this method a series of biological parameters (such as rate of pregnancy, age at sexual maturity, age at recruitment, and natural mortality coefficients before and after recruitment), and their changes against stock size should be ascertained. Further, in the case of sperm whales, age at social maturity as well as the number of females taken care of by a bull are needed in addition to the above (Doi 1971b). In our calculations, however, some of these are assumed because of lack of material; refinement is needed in the future after obtaining more biological evidence. The results of the computations, however, are checked by values obtained from the other sources, for example from sighting, marking, and c.p.u.e., and they are adjusted in the light of present knowledge of the whale population.

Since it was made clear that only one lamina is formed a year, as stated before, recalculation was made of the fin whale ages in the Antarctic (Doi, Ohsumi, Nasu, and Shimadzu 1969). The stock assessment of baleen whales in the North Pacific is also based on this rate

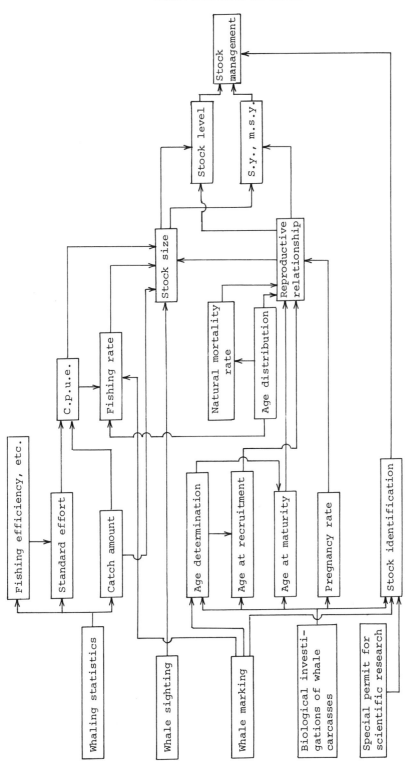

FIGURE 7-1. Arrow diagram for stock assessment of whales.

FIGURE 7-2. Difficulty of getting reproductive relationship directly from catch per unit of effort, and introduction of population model for estimation of reproduction relationship in the whale population assessment. Here relations between population level and several population parameters are needed.
 (a) Reproduction relationship from c.p.u.e.
 (b) Simplified population model.
 (c) Reproduction relationship from population model.
 (d) Sustainable yield curve from population model.
 (e) to (i) Relations between population level of mature females and pregnancy rate, age at sexual maturity, age at recruitment, natural mortality coefficient after recruitment, and natural mortality coefficient before recruitment, respectively.

C - amount of sustainable yield
F - fishing mortality coefficient
M - natural mortality coefficient after recruitment
M' - natural mortality coefficient before recruitment
N - catchable population size

n - population size in each
p - pregnancy rate
R - amount of recruitment
S - population size of matur females
t - age
t_c - age at recruitment
t_m - age at sexual maturity

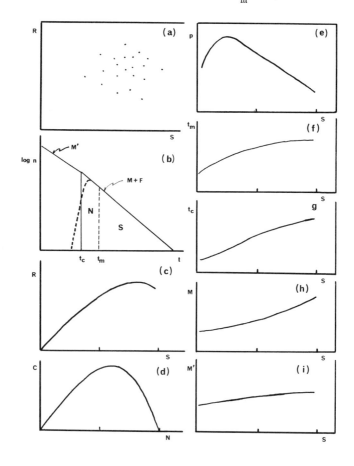

of accumulation (see for example Nemoto, Doi, and Ohsumi 1968).
Our next concern is the idea of the s.y. It is true that the s.y.
is stable when the stock level is unchanged, but if the stock size is
decreasing or increasing, especially in a stock in which the age at
recruitment is comparatively high, the s.y. also changes and hence is
impractical. Therefore, the idea of the actual sustainable yield (a.s.y.)
was introduced instead of s.y. (Doi and Ohsumi 1969; Doi, Ohsumi, Nasu,
and Shimadzu 1969).

Whale Stocks in the North Pacific

In Table 7-1 are shown the results of the assessment of whale
stocks in the North Pacific. The values of a.s.y., m.s.y., and the
corresponding stock size as well as the initial size of the stocks
are shown for fin, sei, blue, and sperm whales. The trends of the
population sizes and the comparison of present stock level against

TABLE 7-1. Stock assessment by Japan of whales in the North Pacific.

	Fin whale	Sei whale	Sperm whale Male	Female	Blue whale
Initial stock size	42,000-45,000	58,000-82,000	167,000	124,000	4,900
Stock size at m.s.y.	26,200-28,200	33,100-47,600			2,370
M.s.y.	1,180-1,270	2,460-3,450			90
Stock size in 1970	13,000-17,600	34,100-58,500			1,340
A.s.y. in 1970	1,270-1,350	3,030-3,200			80
Trend in stock change	±	-	-	-	+
Stock level at m.s.y.	-	±	+	+	-

the m.s.y. level are indicated. In Table 7-2 the corresponding figures
for fin, sei, and minke whales in the Antarctic are shown for reference.

TABLE 7-2.   Stock assessment by Japan of whales in the Antarctic.

	Fin whale	Sei whale	Minke whale
Initial stock size	379,000	150,000	> 70,000
Stock size at m.s.y.	227,000-237,000	52,700-51,400	> 35,000
M.s.y.	9,500-10,600	4,180-6,450	> 4,200
Stock size in 1970	73,300-81,100	82,720	> 70,000
A.s.y. in 1970	3,520-4,350	5,010	
Trend in stock change	+	-	±
Stock level at m.s.y.	-	+	+

The tables are self-explanatory, but brief explanations are given below for each species of whale in the North Pacific.

a. Fin whale

Several subpopulations were identified, as stated before, but for the time being the assessment was carried out dividing the whole population, excepting whales in the East China Sea, into two groups by 180° longitude—American and Asian populations. In the American population the initial size of the stock was estimated to be 25,000 to 27,000 whales, but it dropped below the m.s.y. level (15,600 to 16,900) in the earlier half of the 1960's, and the size in 1970 was 7,900 to 10,100. On the other hand, in the Asian population the initial size was estimated to be 17,000 to 18,000, and the m.s.y. level (10,600 to 11,300) was maintained during the 1940's and 1950's, but after 1952 it fell below the m.s.y. level, and the present size is 5,100 to 7,500. As a whole, the fin whale stock in the North Pacific is below the level of m.s.y., but in recent years the stock has been thought rather stable, due to increased limitation of the catch.

b. Sei whale

Very little is known of the stock units of this species, but the whole population was divided into two groups, as in the case of the fin whale. The sei whale was less exploited by pelagic operations in the years prior to 1963, and the history of exploitation of this species is shorter than that of the fin whale. In addition, the method of age determination by means of the ear plug has not yet been firmly

established. The results of the stock assessment of this species,
therefore, show fairly wide ranges of variation. The initial size of
the stock is estimated to be 30,000 to 50,000 in the American population and 28,000 to 32,000 in the Asian, making a total of 58,000 to
82,000. But as a result of the heavy fishing since 1963, it has
dropped to 34,100 to 58,500 (American 17,400 to 37,900, Asian 16,700
to 20,600) in 1970. This size of the present stock is near or a little above the m.s.y. level.

c. Bryde's whale

Distribution of this species is confined to waters warmer than
roughly $20^{0}C$, and the catch was only made in the waters off Bonin
Island and in the coastal waters of Japan. But catch of two Bryde's
whales by pelagic operations in 1970 was reported by the USSR.

According to a tentative assessment of whales in the Asian side,
the stock size is estimated to be 5,000 to 18,000, from which 300 to
600 whales may safely be taken annually.

d. Sperm whale

The sperm whale is distributed widely in the North Pacific and
is thought to be much more abundant than any of the baleen whales.
In this species, too, very little is known about the stock units,
though possibly several of these exist. The initial catchable stock
of males was estimated to be 167,000, including 134,000 whales which
were distributed in the waters north of $40^{0}N$.

e. Protected species

According to the results of sighting, the stocks of blue, humpback, and right whales are in the stage of recovery, though the population levels are still low. A paper on the stock assessment of the
blue whale in the North Pacific is now under preparation, to be discussed at the Scientific Committee of the International Whaling Commission [cf. Rpt. IWC, 23:32 (1973)—Ed.].

Future Research

There are many problems to be studied in the future, which can
be summarized as follows:

a. Stock unit or subpopulation

Stock assessment should be done by each unit of the stock, but this cannot be done precisely, at least at present. The best approach to this problem is maybe by biochemical studies of the whale.

b. Segregation by sex, sexual condition, or age

This problem was studied in detail for humpback whales in the southern hemisphere. Such study is needed especially for the sei whale in both hemispheres. The constitution of the so-called harem of the sperm whale is also to be studied.

c. Reproduction

Parameters concerning reproduction should be studied in detail. For example, the true rate of pregnancy cannot be derived from the statistics, and continued biological observation on whale carcasses, including those taken by special permit, is needed. Ovaries are now deemed as having less importance from the viewpoint of age, but they should be reevaluated from the standpoint of ecology and life history.

d. Age determination

Age determination by means of ear plugs has not been established yet for the sei whale, and conclusion of this matter is urgently needed.

e. Whale marking

Continued and large-scale whale marking is required, especially for younger whales and for whales in the winter breeding areas.

f. Sighting of whales

A sighting program independent of the industry is needed. This can be done in conjunction with whale marking.

g. True index of abundance or density of population

As stated earlier, standardization of catching effort and of effective overall catching intensity is needed.

REFERENCES

Best, P. B. 1970. The sperm whale (*Physeter catodon*) off the west coast of South Africa. 5. Age, growth and mortality. Investl. Rep. Div. Sea Fish. South Africa, 79:1-27.

Crisp, D. T. 1962. The tonnages of whales taken by Antarctic pelagic operations during twenty seasons and an examination of the blue whale unit. Norsk Hvalf.-Tid., 51:389-393.

Doi, T. 1970. Re-evaluation of population studies by sighting observation of whale. Bull. Tokai Regional Fish. Res. Lab., 63:1-10.
_____ 1971a. A theoretical treatment on reproductive relationship
between recruitment and adult stock. Bull. Tokai Regional Fish.
Res. Lab., 64:39-56.
_____ 1971b. Diagnosis methods of sperm whale population. Bull.
Tokai Regional Fish. Res. Lab., 66:89-143.
_____ 1971c. Further development of sighting theory on whale. Bull.
Tokai Regional Fish. Res. Lab., 68:1-22.
_____ and Ohsumi, S. 1969. A theoretical consideration on the maximum sustainable yield of sei whales in the Antarctic. Bull.
Tokai Regional Fish. Res. Lab., 60:83-93.
_____ _____ Nasu, K., and Shimadzu, Y. 1969. Advanced assessment
of fin whales in the Antarctic. Bull. Tokai Regional Fish. Res.
Lab., 60:95-141.
_____ _____ and Nemoto, T. 1967. Population assessment of sei whales
in the Antarctic. Norsk Hvalf.-Tid., 56:25-41.
Fujino, K. 1960. Immunogenetic and marking approaches to identifying subpopulations of the North Pacific whales. Sci. Rep.
Whales Res. Inst., 15:85-142.
Ichihara, T. 1966. The pygmy blue whale, *Balaenoptera musculus
brevicauda*, a new subspecies from the Antarctic. *In* Whales,
dolphins, and porpoises, ed. K. S. Norris. Berkeley, University
of California Press. Pp. 79-113.
_____ and Doi, T. 1964. Stock assessment of pygmy blue whales in
the Antarctic. Norsk Hvalf.-Tid., 53:145-167.
Mackintosh, N. A., and Wheeler, J. F. G. 1929. Southern blue and fin
whales. Discovery Rpt. 1:257-540.
Nasu, K. 1966. Fishery oceanographic study on the baleen whaling
grounds. Sci. Rep. Whales Res. Inst., 20:157-210.
Nemoto, T. 1959. Food of the baleen whales with reference to whale
movements. Sci. Rep. Whales Res. Inst., 14:149-290.
_____ Doi, T., and Ohsumi, S. 1968. Population assessment of fin
whales in the North Pacific. 1. Catch statistics, biological
information and natural mortality. Bull. Tokai Regional Fish.
Res. Lab., 54:5-52.
Nishiwaki, M., and Kamiya, T. 1958. A beaked whale *Mesoplodon*
stranded at Ōiso beach, Japan. Sci. Rep. Whales Res. Inst.,
13:53-83.
_____ Hibiya, T., and Ohsumi, S. 1958. Age study of sperm whale
based on reading of tooth laminations. Sci. Rep. Whales Res.
Inst., 13:135-153.
Ohsumi, S. 1962. Biological material obtained by Japanese expeditions
from marked fin whale. Norsk Hvalf.-Tid., 51:192-198.
_____ 1964. Examination on age determination of the fin whale.
Sci. Rep. Whales Res. Inst., 18:49-88.
_____ Kasuya, T., and Nishiwaki, M. 1963. The accumulation rate of
dentine growth layers in the maxillary tooth of the sperm whale.
Sci. Rep. Whales Res. Inst., 17:15-35.
Omura, H. 1958. North Pacific right whale. Sci. Rep. Whales Res.
Inst., 13:1-52.

_____ 1963. An improved method for collection of ear plugs from
baleen whales. Norsk Hvalf.-Tid., 52:279-283.
_____ 1966. Bryde's whale in the Northwest Pacific. *In* Whales,
dolphins, and porpoises, ed. K. S. Norris. Berkeley, University
of California Press. Pp. 7-78.
Purves, P. E. 1955. The wax plug in the external auditory meatus of
the Mysticeti. Discovery Rpt. 27:239-302.
Roe, H. S. J. 1967. Seasonal formation of laminae in the ear plug
of the fin whale. Discovery Rpt. 35:1-30.

CHAPTER 8

BALEEN WHALES OFF CONTINENTAL CHILE

Anelio Aguayo L.

This contribution has been prepared to give the student all the information we have gathered on baleen whales in Chile. Whale research in Chile is conducted jointly by the University of Chile (Departamento de Oceanología) and the Ministry of Agriculture (División de Pesca y Caza).

Whaling in Chile is regulated by the Permanent Commission for the Exploitation and Conservation of the South Pacific Marine Resources, set up in 1952 by agreement between the governments of Chile, Ecuador, and Peru.

Clarke, Aguayo, and Paliza (1968) state: "Whalebone whales comprising humpback, blue, fin and sei whales (Clarke 1962) and Bryde's whales (Clarke and Aguayo 1965), and the protected southern right whales (Clarke 1965), are all to be found in the region, but for technical and economic reasons the whaling companies of Perú confined their catch to sperm whales until 1964, whilst in Chile whalebone whales were only a small fraction of the catch until 1965."

The Japanese whaling companies (the Nitto Whaling Co. Ltd.) operated in Chile from October 1964 to March 1968, increasing significantly the baleen whale captures in Chile. The annual average baleen whale catch from 1964 to 1967 was 568 animals; the average for the previous ten years (1954 to 1963) was only 266 animals (Table 8-1).

Blue Whale, *Balaenoptera musculus* (Linné 1758)

The color of the blue whales we examined and observed in Chile was dark grayish blue except for patches of paler pigmentation on the ventral and lateral surfaces of the body.

Our observations at sea indicate that blue whales are not

TABLE 8-1. Baleen whales captured in Chile between 1929 and 1970.[a]

Year	Blue whales	Fin whales	Sei whales[c]	Humpback whales	Right whales	Total
1929	150	85	–	23	9	267
1930	97	62	–	34	1	194
1931	23	1	–	47	–	71
1932	29	14	–	20	22	85
1933	16	44	–	11	11	82
1934[b]	– ?	– ?	–	1	15	124
1935	31	42	–	15	27	115
1936	39	9	–	14	1	63
1937	18	25	13	16	2	74
1938	15	56	44	6	14	135
1939	2	99	15	7	5	128
1940	–	–	–	–	–	–
1941	–	–	–	–	–	–
1942	–	–	–	–	–	–
1943	2	13	–	–	1	16
1944	3	49	–	3	–	55
1945	60	64	1	11	–	136
1946	11	228	13	15	–	267
1947	24	88	2	17	–	131
1948	85	289	6	5	–	385
1949	35	219	–	6	–	260
1950	45	274	–	5	–	324
1951	77	279	2	3	–	361
1952	142	423	10	15	–	590
1953	172	301	27	5	–	505
1954	70	434	26	–	–	530
1955	149	359	33	5	6	552
1956	207	203	47	3	–	460
1957	100	69	39	5	–	213
1958	165	74	16	–	–	255
1959	80	70	17	4	1	172
1960	131	52	13	2	–	198
1961	142	16	13	3	–	174
1962	11	34	9	4	–	58
1963	31	11	6	1	1	50
1964	112	136	47	–	–	295
1965	385	266	487	6	2	1,146

TABLE 8-1 (cont'd.)

Year	Blue whales	Fin whales	Sei whales	Humpback whales	Right whales	Total
1966	128	84	210	7	1	430
1967	65	7	330	-	-	402
1968	-	24	70	-	-	94
1969	-	-	81	-	-	81
1970	-	-	19	-	-	19
Total	2,852	4,503	1,596	319	119	9,497

[a] Captures recorded by the Ministry of Agriculture and the University of Chile.

[b] Captures recorded by the International Whaling Statistics and Macaya Hnos. y Compañía.

[c] Includes Bryde's whales.

concentrated in schools, but are usually found single or in small groups of two or three animals. In 1966 we found five blue whales at 43°S, 75°W (March) and four at 45°-46°S, 75°-76°W (December), see Table 8-2. Two of these whales were marked at 44°56'S, 75°27'W (mark numbers 22709 and 22723, 17 December 1966[1]).

TABLE 8-2. Baleen whales observed off the coast of Chile in 1966.

Date	Area	Sei	Fin	Blue	Right	Minke	Hump
March (25)-April (3) 9 days	33°S-46°S 72°W-78°W	344	-	5	-	-	-
October (9-20) 11 days	33°S-44°S 72°W-78°W	399	-	-	2	1	-
December (13-23) 10 days	33°S-48°S 72°W-78°W	142	9	4	-	-	-

We have records in Chile of the catch from 1929 to 1970. Of the baleen whales captured, 30.9% were blue whales. We also have data on 168 blue whales examined, the size range being 16.2 to 25.7 m.

---

[1] Marks and marking guns were supplied by the National Institute of Oceanography, England.

Pygmy Blue Whale, *Balaenoptera musculus brevicauda*
Zemsky and Boronin 1964

In Quintay ($33^{\circ}11$'S, $71^{\circ}42$'W), during the seasons 1965/66 and
1966/67, we identified, among 168 blue whales examined, 10 specimens
of the subspecies pygmy blue whale (Aguayo et al., in prep.).

The subspecies *brevicauda* was proposed by Ichihara in 1963, but
his paper was not published until 1966. Meanwhile, Zemsky and Boronin
(1964) published the denomination *brevicauda*.

Fin Whale, *Balaenoptera physalus* (Linné 1758)

The color of the fin whales we examined and observed in Chile was
gray above and white below. The asymmetry of color is especially observed on the head, and the baleen reflects the asymmetry.

Part of the Chilean catch of fin whales are migrants traveling
between summer feeding grounds in high latitudes and winter breeding
grounds in low latitudes. Clarke (1962) said: "Four fin whales,
from eleven marked off the coast of Chile in November 1958, have been
recaptured by pelagic whaling expeditions in the western part of Area
II, the Atlantic sector of the Antarctic, showing that at least some
fin whales off the coast of Chile in spring are migrants belonging to
a stock frequenting the Antarctic in summer, and that fin whales
migrating southwards off the Atlantic and Pacific coasts of South
America mingle on feeding grounds south of Cape Horn."

Among baleen whales, the fin whale was the most abundant species
of the Chilean catch from 1937 to 1955 (see Table 8-1).

We have some biological data which are being analyzed.

Sei Whale, *Balaenoptera borealis* Lesson 1828

In recent years, with the declining abundance of other baleen
whales, the sei whale has become the most important baleen whale species in the Chilean catch; in Table 8-1 sei whales include sei and
Bryde's whales.

In the last three observation cruises made off the coast of
Chile, the most abundant species was the sei whale (Table 8-2). In

March, the concentration of sei whales (286 animals) was found between 43°S and 45°S, and at 60 to 70 miles from the coast.   In October, the concentration of sei whales (345 animals) was between 39°S and 41°S, and at 60 to 120 miles from the coast.   In December, the concentration of sei whales (114 animals) was between 46°S and 48°S, and at 20 to 60 miles from the coast (Aguayo 1966).

We have some biological data from captures off Valparaíso and Talcahuano which are being analyzed.

Bryde's Whale, *Balaenoptera edeni* Anderson 1879

This species appears in the Chilean statistics included with the sei whale (Table 8-1). We have identified it in the captures from Iquique and Valparaíso, but not in Talcahuano. Clarke and Aguayo (1965), comparing the first identified specimen from Iquique with the southern sei whale, say: "The whale from Iquique had a smaller dorsal fin set farther back, the ventral grooves reached beyond the umbilicus, and the baleen had fewer and shorter plates which themselves were thicker, wider in proportion to their length, and with thicker and longer bristles than those of the sei whale."

We have some biological data from captures off Valparaíso which are being analyzed.

Humpback Whale, *Megaptera novaeangliae* (Borowski 1781)

This species was formerly abundant along the coast of Chile, but now it is rarely found. In 1966 we made three cruises to observe whales, in March, October, and December, and we could not see any humpback whales between 33°S and 48°S (Table 8-2).

Among baleen whales, the percentage of humpback whales caught in Chile from 1929 to 1970 is 3.5%.

We do not have biological data on this species in Chile.

Southern Right Whale, *Eubalaena australis* (Desmoulins 1822)

This whale was captured at the beginning and middle of the century in the Chilean central and southern areas, especially off Concepción, Arauco, Valdivia, and Chiloe (Table 8-3). Clarke (1965) observed it

TABLE 8-3.   Right whales captured off the coast of Chile from 1929 to 1970.[a]

Year	No. of whales	Locality
1929	9	
1930	1	
1931	–	
1932	22	21 from Corral (39°53'S) and 1 from Isla Santa María (37°01'S)
1933	11	
1934	15	Without locality (I.W.S.)
1935	27	1 from Isla Santa María
1936	1	Isla Santa María
1937	2	
1938	14	
1939	5	
1940	–	
1941	–	
1942	–	
1943	1	Isla Santa María
1944	–	
1945	–	
1946	–	
1947	–	
1948	–	
1949	–	
1950	–	
1951	–	
1952	–	
1953	–	
1954	–	
1955	6	4 from Talcahuano (36°45'S) and 2 from Quintay, which were caught in 35°07'S
1956	–	
1957	–	
1958	–	
1959	1	Talcahuano
1960	–	
1961	–	
1962	–	
1963	1	Talcahuano
1964	–	
1965	2	Quintay (33°11'S)

TABLE 8-3 (cont'd.)

Year	No. of whales	Locality
1966	1	Quintay
1967	-	
1968	-	
1969	-	
1970	-	
1929-1970	119	33°S-40°S

a This table is taken from Clarke 1965, pp. 25-26.

at Cartagena ($33^{\circ}32$'S) and Aguayo (unpublished data) off Isla de Chiloe ($41^{\circ}58$'S) in October 1966, marking the two whales just mentioned (marks 23572 and 23580).

The right whale was abundant off the coast of Chile and Argentina, but these stocks were devastated by whaling in the nineteenth century.

Clarke (1965) notes, "Right whaling in the South Pacific began after 1790 when the British whaleship *Amelia* had returned from her pioneer voyage after sperm whales round Cape Horn. Vessels which followed the *Amelia* to look for sperm whales in that ocean found plenty of right whales as well, and a favourite right whaling ground was the coast of Chile. In Townsend's chart (1935) which shows the positions of 6,262 southern right whales captured by American whaleships between 1785 and 1913, the coast of Chile ground is seen to have extended from 30° to 50°S, with most captures concentrated near the coast, although between 40°S and 50°S there were outliers extending westwards to 600 miles from the coast. On the Argentine side the whaling stretched from Cape Horn eastward and northward all along the coast to 30°S, so that the absence of captures on the Chile side between Cape Horn and 50°S was probably due not to a lack of whales but to the prevailing heavy weather which makes that far southern part of the west coast a dangerous lee shore at all times of the year."

It is interesting to quote here what Best (1970) said when referring to seasonal movements off the west coast of South Africa: "If this is actually the case, then the difference in size between the

two peaks suggests that during their northward migration the majority
of right whales remain outside coastal waters, but that on their southward leg when many females are accompanied by young calves they tend
to travel closer to the coastline. This may also account for the slower
rate of migration during the southern leg."

We have records of the catch since 1929 (see Table 8-3).

Pygmy Right Whale, *Caperea marginata* (Gray 1846)

This whale has always been considered to be a rare species. It
has been recorded from South America (Norman and Fraser 1937), from
Chile (Cabrera and Yepes 1940, Yañez 1948), and from Argentina (Cabrera
and Yepes 1940). I was not able to find in their papers any data of
capture, sightings, or strandings from Chile, or any actual records
from 1958 to 1970. At present I do not know if this species can be
cited or not for our country.

Minke Whale, *Balaenoptera acutorostrata* Lacépède 1804

This species is absent from the Chilean catch to date (Table 8-1).
It is probable that in the near future the minke whale will become part
of Chilean whaling. I know only three reliable records of minke whale
in Chilean waters: Clarke in 1958 at $33^{\circ}14'S$, $73^{\circ}15'W$ (two animals),
and Aguayo in 1966 at $41^{\circ}40'S$, $75^{\circ}45'W$ (one animal) (Table 8-2). All
these animals had a white band across the flipper.

Summary

To date, the following baleen whale species from Chile have been
certainly recorded: blue, pygmy blue, fin, sei, Bryde's, humpback,
right, and minke whales.

Acknowledgments

I wish to thank Dr. Rene Maturana of the Ministry of Agriculture,
Valparaíso, for his help in preparation of the tables, and Miss Sara
Soudy of the University of Chile, Valparaíso, for her valuable sug-
gestions in my English writing.

REFERENCES

Aguayo L., A. 1966. Observaciones de cetáceos frente a la costa de Chile durante el año 1966. Montemar, Informe enviado al Ministerio de Agricultura (Departamento de Pesca y Caza). No publicado.

Best, P. B. 1970. Exploitation and recovery of right whales *Eubalaena australis* off the Cape Province. Investigational Rpt., Div. Sea Fisheries South Africa, 80:1-20.

Cabrera, A., y Yepes, J. 1940. Mamíferos sudamericanos: Vida, costumbre y descripción. Buenos Aires, Historia Natural Ediar. Pp. 291-318.

Clarke, R. 1962. Whale observation and whale marking off the coast of Chile in 1958 and from Ecuador towards and beyond the Galápagos Islands in 1959. Norsk Hvalf.-Tid., 51:265-287.

_____ 1965. Southern right whales on the coast of Chile. Norsk Hvalf.-Tid., 54:121-128.

_____ and Aguayo L., A. 1965. Bryde's whale in the southeast Pacific. Norsk Hvalf.-Tid., 54:141-148.

_____ _____ and Paliza G., O. 1968. Sperm whales of the southeast Pacific. Pts. 1 and 2. Oslo, Hvalrådets Skrifter, 51:1-80.

Ichihara, T. 1966. The pygmy blue whale, *Balaenoptera musculus brevicauda*, a new subspecies from the Antarctic. *In* Whales, dolphins, and porpoises, ed. K. S. Norris. Berkeley, University of California Press. Pp. 79-113.

Norman, J. R., and Fraser, F. C. 1937. Giant fishes, whales and dolphins. London, Putnam.

Yañez, P. 1948. Vertebrados marinos chilenos: 1. Mamíferos. Rev. Biol. Mar. Valparaíso, 1:103-123.

Zemsky, V. A., and Boronin, V. A. 1964. On the question of the pygmy blue whale taxonomic position. Norsk Hvalf.-Tid., 53:306-311.

CHAPTER 9

STATUS OF ANTARCTIC RORQUAL STOCKS

D. G. Chapman

As is well known, the discovery of the explosive harpoon and the
development of faster ships made possible worldwide exploitation of
the rorqual whales—blue, humpback, fin, and sei—and especially in
the Antarctic. Until the further development of the factory ship with
its attendant fleet of catcher-boats, Antarctic whaling was rather
small. In the late 1920's it expanded rapidly and, except for a brief
recession in 1931/32 caused in part by the collapse of the whale oil
market because of the worldwide depression, this expansion continued
through the 1930's.

While early whalers had learned quite a lot about the right and
sperm whales they hunted, the new era in whaling gave similar and
greater opportunities to learn about the rorquals, which had been primarily studied only from their occasional strandings.

Research on southern whales based on studies of animals captured
in whaling operations began at South Georgia in 1913, but may be said
to have reached maturity when Mackintosh and Wheeler (1929) published
their work on blue and fin whales. Research proceeded at an acceler-
ated rate in the 1930's, not only at whaling stations and by biologists on board factory ships, but also through the Discovery expeditions of the United Kingdom. These expeditions did an extensive
amount of whale marking and also made studies of whale density from
sighting data by areas.

At the same time that the British were undertaking whale marking,
the Norwegians were also undertaking their own research program with a
special interest in analysis of the statistics of catch and other data
obtained from the factory ships. Successive reports on this analysis
were published through the 1930's (Hjort, Lie, and Ruud 1932-1938).

The early biological work gave information on the breeding cycle
and the growth pattern, though this work was handicapped by the absence
of a reliable indicator of age. Nevertheless, much was learned about
the annual migration of whales from their feeding areas in the Antarctic to temperate and subtropical parts of the ocean where breeding
takes place and where the females give birth. In the case of humpback whales, migration patterns are reasonably known, because this
species stays close to coasts both in its travels and during the breeding season. The other rorquals appear to disperse more widely in
their winter habitat; few breeding concentrations have been found, so
that the actual location and timing of activities is largely a matter
of conjecture.

Independently of the marking activities of the Discovery expeditions, Hjort, Lie, and Ruud used the catch data to divide the Antarctic into broad areas. For a number of years Area I included the Antarctic sectors from longitudes 60°W to 170°W, the major part of which
was closed to whaling (the so-called Sanctuary area included the sectors 70°W to 160°W). In 1955 the Sanctuary area was opened to whaling
and thereafter the statistical areas were redefined as follows:

Area I	120°W-60°W
Area II	60°W-0°
Area III	0° -70°E
Area IV	70°E-130°E
Area V	130°E-170°W
Area VI	170°W-120°W

It is convenient to regard the areas as bounded on the north by
40°S latitude because factory ships have not in postwar years been
permitted to take baleen whales north of this boundary. These are
part of a variety of restrictions upon whaling established by the
treaty that subsequently created the International Whaling Commission.

While pelagic operations for baleen whales are banned north of
40°S, land stations do operate further north. From such catches and
from long series of observations at sea, it is known that, in general,
these whales migrate to the north of this line in the southern winter.
The sighting data analyzed by Mackintosh and Brown (1956) show that

some whales are to be found in the Antarctic Ocean at all times of the
year, though the numbers present reach a minimum in the winter months.
Conversely, sightings in the warmer waters show that there are some
whales of these species present in these latitudes throughout the year.

These observations have been interpreted by some to imply that
some whales fail to migrate each year. This may be possible, but
another explanation is that there is a large dispersion in the timing
of migration so that the last departures from any region overlap the
arrivals of the earliest animals of the next season. Another pattern
of movement has been suggested for North Atlantic baleen whales by
Mitchell: a rather restricted movement. He suggests that whales
feeding off Nova Scotia in the summer move south in the winter to be
replaced by another population that had fed further north, possibly
in Denmark Strait or off Iceland. Stomach analyses, however, have
shown that North Atlantic fin whales have a greater variety in their
diet than the southern hemisphere members of the same species, which
concentrate much more on the euphausid *E. superba*. Thus, whether
this pattern of limited movement might also apply in the southern
hemisphere is very conjectural. There is no evidence that northern
and southern rorqual stocks intermingle—no Antarctic mark has been
recovered north of the equator and no North Pacific mark in the
southern oceans.

Separation of Stocks

Humpback Whale, Megaptera novaeangliae

The best case for the separation of stocks and the identity of
the subgroups can be made for the humpback whales. Mackintosh (1965)
identifies six groups, one associated with each coast of the southern
continents (South America, Africa, and Australia). There is some
question whether the eastern Australian stock is partially separate
from the stock that formerly migrated along the New Zealand coast and
was also found in the islands of the southwest Pacific Ocean. It has
also been assumed that these stocks feed in the Antarctic in regions
immediately to the south of their migration routes, but some intermingling is possible.

Thus it is quite likely that the humpback stock off the west coast of South America might move eastward to feed in the richer waters of the South Atlantic (Area II) rather than migrating due southward to Area I. It is also probable that the separate breeding stocks off each coast of Africa intermingle when feeding in Area III, with occasional movement west to Area II or eastward to Area IV. "Intermingle" as used in this paper should be taken in a very broad sense—it is not meant to imply that animals from different stocks mix in common groups or pools, rather that they inhabit the same extremely large statistical area (which for Area III is 70^{0} wide, so that these "intermingling stocks" may be hundreds of miles apart).

Finally, recovery in Area IV of a whale mark placed in a humpback on the east coast of Australia demonstrates that the eastern and western Australian stocks may overlap in their feeding areas. However, for purposes of analysis it has been assumed that Area IV catches are from the western Australian stock and Area V catches are to be associated with catches off eastern Australia and New Zealand and identified with the so-called eastern Australian stock. The limited degree of intermingling in feeding areas serves to restrict the amount of interbreeding to very low levels, if any. However, if there is complete genetic separation, it has not been in existence for a sufficiently long period to show up in distinguishable characteristics of these stocks.

Finback Whale, Balaenoptera physalus

The identification of fin whale stocks has been based for the most part on mark recoveries, though Lund (1950, 1951) studied iodine levels of fin whale oil, while Laws (1960) and Ohsumi and Shimadzu (MS) have attempted to distinguish between stocks by morphometric measurements. Fujino (1964) used serological methods to distinguish subpopulations.

The most complete analysis of migrations and movements from mark recoveries is found in several papers by Brown (1962a, 1962b, MS). Through 1970, fourteen marks have been recovered that show north-south movement. These demonstrate southward movement from about latitude

30°S on both coasts of South America into Area II (0°-60°W) and northward movement from Area III (0°-70°E) to both coasts of South Africa
(also to about latitude 30°S). The mark recoveries in temperate and
subtropical waters are of course dependent on the location of whaling
stations, and it needs to be remembered that little marking has taken
place in such waters and that catches from land stations have been
only a small fraction of Antarctic pelagic catches.

Brown has also analyzed the east-west movement of this species
from marks fired and recovered in the Antarctic. Of 466 marks recovered up to 1970, 82 were recaptured in areas different from that in
which the mark was placed, but in all 82 cases the area of recovery
was adjacent to that in which the mark was fired. Moreover, in a considerable number of these cases, the location of placement and of recovery were quite close—only a few degrees of longitude apart, but
on opposite sides of the convenient but arbitrary boundaries defined
by the Bureau of International Whaling Statistics.

Because fin whales appear to be more dispersed in their breeding
areas and because they are found further from land than are humpback
whales, it is possible that each southern ocean contains only one
breeding stock. On the other hand, a subdivision similar to that for
humpbacks is more likely. A pattern of this type has been proposed
by Ivashin (MS). Table 9-1 is based on his presentation at a meeting
on Antarctic fin whale stock assessment held in Honolulu in March 1970.

This table would suggest that two breeding stocks are feeding in
each of Areas II and III (with some overlap into IV). The possibility
of two stocks in III was suggested by Fujino (1964) on the basis of
blood-typing. His analysis suggested different blood-type groups
north and south of 50°S. This suggests a division along the lines
Mitchell proposed for the North Atlantic, but it is possible that one
of the two South Atlantic breeding stocks identified by Ivashin moves
into more southerly waters and the other remains north of 50°S to
feed. Certainly Areas II and III have been the most productive regions
for whale catches. This has been in part attributed to the greater
primary productivity which, of course, could support additional whale
stocks. If in fact there are two stocks being exploited in Areas II

TABLE 9-1. Distribution of stocks of fin whales in the southern hemisphere (after Ivashin MS).

Stock	Wintering area	Summer distribution
Chile-Peruvian	West of northern Chile and Peru	From 100° (110°) W to Drake Passage and S. Shetland (about 25°W)
South Georgian	East of Brazil	From eastern South America as far east as S. Sandwich Islands (about 25°W long.) and about as far south as the S. Orkney Islands (about 60°S lat.)
West African	West of South Africa, Angola, and Congo	From the S. Sandwich Islands to Zero Meridian
East African	Off eastern Africa and Madagascar	From Zero Meridian to 40°E
Crozet-Kerguelen	East of Madagascar	40°-80°E
West Australian	Northwest of Western Australia	80°-110° (120°) E
East Australian	Coral Sea	140°-170°E (including Balleny Islands)
New Zealand	Fiji Sea and the adjacent waters eastwards	170°E-145°W

and III, the degree to which each has been taxed is unknown. Even if segregation existed between the stocks prior to exploitation, it is possible that the pattern of segregation has been altered by exploitation. If this exploitation drastically reduced the stocks that migrate to this southerly area, the vacated feeding areas could have been taken over by the stock formerly farther north. In the absence of further information, it is convenient to retain the basic statistical subdivisions as appropriate to separate stocks.

Blue Whale, Balaenoptera musculus

Less is known about blue and sei whale stock subdivisions than for fin whales. It is true that the original subdivision of the southern oceans was in fact based largely on blue whale catches rather

than on fin whale catches. Blue whale mark recoveries have been much
fewer than fin whale recoveries, since the number marked has been much
smaller. However, with this reservation, it must be noted that the
pattern of mark recoveries is similar—some movement into adjacent
areas, but most recoveries are made in the same area that the mark
was placed. There is one recovery of a blue whale mark in Area IV
that was placed in Area II.

Sei Whale, Balaenoptera borealis

Because sei whales were of little interest to pelagic whaling
fleets until the late 1950's and because almost none were marked until
the 1960's, stock identifications for this species are even more tenu-
ous. It was formerly assumed that this species did not migrate into
Antarctic waters, and it is true that the heaviest catches have occurred
between 40^{0} and 50^{0}S. Clearly if the species breeds and spends some
time feeding in temperate or subtropical latitudes, it too must have
at least one breeding stock in each of the southern oceans.

The patterns of exploitation and response in the early 1960's suggested to FAO analysts (Gulland and Boerema) the presence of two stocks
in Area II. These could be stocks breeding off the east coast of South
America and the west coast of Africa parallel to the identification
suggested by Ivashin for fin whales, though western Area II catches
could include some whales from the west coast of South America also.

Recent Research

The active research programs carried on by Norway and the United
Kingdom prior to World War II were, of course, interrupted during the
war years. However, shortly after the resumption of whaling, research
programs were also resumed under Ruud in Norway, Slijper in the Netherlands, and at the National Institute of Oceanography in Great Britain.
In Japan the Whales Research Institute was established in 1946; for
much of its history the Institute has been under the leadership of H.
Omura. Scientists of this Institute and of the Japanese fisheries
agency have played a large role in data collection and biological research since that time.

Some of the work of these groups involved the marking programs already discussed under separation of stocks. Some has to do with basic physiology of the cetaceans, and for the results of this research (not only for southern baleen whales but whales in general) reference is made to Slijper's book on the subject (1962) and to chapters in *The Whale*, edited by Matthews (1968), and in *The Biology of Marine Mammals*, edited by Andersen (1969).

Proper management of the whale stocks requires an understanding of their age and growth and of their reproductive cycles, and of how they are affected by stock manipulation. There have been many contributors to such knowledge, but we refer particularly to two monographs, by Laws (1961) on fin whales and by Gambell (1968) on sei whales.

Improvements in Aging Techniques

Purves (1955) and Laws and Purves (1956) showed that alternate dark and light layers are laid down in the ear plug. It was first thought that two such pairs of layers are laid down per year. It became clear that this is not always the case because marked whales were recovered with fewer than the layers indicated by the lapse of time between marking and recovery. In the case of fin whales the matter was resolved by Roe (1967), who read samples taken throughout the different months of the year to show that, in fact, one pair of alternate dark and light layers is indeed formed per year; he also showed that there are a number of "false" laminae, particularly at young ages, which make age determination difficult in some cases.

The ear plug method has been used with all the large rorquals with which we are concerned here. In fact, recoveries of humpback whales marked as calves had seemed to confirm the two layers per year theory (Chittleborough 1960) for this species. In the case of sei whales, it has been very difficult to read the ear plugs, and there were wide variations in age determinations by different scientists. A new method of treatment developed by Lockyer (MS) appears to show promise of remedying this problem.

The number of layers laid down per year has not been completely resolved for baleen whales other than fin whales, but most scientists

working with this material have taken one layer per year as a useful
working rule. The present author believes that this is indeed valid,
though reading is complicated by the events of early life (weaning,
sexual maturity, and possibly departures from the adult migration pattern) which cause alterations in the ear plug growth pattern and hence
give the appearance of additional layers.

Population Estimates

Humpback Whale, Megaptera novaeangliae

The only thorough analysis of humpback whale stocks in the southern hemisphere has been made for Area IV and Area V stocks. These
analyses were published by Chapman, Allen, and Holt (1964) and by
Chittleborough (1965). In addition, we have records of the catch by
pelagic operations and from land stations. Since humpback whales are
quite vulnerable to land stations on their near-shore migrations, the
catches of land stations have been of considerable importance. Table
9-2 shows these land-station catches with the catches of the associated
Antarctic areas, except for the Australian stocks. While this table
shows the levels of exploitation since 1934 (1933/34 in the Antarctic),
it is to be noted that there was a brief period of heavy exploitation
at South Georgia and at the South Sandwich Islands much earlier (8,294
taken in 1910/11), and in tropical waters off Africa at about the same
period (5,649 in 1912). Exploitation off Madagascar, Angola, and the
Congo delta has been erratic, which explains in part the fluctuations
of catches off Africa. The kill of slightly more than 30,000 in Area
II from 1909/10 to 1932/33 undoubtedly accounts for the low catches
since that time off both coasts of South America and in adjacent Antarctic areas. In fact, during the seven seasons 1909/10 to 1915/16
(International Whaling Statistics 16, pp. 78-79) 27,016 were taken in
Area II. These clearly depleted the stock, for catches have averaged
only a few hundred or less ever since. This suggests that the original
Area II stock numbered 25,000 to 35,000. The original Area III stock
was not greater than this and probably less. Analysis is complicated
by the catches off Africa in the period prior to 1934. These are

TABLE 9-2. Recent humpback kill in the southern hemisphere by Antarctic areas and at land stations. From published tables of the Bureau
of International Whaling Statistics, Oslo.

Year	West coast of South America	Area I	East coast of South America	Area II and South Georgia	West coast of Africa	East coast of Africa	Area III
1934	12	0		205	724	514	8€
1935	29	0		94	1,241	418	53§
1936	18	0		329	67	301	1,88£
1937	18	0		259	326	1,463	2,78(
1938	6	6		375	0	1,927	82§
1939	7	7		0	0	200	(
1940	0	0		0	0	0	(
1941	0	0		231	0	0	4£
1942	0	0		16	0	0	(
1943	0	0		0	0	0	(
1944	0	0		4	0	0	(
1945	0	·0		60	0	0	(
1946	15	0		238	5	0	(
1947	14	11		29	14	0	(
1948	5	21		25	1,348	0	(
1949	6	15		22	718	1,523	§
1950	5	24		198	9	862	25]
1951	26	262	28	10	15	103	2(
1952	27		10	43	9	111	18§
1953	27		8	23	0	89	21£
1954	1		18	24	0	27	17ʔ
1955	7		9	4	0	49	14ʔ
1956	10	14	14	0	0	36	6∠
1957	5	655	0	0	3	34	£
1958	0	90	5	0	2	39	11€
1959	3	0	8	0	7	38	9ʃ
1960	2	179	13	0	4	36	15€
1961	3	82	13	3	4	36	11ʔ
1962	4	125	11	3	9	37	3ʔ
1963	1	49	12	0	3	37	3ʔ
1964	0	-	0	0	-	-	
1965	35	-	0	0	-	-	
Total 1934 to 1965	286	1,456	233	2,195	4,508	7,880	7,79ʔ

1,742   2,428   20,180

estimated to be about 20,000, but were not fully recorded. As catches
since that time have been almost 20,000, the 1910-1933 removals did not
decimate the stock as in Area II. However, catches were much more
spread out over time, so that it was possible for recruitment to play
a greater role.

The catches from the western and eastern Australian stocks have
been listed by Chittleborough (1965) and are reproduced in Table 9-3.

TABLE 9-3. Humpback whale catches from western and eastern Australian
stocks, 1949 to 1962.

Year	Western Australian stock Area IV	Eastern Australian stock Area V
1949	190	141
1950	1,167	982
1951	2,336	273
1952	2,314	868
1953	1,496	1,313
1954	1,578	898
1955	1,154	1,929
1956	1,951	1,207
1957	1,120	1,025
1958	967	1,023
1959	2,113	2,163
1960	611	2,272
1961	584	1,274
1962	599	209[a]
Totals	18,180	15,577

[a] Chittleborough (1965, p. 112) gives evidence of an illegal catch of
humpback whales in Antarctic Area V in the 1961/62 season in addition
to the number shown here.

The Area IV population was hunted along the west coast of Australia from 1912 to 1916, from 1925 to 1928, and from 1936 to 1938
prior to the exploitation shown in Table 9-3 (Ruud 1952). From 1935
to 1938 at least 12,673 whales were taken from this population off
Australia and in the Antarctic. On the other hand, the Area V population was exploited very lightly prior to 1950.

Chittleborough gives the following estimates for these populations at various times:

TABLE 9-4. Humpback population estimates (after Chittleborough 1965).

| | Population size | |
Year	Area IV	Area V
1934	12,000-16,000	10,000
1950	9,400	10,000
End of 1962	800	200[a]

[a] Chittleborough estimates 500, taking into account the suspected illegal kill referred to in Table 9-3.

To some extent these estimates may be in error, because they depend in part on age determinations based on the assumption that two layers are laid down in the ear plug each year. Reanalysis of fin whale data when this assumption was shown to be invalid for that species led to somewhat higher population estimates, particularly of the initial stocks, though much lower rates of recruitment. However, the final estimates of Chittleborough are confirmed by recent Japanese sighting data. These are analyzed by two methods by Ohsumi and Masaki (1972) and by Doi, Ohsumi, and Shimadzu (1971). These data give the average population throughout the entire Antarctic for the period 1965/1966 to 1970/1971 as either 1,700 or 2,800 (using two different methods to convert sighting data to absolute numbers). Since humpback whales have been protected since 1963, it is certainly true that in that year all populations were of the order of a few hundred. Chittleborough estimates that it will take 49 years for the Area IV stock to reach the level of 50% of the original unexploited stock size; for Area V stocks the required time is estimated to be 63 years. The 50% level of the unexploited stock is an approximation to the level that might provide maximum sustainable yield. These estimates are predicated upon the assumption that reproduction is not impaired by the low levels to which this population has been reduced.

Finback Whale, Balaenoptera physalus

The fin whale population of the southern oceans has been analyzed more carefully than any other whale population. Some analysis had been made by several authors prior to the special study of the Committee of Three authorized by the International Whaling Commission in 1960. This study is cited as Chapman, Allen, and Holt (1964). Subsequent updating of this study was carried out by FAO scientists and reported to the Commission. These reports may be found in the annual reports (Sixteenth to Nineteenth). Beginning in 1969 additional studies were made by Allen (1971), Chapman (1971), and by Doi, Ohsumi, and Shimadzu (1971). In addition, a special meeting of the Scientific Committee of the Commission was held in Honolulu in March 1970 to study further the assessment of this stock (Anon. 1971).

These studies have used a variety of methods, and while the different methods give slightly different estimates, there is general agreement on the present population size. Further, because of the detailed data available, it is possible to estimate this population by areas. Particularly good data are available for the period 1954/55 to 1962/63, when this stock was overwhelmingly the target of the industry. The basic catch data are shown in Table 9-5.

Estimates of the size of these stocks at two periods by the present author and by Allen are shown in Table 9-6.

The estimates given in the first two lines of this table are updated from those given by Chapman (1971, p. 71), except for Area I where a new estimate based on a lower gross recruitment rate has been used. This seems more reasonable in view of the exploitation history of this area. Allen's estimate (not yet published) was presented to the meeting of the Scientific Committee of the Commission at its 23rd meeting in Washington in 1971. Doi, Ohsumi, and Shimadzu (1971) estimated the 1970/71 population to be 73.3 to 81.1 thousand (average 77.2), so that these several estimates are in close agreement. It should be noted that all estimates apply to the beginning of the seasons indicated (that is, in December of the first year of the pair).

The size of the Antarctic fin whale stock in its unexploited state is believed to have been between 350 and 400 thousand, and the level

TABLE 9-5. Postwar catches of fin whales by area (including catch at
South Georgia in Area II). From published tables of the Bureau of
International Whaling Statistics, Oslo.

Season	I	II	III	IV	V	VI	Total
1946/47	0	6,345	4,382	3,355	478	0	14,560
1947/48	0	8,278	6,483	4,654	1,591	0	21,006
1948/49	0	6,917	6 457	3,068	2,489	72	19,003
1949/50	0	9,149	5,117	2,940	1,664	1,005	19,875
1950/51	462	7,003	4,133	4,210	2,198	1,212	19,218
1951/52	0	7,984	7,122	2,141	2,768	2,304	22,319
1952/53	0	6,221	11,364	1,449	1,709	1,891	22,634
1953/54	98	10,326	11,459	2,195	654	2,616	27,348
1954/55	0	7,709	13,607	2,790	3,021	1,227	28,354
1955/56	4,108	9,288	7,088	3,208	1,494	2,585	27,771
1956/57	5,615	12,754	3,812	291	23	5,064	27,559
1957/58	1,843	7,462	9,197	3,579	1,130	4,107	27,318
1958/59	0	5,589	9,385	7,277	2,625	2,102	26,978
1959/60	412	6,383	9,307	6,501	4,730	401	27,734
1960/61	373	6,610	13,269	2,086	3,174	3,174	28,686
1961/62	2,520	7,311	11,847	3,129	1,098	1,120	27,025
1962/63	1,373	5,570	8,977	1,725	645	346	18,636
1963/64	34	7,871	4,753	603	1,144	0	14,405
1964/65	66	5,031	1,199	766	747	0	7,809
1965/66	17	854	1,008	64	385	204	2,532
1966/67	44	81	1,554	372	304	530	2,885
1967/68	0	173	780	749	223	227	2,152
1968/69	130	32	552	1,627	413	260	3,014
1969/70	0	32	1,550	1,064	321	33	3,000
1970/71	0	307	1,710	657	115	99	2,888

that would provide maximum sustainable yield would be close to or
slightly larger than 200 thousand. The maximum sustainable yield at
such a population level has been estimated as 8 to 10 thousand animals
per year. Also closely estimated for this stock is the annual mor-

TABLE 9-6. Estimates of fin whale stocks by areas (all figures in thousands).

	Area						
	I	II	III	IV	V	VI	Total
1961/62	6.3	28.7	37.4	14.4	7.0	5.3	99.1
1970/71	4.3	20.0	29.0	12.5	7.4	6.4	79.6
Allen's least square estimate 1970/71	(4.4)	18.6	34.9	7.2	2.8	6.9	74.8

tality rate, which is 0.04 on the average for the adults; the rate for females is slightly higher than this and for males slightly lower, according to the best evidence.

The effect of stock density changes on the rate of recruitment has been observed through changes in the pregnancy rate and in a lowered age of sexual maturity with decreases in density. It is also possible that at lower densities the juvenile mortality rate is decreased, but unfortunately there is no direct evidence to support this conjecture. Thus, there is no adequate estimate of the relation of recruitment of the adult stock to stock size, though Doi, Ohsumi, and Shimadzu (1971) have developed some theoretical models for this relationship. From actual observations it has been shown that the average rate of recruitment to the exploited stock is approximately 8% of the parent exploited stock from which this recruitment is derived, and the median age of recruitment is at or slightly below 5 years of age.

Blue Whale, Balaenoptera musculus

Some analyses of blue whale stocks were also carried out by the Committee of Three (later Four) and reported by Chapman, Allen, and Holt (1964). The blue whale analysis has been complicated by the absence of adequate age data; catches of blue whales were very small by the time the ear plug method of age determination was fully usable. However, the present author has combined a variety of methods, including modified DeLury methods, to estimate the size of this population

at different periods. The different methods yield quite comparable
results and agree with the limited mark-recapture data. The author
has developed a complete estimated reconstruction of the blue whale
stock in Antarctic Areas II to V on the assumption that recruitment
and mortality rates for this population are the same as for the fin
whale. This reconstruction, shown in Table 9-7, has been compared

TABLE 9-7. Reconstruction of Area II-V blue whale stock, 1933/34 to
1957/58 (all figures in thousands).

Season	Initial stock size	Catch	End-of-year stock size	Recruit- ment	Stock size beginning of next season
1934	100.0	17.3	79.4	8.0	87.4
1935	87.4	16.5	68.1	8.0	76.1
1936	76.1	17.7	56.1	8.0	64.1
1937	64.1	14.3	47.8	8.0	55.8
1938	55.8	14.9	39.3	8.0	47.3
1939	47.3	14.1	31.9	7.0	38.9
1940	38.9	11.5	26.3	6.1	32.4
1941	32.4	4.9	26.4	5.1	31.6
1942	31.6	0.1	30.2	4.5	34.7
1943	34.7	0.1	33.2	3.8	37.0
1944	37.0	0.3	35.2	3.1	38.3
1945	38.3	1.0	35.8	2.6	38.0
1946	38.0	3.6	33.0	2.5	35.5
1947	35.5	9.2	25.2	2.8	28.0
1948	28.0	6.9	20.2	3.0	23.2
1949	23.2	7.6	15.0	3.1	10.1
1950	18.1	5.0	12.6	3.0	15.6
1951	15.6	6.3	9.3	2.8	12.1
1952	12.1	5.0	6.8	2.2	9.0
1953	9.0	3.6	5.2	1.9	7.1
1954	7.1	2.6	4.3	1.5	5.8
1955	5.8	2.0	3.6	1.2	4.8
1956	4.8	1.1	3.6	1.0	4.6
1957	4.6	0.7	3.7	0.7	4.4
1958	4.4	1.1	3.2	0.6	3.8

with independent estimates of the stock size in 1946/47, 1953/54, and
1957/58. The differences are +1.1, -2.0, and +1.2 thousand, providing considerable support for the reconstruction.

The estimated reconstruction has not been carried further than
the 1957/58 season because of the difficulty of separating the pygmy
blue whale catch from that of the blue whales. From Japanese sighting data of blue whales using fin whales as an index, Ohsumi and
Masaki (1972) estimate the 1965/66 to 1969/70 blue whale population
as 6,400; a different direct model yields a slightly higher estimated
average for this period plus 1970/71; this is 10,600. These totals
apply to the total Antarctic except for Area I, and hence refer essentially to the same stock as in Table 9-7. Thus the Table 9-7 reconstruction is quite in agreement with these recent estimates.

If the population in 1933/34 was about 100,000 after several years
of exploitation, the level of the unexploited population was perhaps
in the neighborhood of 150,000 and the level that would provide maximum sustainable catches would be between 75,000 and 100,000.

Sei Whale, Balaenoptera borealis

Sei whale stocks have been exploited intensively only for the past
decade. The record of catches by areas is shown in Table 9-8.

The interest in sei whales developed after blue whales had been
depleted to the level where catches were negligibly small, and at the
time that fin whales were becoming increasingly scarce. However, some
change in utilization of whale products further increased the interest
in sei whales. It is also possible that to some extent sei whales
took over the feeding areas of the other larger rorquals as these became depleted. Previously, sei whales had been regarded as temperate-
water species; it is true that the largest fraction of the catches are
obtained in the zone 40°S to 50°S (the most northerly zone in which
factory ships can take baleen whales). It is also possible that the
sei whale populations have actually increased with the additional food
supply available to them after the elimination of most of the blue and
the larger fraction of the fin whales. Unfortunately, there is no way
of proving or disproving such a conjecture.

TABLE 9-8. Sei whale catches in the Antarctic by Areas (1959/60 to
1970/71). From published tables of the Bureau of International Whaling Statistics, Oslo.

Season	I	II	III	IV	V	VI	Total
				Area			
1959/60	159	1,498	230	526	1,649	232	4,294
1960/61	102	1,938	336	103	563	2,030	5,072
1961/62	1,629	1,696	427	633	409	369	5,163
1962/63	807	1,812	1,457	631	430	345	5,482
1963/64	28	4,459	1,984	274	1,820	-	8,565
1964/65	40	16,076	443	1,564	2,207	-	20,330
1965/66	32	12,722	2,724	436	1,014	599	17,527
1966/67	-	1,540	6,865	2,826	717	402	12,350
1967/68	-	195	2,352	2,271	3,327	2,207	10,352
1968/69	73	188	1,771	1,030	2,156	552	5,770
1969/70	-	1,278	1,997	1,925	474	156	5,830
1970/71	-	640	1,065	3,967	285	194	6,151
Total	2,870	44,042	21,651	15,186	15,051	7,086	106,886

Gulland (MS) has developed estimates of the sei whale population
by areas. These have been extrapolated to December 1971 and are presented here (figures in thousands):

I	II	III	IV	V	VI	Total
11.5	25.4	11.7	2.4	12.2	9.7	72.9

These estimates, which apply to December 1971, are calculated assuming
the recruitment rate to be 0.07 and the natural mortality rate 0.04.
Recent catch and sighting data suggest that these estimates are not
necessarily in the correct relative order. For example, it is unlikely
that Area IV has this small a population; in fact, Area IV may well
have the largest population of any of the areas. Additional research
is necessary to provide reliable estimates. Nevertheless, the total
may not be unreasonable: Ohsumi and Masaki (1972), using a different
method, suggest that the total population is 82,000 and that the original population was 150,000. This refers to the level of the recruited

population. There remains, however, some doubt about the median age
of recruitment and the fraction of unrecruited stock in the Antarctic
and the fraction north of 40°S. Hopefully within the next year some
of the biological questions regarding this species will be answered,
so that it will be possible to develop better population estimates
and better estimates of the level that provides maximum sustainable
yield. It is important that this be done so that the Commission can
fix proper quotas and prevent the overexploitation that occurred with
the other large rorquals.

Summary

The table below summarizes the estimates given above for the Antarctic of original and present stock sizes and of present and maximum
sustainable yields.

Species	Original stock size	Present stock size	Present sustainable yield	Maximum sustainable yield
Blue (*Balaenoptera musculus*)	150,000	5,000–10,000	Near zero	3,000–4,000
Fin (*B. physalus*)	350,000–400,000	70,000–80,000	2,000–4,000	8,000–12,000
Sei (*B. borealis*)	150,000	70,000–80,000	5,000	5,000
Humpback (*Megaptera novaeangliae*)	90,000–100,000	1,700–2,800	Near zero	2,000–4,000

REFERENCES

Most of the analyses of Antarctic whale populations are included
in reports of the International Whaling Commission in recent years.
These may be included as annexes to the Scientific Committee Report or
as appendices to the main report. The reports of the Scientific Committee, the special reports, and the analyses by the FAO assessment
group should be consulted in the IWC reports from 1965 (Fourteenth
Report) to 1972 (Twenty-Second Report). However, some of the reports
or papers are cited in the body of this paper, and these are noted
below.

Allen, K. R. 1971. Notes on the assessment of Antarctic fin whale stocks. Rpt. IWC, 21:58-63.

Andersen, Harald T., ed. 1969. The biology of marine mammals. New York, Academic Press.

Anonymous. 1971. Report of the special meeting on Antarctic fin whale stock assessment, Honolulu, Hawaii, 13th-25th March 1970. Rpt. IWC, 21:34-39.

Brown, S. G. 1962a. A note on migration in fin whales. Norsk Hvalf.-Tid., 51:13-16.

_____ 1962b. The movements of fin and blue whales within the Antarctic zone. Discovery Rpt. 33:1-54.

_____ MS. A note on the migrations and movements of fin whales in the southern hemisphere as revealed by whale mark recoveries. Unpublished MS for the special meeting on Antarctic fin whale stock assessment, Honolulu, Hawaii, 13th-25th March 1970.

Chapman, D. G. 1971. Analysis of the 1969/70 catch and effort data for Antarctic baleen whale stocks. Rpt. IWC, 21:67-75.

_____ Allen, K. R., and Holt, S. J. 1964. Reports of the Committee of Three Scientists on the special scientific investigation of the Antarctic whale stocks. Rpt. IWC, 14:32-106.

Chittleborough, R. G. 1960. Marked humpback whales of known age. Nature (London), 187:164.

_____ 1965. Dynamics of two populations of the humpback whale, Megaptera novaeangliae (Borowski). Australian Journal Mar. and Freshw. Res., 16:33-128.

Doi, T., Ohsumi, S., and Shimadzu, Y. 1971. Status of stocks of baleen whales in the Antarctic, 1970/71. Rpt. IWC, 21:90-99.

Fujino, K. 1964. Fin whale subpopulations in the Antarctic whaling areas II, III and IV. Sci. Rpts. Whales Res. Inst., 18:1-27.

Gambell, R. 1968. Seasonal cycles and reproduction in sei whales of the southern hemisphere. Discovery Rpt. 35:31-134.

Gulland, J. 1969 (MS). The exploitation of the sei whale stocks of the Antarctic. Rome, FAO.

Hjort, J., Lie, J., and Ruud, J. T. 1932 to 1938. Norwegian pelagic whaling in the Antarctic [Pts.] 1-7. Hvalrådets Skr., 3:1-37; 7:128-152; 8:1-59; 12:1-52; 14:1-45; 18:1-44.

International Whaling Statistics. 1942. Oslo. Vol. 16.

Ivashin, M. V. MS. Locality of some commercial species of whales in the southern hemisphere. Unpublished MS for the special meeting on Antarctic fin whale stock assessment, Honolulu, Hawaii, 13th-25th March 1970.

Laws, R. M. 1960. Problems of whale conservation. Trans. N. Amer. Wildlife Conf., pp. 304-319.

_____ 1961. Reproduction, growth and age of southern fin whales. Discovery Rpt. 31:327-486.

_____ and Purves, P. E. 1956. The ear plug of the Mysticeti as an indication of age with special reference to the North Atlantic fin whale (Balaenoptera physalus Linn.). Norsk Hvalf.-Tid., 45:413-425.

Lockyer, C. MS. A method of bleaching earplugs of the sei whale (Balaenoptera borealis) in preparation for the counting of growth layers. London, Whale Research Unit, National Inst. of Oceanography.

Lund, J. 1950. Charting of whale stocks in the season 1949/50 on
the basis of iodine values. Norsk Hvalf.-Tid., 39:53-60, 298-305.
_____ 1951. Charting of whale stocks in the season 1950/51 on the
basis of iodine values. Norsk Hvalf.-Tid., 40:384-386.
Mackintosh, N. A. 1965. The stocks of whales. London, Fishing News
(Books), Ltd.
_____ and Brown, S. G. 1956. Preliminary estimates of the southern
populations of the larger baleen whales. Norsk Hvalf.-Tid.,
45:467-480.
_____ and Wheeler, J. F. G. 1929. Southern blue and fin whales.
Discovery Rpt. 1:257-540.
Matthews, L. H., and others. 1968. The whale. New York, Simon and
Schuster.
Ohsumi, S., and Masaki, Y. 1972. Status of stocks of baleen whales
in the Antarctic, 1971/72. Rpt. IWC, 22:60-68.
_____ and Shimadzu, Y. MS. Comparison of growth of fin whales among
various areas of the Antarctic Ocean. Shimizu, Japan, Far Seas
Fisheries Research Laboratory.
Purves, P. E. 1955. The wax plug in the external auditory meatus of
the Mysticeti. Discovery Rpt. 27:293-302.
Roe, H. S. J. 1967. Seasonal formation of laminae in the earplug of
the fin whale. Discovery Rpt. 35:1-30.
Ruud, J. T. 1952. Modern whaling and its prospects. FAO Fish. Bull.
No. 5, 165-183.
Slijper, E. J. 1962. Whales. New York, Basic Books, Inc.

CHAPTER 10

WHALE POPULATIONS AND CURRENT RESEARCH OFF WESTERN AUSTRALIA

J. L. Bannister

Only two whale species, the humpback (*Megaptera novaeangliae*) and
the sperm whale (*Physeter catodon*), have been commercially important
off Western Australia in recent years. A third species, the southern
right whale (*Eubalaena australis*), was taken by pelagic and coastal
whalers in the nineteenth century, while small catches of blue (*Balaenoptera musculus*), sei (*B. borealis*), Bryde's (*B. edeni*), and fin (*B. physalus*) have been obtained from time to time in this century.

I shall therefore consider two species, the humpback and the sperm,
and of those I shall concentrate on the sperm whale. That is the species I have been mainly concerned with myself, while the humpback situation has been comprehensively analyzed and described in a series of
publications by Chittleborough, culminating in his major paper (1965)
on the dynamics of the two "Australian" populations, those of Group IV
and Group V with breeding grounds off the west and east coasts of
Australia respectively.

The Humpback *Megaptera novaeangliae*

Here I would like to recapitulate the humpback story very briefly.
Its lessons are plain and emphatic and already well-known, the results
of overexploitation and lack of effective control. Even so, I shall
recapitulate some of the main points, taken from Chittleborough (1965),
as they apply to Western Australia.

Since 1900 the Group IV (west coast of Australia and Antarctic
Area IV) humpback population has been fished in two main cycles, 1934
to 1939 and 1949 to 1963. In the first period, over 12,500 humpbacks
were taken from the population both off the Western Australian coast

and in the Antarctic; the catch per unit effort declined over that
period, catching was concentrated on small whales, and the population
obviously became severely depleted. After full protection in the intervening years, catching began again in 1949. By 1963 a total of 18,180
whales had been taken from the population. A steady decline in abundance began in 1954. Despite attempts at selection of the large animals, the mean length of both sexes declined progressively after 1956
until 1961 when the minimum legal length was more strictly enforced.
The catch per unit effort declined by 90% over the period, from 0.48
to 0.05 whales per steaming hour. Chittleborough has calculated that
the Group IV stock consisted initially of 12,000 to 17,000 animals in
the unfished state, of about 10,000 in 1949, and of no more than 800
in 1962. In that state, Group IV could have given a sustainable yield
of some 18 whales. After regeneration the maximum sustainable yield
would be 390 whales per year; that state would not be reached before
the year 1991, and perhaps not until 2012, assuming of course complete
protection in the interim.

The Sperm Whale *Physeter catodon*

Sperm whales have been fished off Western Australia, as elsewhere
in the southern hemisphere, since at least early in the nineteenth
century. Townsend's (1935) charts of the positions of American whalers
on days when sperm whales were caught show two main grounds off Western
Australia, one centered some 300 to 400 miles off Carnarvon between
19 to 30°S and 102 to 112°E, the other extending from around Cape
Leeuwin to just east of Eyre in the Great Australian Bight—that is,
from 113° to 127°E and much closer to the coast. Both grounds were
visited in the warmer half of the year, while in the colder months
there was a considerable concentration of effort around the west coast
ground. That ground differed from most others away from the equator
in not being predominantly the site of a summer fishery. Further
east, and around Tasmania, Townsend's charts indicate very little
activity on sperm whales, possibly because it would have been a dangerous lee shore to sailing vessels. There are references to foreign

whaling vessels taking sperm whales there early in the nineteenth century (Lord and Scott 1924) and to local sperm whaling there from 1830 to 1850 and again in the 1870's (Crowther 1920).

In the twentieth century sperm whaling has been carried out from one main center, Albany on the south coast. Here there was some activity from 1912 to 1916, when the season was a summer one (Dakin 1938), and catches were fairly small (205 whales being recorded as taken in 1912/13, for example). But the main activity has been since 1955, with fishing mainly in the colder months. Operations were influenced by the taking of humpbacks during their northern winter migration; sperm whaling occurred at first after the humpback season. No humpbacks have been taken since 1963, and recently the sperm whale season has extended from March until December. Sperm whales are taken off Albany in a relatively restricted area about 25 miles wide and up to 60 miles long, off the continental shelf which is itself quite narrow off Albany, extending about 18 to 20 miles offshore. Sperm whale catches have averaged over 600 in recent years (Table 10-1). Very few females are taken; for example, only 7% of the catch was female from 1962 to 1967.

TABLE 10-1. Sperm whale catches in coastal whaling off Albany, Western Australia, 1962 to 1970.

Season	Males	Females	Total[a]
1962	555	12	567
1963	575	20	595
1964	716	85	801
1965	647	105	752
1966	595	11	606
1967	560	26	586
1968	583	75	658
1969	636	42	678
1970	776	23	799

[a] Including catches of small whales taken under permit for research, as follows: 1962—2 whales; 1963—21; 1964—89; and 1965—84.

Just as the modern shore-based fishery off Albany is sited close
to the center of an old southern pelagic ground, the only other shore
station to take sperm, at Carnarvon, operated close to the old western
ground. But the Carnarvon company concentrated on humpbacks migrating
north and south close to the coast. Sperm whales were only taken immediately before or after the humpback season; they were found much
farther from the coast than off Albany—a result of the greater width
of the continental shelf and more seaward dispersal of sperm whales
off Carnarvon.

In the past eight or nine years modern pelagic factories have
taken large numbers of sperm whales enroute to and from Antarctic
whaling grounds (Table 10-2). In the southern Indian Ocean these
catches were concentrated at first around Amsterdam and St. Paul Islands and eastward, but in 1965 and 1966 there were large catches
close to Western Australia, an area almost entirely neglected earlier
by modern pelagic factories. Large catches have also been taken off
Tasmania and New Zealand. One feature of these catches, in which they
differ from those taken by the Albany company, has been the high number of whales reported at or just above the minimum legal length (see
below) and the high proportion of females taken. The length limit of
35 ft (10.7 m) at Albany has resulted in only about 7% of the catch being
female, while in pelagic operations in the southeast Indian Ocean,
north of $40°S$, the proportion has been much higher—for example, 34%
in 1962-1966 (and even higher, 58%, in the southwest Pacific in the
same period).

Prior to 1962 pelagic sperm whale catches were restricted to
waters south of $40°S$. In Antarctic Area IV ($70°E$ to $130°E$) an average
of about 900 per year were taken each year, and rather more in Area V
($130°E$ to $170°W$).

Research

A comprehensive program aimed at stock assessment was begun in
1964, using data mainly from the Albany catches. Some material had
been obtained from both Albany and Carnarvon in previous years. A

TABLE 10-2. Sperm-whale catches in and north of Antarctic Areas IV
and V in pelagic whaling, 1962 to 1970.

Season	Sex	North of Area IV	Area IV	Total Males	North of Area V	Area V	Total Males
1962/63	Males	133	797	930	60	732	792
	Females	208			104		
1963/64	Males	618	1,400	2,018	382	2,434	2,816
	Females	904			839		
1964/65	Males	1,835	827	2,662	1,027	1,503	2,530
	Females	477			1,121		
1965/66	Males	960	617	1,577	42	2,209	2,251
	Females	198			26		
1966/67	Males	655	777	1,432	29	2,091	2,120
	Females	83			7		
1967/68	Males	526	376	902	9	1,430	1,439
	Females	35			–		
1968/69	Males	105	587	692	–	809	809
	Females	20			–		
1969/70	Males	515	887	1,402	96	1,085	1,181
	Females	63			70		

two-year aerial survey, begun off the Western Australian coast in 1963,
gave information on distribution, direction of movement, and seasonal
abundance from north of Carnarvon around the coast to east of Albany.
Each month, legs were flown parallel to the edge of the continental
shelf and out to sea, up to 150 miles from the shelf, from various
points (Bannister 1968). Since 1967, when full-time work on collection
of data and analysis was terminated, a monitoring program has continued
from Albany, with pelagic catch and effort statistics being obtained
from the Bureau of International Whaling Statistics. Certain biological collections were made from Albany in 1971 and it is expected that
with continued government support the work will continue.

The "Western Australian" Stock

    Stock identity. There is at present no direct evidence on the
limits of the stock fished off Western Australia. The presence of a

film of an Antarctic diatom (*Cocconeis*) on males caught off Albany suggests very strongly that some proportion, at present unknown, of the
bulls returns to the breeding stocks from colder waters. So far only
two whale marks have been recovered, both from females and by USSR
pelagic factories, from over 400 fired into sperm whales off the coast.
One was recovered after two years, close to where it was fired; the
other, about 200 miles out into the Indian Ocean 88 days after marking
off Albany. The aerial survey results (see Bannister 1968) gave no
reason to suppose that sperm whales off the west coast were separate
from those off the south coast, and in summer months there was a predominantly southward movement, from warmer waters, along that coast.
Unfortunately there were very few winter data for comparison and no
evidence of a reversal in the trend in colder months.

Sperm whales certainly seem to be concentrated along the continental slope off Albany, more so than at any other place on the coast
at least as far north as beyond Carnarvon. This may be due to a combination of two factors, the first the presence of food on the steeply
sloping edge of the shelf at that point, possibly enhanced by the
presence of submarine canyons, and the second the concentrating effect
of the east-west coastline on whales heading northward from colder
waters. A remarkable feature of the Albany spotter data is the consistently westward or southwestward movement of whales observed throughout the season (in 1963 and 1964, for example, 84% of over 800 observations were of whales heading in that direction). Pelagic activities
east of Albany in March 1965 disrupted the normal Albany operations to
such an extent that one can only conclude that at least at that time
of year whales were moving along the coast from some distance to the
east.

As a working hypothesis, then, I have assumed that the sperm
whales off Western Australia belong to a single stock, the males
traveling to and from the Antarctic and whales off the south coast
traveling from (because I have no evidence to the contrary) possibly
as far east as Tasmania, and out into the Indian Ocean. How far into
the Indian Ocean is again a matter for conjecture, although from plots
of recent catches and catch rates there seems to be some concentration

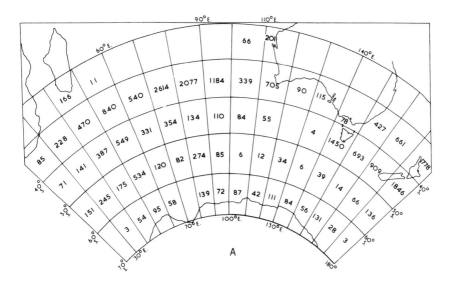

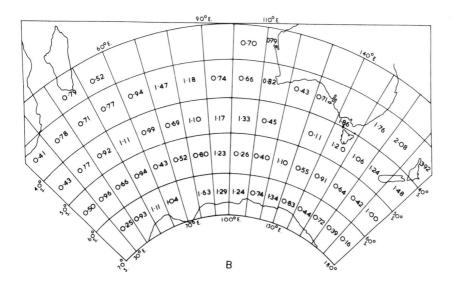

FIGURE 10-1. Number of sperm whales caught in pelagic whaling, 1962 to 1970, south of 30°S and between longitudes 20°E and 180°E, from data provided by the Bureau of International Whaling Statistics.
  (a) Catches (both sexes combined) by 10° square.
  (b) Catch rate (catch per net catcher day) per 10° square.

of whales in the St. Paul/Amsterdam Islands area, and therefore the possibility of a central oceanic stock (Fig. 10-1). On the assumption that few females move into cold waters, I would expect rather little link of breeding stocks south of Tasmania. It may be that we should consider the whole Indian Ocean, at least as far north as the equator and possibly farther, as containing one stock, but for the present I have restricted my hypothesis to a stock extending not much farther east than 95^{0}E, about halfway between the Western Australian coast and St. Paul/Amsterdam Islands, and therefore including the eastern part of Antarctic Area IV (and warmer waters north of it) as well as some of Area V.

Life history. From the commercial catches at Albany, particularly between 1964 and 1967, a considerable amount of biological material was collected; a small amount had been obtained earlier from Albany and Carnarvon. The usual biological collections of gonad material and stomach contents were obtained as well as teeth for age determination, both from these and from special "research" catches of small whales taken under permit, particularly off Albany in 1964 and 1965 (163 whales taken) and off Carnarvon in 1962.

Analysis of these data (summarized in Bannister 1969) has shown that the sperm whales off Albany do not differ greatly from those examined elsewhere, particularly off South Africa. For example, they feed on much the same food as at Durban—mainly histioteuthid and ommastrephid squid, although deep-sea fish (e.g. the angler *Himantolophus*) are obtained occasionally. Dr. M. R. Clarke, who has examined the squid, tells me that species typical of Antarctic waters are present at least in some large males.

From counts of laminae in mandibular teeth (made easier by etching in dilute formic acid, a technique developed in Australia by Bow and Purday 1966), there seems to be little difference in the age at sexual maturity or in ovulation rate between whales sampled at Albany and elsewhere, assuming one lamina is formed each year. From information on fetuses and recently born calves, gestation seems to last about $15\frac{3}{4}$ months, with most conceptions in midsummer (December to January) and most births in early winter (April to May). More mature males

had larger seminiferous tubule diameters in the testis in early summer
(October to December) than in winter (May to July), and there is some
evidence of an increase in ovarian follicle diameter in early summer.
All this points to a main pairing period in early summer off the Western
Australian south coast.

The annual pregnancy rate is calculated at about 25%, a little
higher than in some other areas. From adult females sampled between
June and October (to avoid complications due to the $15\frac{3}{4}$ -month gestation period) and basing results on the proportions of adult females
in the various stages of the breeding cycle, it seems that lactation
might last rather a short time (of the order of four to six months)
and the resting stage a very long time (over a year at least) in a
three- to four-year cycle. These last two are the most strikingly
different results by comparison with those from elsewhere, possibly
because of the small size of the sample. In South African data, for
example, lactation has been estimated at 24 to 25 months and the resting stage at 8 to 9 months (Anon. 1971). But with the relatively
small numbers of females examined, it has not been possible to analyze
the various stages of ovarian corpora regression as some other workers
have done and so arrive at independent estimates of the duration of
the breeding cycle and the rate of ovulation.

Population structure. The Albany fetal sex ratio (from 55 records)
was very close to 50% male. The small number of females relative to
males in the Albany catch (135 females compared to 4,500 males from
1955 to 1966) has already been mentioned. Of the research catches,
70 to 80% were female.

The Albany catch has been predominantly males 39 to 47 ft (11.9
to 14.3 m) long, with subsidiary groups less than 39 ft (11.9 m) and
greater than 47 ft (14.3 m) (Fig. 10-2). There have been changes in
the length distribution from year to year, but the average length has
varied little, between just under 43 ft (13.1 m) and almost 44 ft
(13.4 m), increasing somewhat since 1968 (Table 10-3). Using an age-
length key prepared for males from teeth collected in 1964/65, the
calculated age distributions show that most catching is on males of
18 to 30 tooth laminae (17 to 29 years), being fully recruited from

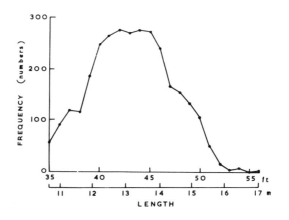

FIGURE 10-2. Length frequency distribution of the catch of male sperm
whales caught off Albany, Western Australia (combined data for 1962
to 1966).

TABLE 10-3. Annual mean length of sperm whales caught off Albany,
Western Australia, from 1962 to 1970. The figures are for males ≥
35 ft (10.7 m) in length.

| | Mean length | | |
Year	(ft)	(m)	Catch
1962	43.6	13.3	555
1963	43.2	13.2	570
1964	42.7	13.0	697
1965	42.6	13.0	633
1966	43.6	13.3	592
1967	43.3	13.2	559
1968	42.9	13.1	570
1969	43.3	13.2	630
1970	43.9	13.4	777

21 to 25 years (Fig. 10-3). There is a remarkable spread in the age

distribution of females (Fig. 10-4), and no age-length key has been

used for female catches.

The modal length and age of the Albany catches are higher than

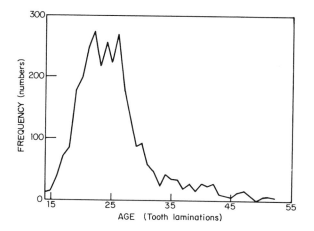

FIGURE 10-3. Calculated age distribution of the catch of male sperm whales off Albany, Western Australia (combined data 1962 to 1966).

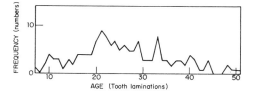

FIGURE 10-4. Age distribution of the catch of female sperm whales off Albany, Western Australia (1964/65 data).

those of catches in similar latitudes, such as at Saldanha Bay where the modal age of the male catch is 18 years (Best 1970). They are also rather higher than in pelagic catches for similar latitudes, but those data are affected by the large numbers recorded at the minimum legal length, as shown in Fig. 10-5. The modal length for the Albany male catch is much closer to the pelagic figure for latitudes 40° to 49°S than for 30° to 39°S. This may also be due in part to the coastline near Albany acting as a concentration area for males traveling north from colder waters.

Population density. The relative abundance of legally catchable bulls— $\geq$ 35 ft (10.7 m)—declined at Albany over the period 1962 to 1966, by a total of between 20 and 40%. The annual catch rates, adjusted for fishing power and catchability, and the sighting rates from the

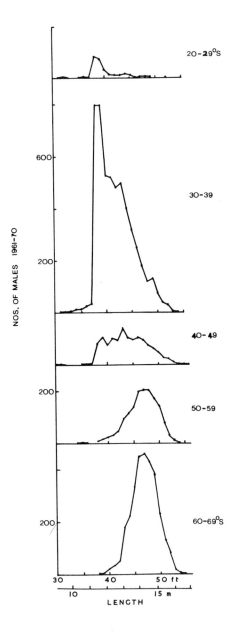

FIGURE 10-5. Length frequency distribution of male sperm whales, pelagic catches, 95°E to 180°E, by 10° latitude bands; combined data 1962/63 to 1965/66.

spotter aircraft, showed much the same trend. But since 1966 there
have been changes in aircraft and catchers, as well as the introduction of asdic on one catcher in 1969 and on another in 1970, all increasing the total efficiency of the operation in a way which so far
it has not been possible to quantify exactly. This increase in efficiency must have been responsible, for example, for much of the almost
40% increase in catch rate in 1970 relative to 1969, since the sighting
rate from the aircraft showed no increase in that period, without, as
far as is known, a change in aircraft operation. The unadjusted catch
rates and the aircraft sighting rates are shown in Table 10-4.

TABLE 10-4. Annual catch and sighting rates of sperm whales off Albany,
Western Australia, 1962 to 1970.

	1962	1963	1964	1965	1966	1967	1968	1969	1970
Catch (males)[a] per hunting hour	0.21	0.17	0.17	0.13	0.13	0.14	0.16	0.15	0.20
Catchable males[a] seen per effective searching hour	2.9	2.5	2.5	2.4	1.8	2.3	2.4	2.1	1.9

[a] Males $\geq$ 35 ft (10.7 m) in length.

Catchable bulls are most abundant in spring and early summer
(Fig. 10-6). Juveniles and other undersized whales are most abundant
in autumn, but they tend to be in the area again in spring and early
summer. Small bulls $\leq$ 38 ft (11.6 m) are abundant in April and August;
larger bulls, however, tend to appear more frequently in early summer,
with those $\geq$ 44 ft (13.4 m) appearing about a month later than the
medium-sized bulls, 39-43 ft (11.9-13.1 m), with the mode in July to
August. Unfortunately these peaks of abundance cannot be related to
changes in direction of movement, although the appearance of bulls in
late winter suggests that they may be returning north to join the
breeding herds at that time of year.

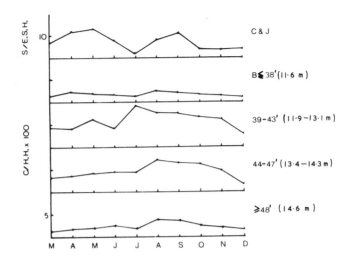

FIGURE 10-6. Monthly abundance through the season of various classes of the catch of sperm whales off Albany, Western Australia: cows and juveniles per aircraft spotter effective searching hour, bulls per 100 catcher hunting hours.

Present status. The downward trend in Albany catch and sighting rates from 1962 to 1966 seems not to have continued since then. Coupled with the recent increase in mean length, this indicates that there seems no immediate cause for alarm, though the situation needs to be watched closely, particularly as the effects of large catches may not be felt for some years.

An attempt to estimate stock sizes and sustainable yields, using a rather incomplete mathematical model, was made by Bannister (1969). The results suggested a rather lower yield than that likely from models derived by the IWC Scientific Committee (Anon. 1969). Some preliminary analyses at the special meeting of the Scientific Committee in Honolulu 1970 suggested that "Area IV plus Australia" female stock might be on the order of 47,000. That size female stock, on the basis of the models derived earlier, could give a sustainable yield of approximately 2,000 males. Recently the total male catches from that region have been close to that figure (Table 10-2), so that if the preliminary calculations are not greatly in error, there should be no reason to believe the stock there at least is being very badly

depleted. But, as pointed out in the Scientific Committee's report
of the special meeting, the estimates are only rough; they are based
on assumptions of stock units, of the relation between males fished
by pelagic and shore-station operations, and of the effects of migration between cold and warm waters—all of which urgently need investigation.

Future Research

Records of humpback sightings, collated as observations are reported, are so far not comprehensive or comparable enough to give
positive evidence on whether the stocks off Western Australia are
showing signs of recovery. The same may not be quite so true for
right whale sightings, none of which were reported off Western Australia in this century before 1955 (Chittleborough 1956) but which are
now made two or three times each year, particularly in late winter
when females come inshore to give birth. Over the years these sighting data, together with observations of blue, fin, and sei whales,
should provide a body of evidence to give indications of the health
of the various stocks.

Monitoring of sperm whale data from Albany will proceed, assuming continuation of present government support. At the same time,
there is an urgent need for investigation of the link between Antarctic and warmer-water catches. Albany seems to be in a unique position
to provide evidence of this link, and at the time of this writing collections of diatom film and stomach contents are being made (the latter in cooperation with Japanese workers) for this purpose. With
detailed analyses of pelagic catch records it is hoped that this
biological evidence will give results useful in refining some of the
approximations currently used in estimating sustainable yields.

REFERENCES

Anonymous. 1969. Report of the IWC-FAO working group on sperm whale
    stock assessment. Rpt. IWC, 19:39-83.

_____ 1971. Report of the special meeting on sperm whale biology and stock assessments, Honolulu, Hawaii, 13th-24th March 1970. Rpt. IWC, 21:40-50.

Bannister, J. L. 1968. An aerial survey for sperm whales off the coast of Western Australia 1963-65. Aust. J. Mar. Freshw. Res., 19:31-51.

_____ 1969. The biology and status of the sperm whale off Western Australia—an extended summary of results of recent work. Rpt. IWC, 19:70-76.

Best, P. B. 1970. The sperm whale (*Physeter catodon*) off the west coast of South Africa. 5. Age, growth and mortality. Investl. Rpt. Div. Sea Fish. S. Afr., 79:1-27.

Bow, J. E., and Purday, C. 1966. A method of preparing sperm whale teeth for age determination. Nature, Lond., 210:437-438.

Chittleborough, R. G. 1956. Southern right whale in Australian waters. J. Mammal., 37:456-457.

_____ 1965. Dynamics of two populations of the humpback whale *Megaptera novaeangliae* (Borowski). Aust. J. Mar. Freshw. Res., 16:33-128.

Crowther, W. L. 1920. Notes on Tasmanian whaling. Proc. Roy. Soc. Tasmania:130-151.

Dakin, W. J. 1938. Whalemen adventurers. Rev. ed. Sydney, Sirius Books.

Lord, C. E., and Scott, M. H. 1924. A synopsis of the vertebrate animals of Tasmania. Hobart. Oldham, Beddome, and Meredith.

Townsend, C. H. 1935. The distribution of certain whales as shown by logbook records of American whaleships. Zoologica (N.Y.), 19:1-50.

PART THREE
Biology

CHAPTER 11

THE BIOLOGY OF THE SPERM WHALE AS IT RELATES TO STOCK MANAGEMENT

Peter B. Best

The annual world catch of sperm whales from 1910 to 1946 did not exceed 7,500 animals, but postwar catches have increased to nearly 30,000 animals a year, about two-thirds of which are males. With the recent high prices being paid for sperm oil, and the decreasing stocks of baleen whales in general, the species must now be considered a major resource for the whaling industry. Peculiarities of the behavior of the sperm whale make it necessary for new models for stock assessment to be created, but unfortunately our knowledge of the biology of the species has lagged behind that of baleen whales.

This paper is intended as a summary of present information on those aspects of sperm whale biology that are significant as parameters in stock assessment procedures. The most reasonable estimates for these parameters are then applied to existing assessment models to obtain estimates of the possible yield for a given population size.

In compiling this paper, I am indebted to the following people who had no objection to my quoting from their unpublished work:

J. L. Bannister, Western Australian Museum, Perth

L. K. Boerema, Fish Stock Evaluation Branch, Food and Agriculture Organization, Rome

Y. Masaki
and   } Far Seas Fisheries Research Laboratory, Shimizu, Japan
S. Ohsumi

D. W. Rice, National Marine Fisheries Service, La Jolla, California, U.S.A.

Y. Shimadzu, Far Seas Fisheries Research Laboratory, Shimizu, Japan

Stock Units

The area for which there is the best information available on

stock identities of sperm whales is the North Pacific. An analysis
of mark recoveries suggests that there are migration routes north and
south on either side of the Pacific, though whales from either route
intermingle in latitudes above $40°N$ (Anon. 1969b; Masaki 1970a). From
the incidence of blood types, Fujino (1963) was able to distinguish
between two stocks of sperm whales divided approximately into east and
west by the $170°E$ line, the two populations mixing around the Kurile
Islands and the east coast of Kamchatka. The density of sperm whale
distribution, as measured by sightings from survey vessels, indicates
three areas of high density—one west of $170°E$, one between $170°E$ and
$160°W$, and one east of $160°W$. This evidence, in conjunction with differences in the seasonal size composition, has led Masaki (1970a) to
conclude that there are basically three stocks in the North Pacific.
One occurs in the western part (west of $170°E$) and includes whales
from the Bonin Islands, Japanese coastal waters, and the east coasts
of the Kurile Islands and Kamchatka. A second stock occurs in the
eastern part of the Pacific (east of $150°W$) and includes Alaska Bay,
while the third stock is a central one (between $170°E$ and $150°W$), including waters around the Aleutian Islands. Considerable intermingling
may occur at the boundaries of these areas and also north of $40°N$.

Elsewhere in the world there is much less evidence for stock identification. Various morphological features such as body color, tooth
counts, and the number of dorsal humps, which have some inherent variation, apparently differ little between either coast of South Africa,
the North Pacific, the Antarctic, the Azores, or the southeast Pacific
(Best and Gambell 1968). Similar results were obtained in an analysis
of the external characteristics and number of teeth in sperm whales
from four localities on the west coast of South America (Clarke,
Aguayo, and Paliza 1968).

Although marking of sperm whales in the southern hemisphere has
been much less intensive than marking of baleen whales (according to
information available to the author, about 1,500 sperm whales have
been marked south of the equator to date), certain recoveriees have
been of great interest. Two whales marked by the Soviets off the
Congo ($6°S$) and off Cap Blanc ($21°$ to $22°N$) were killed off Saldanha

Bay (33°S) on the west coast of South Africa. A third whale, marked
by the Soviets in the western half of Antarctic Area III (ca. 62°S),
was killed off Durban (30°S) on the east coast of South Africa (Best
1969b). These recoveries provide direct evidence of extensive north-
south movements, suggesting that the separation of stocks should be
considered on a longitudinal rather than a latitudinal basis. Although one Soviet mark recovery also implies that intermingling with
northern hemisphere animals may occur across the equator, the fact
that the female breeding seasons in the two hemispheres are about six
months apart would seem to preclude their interbreeding to any extent.

Additional confirmation of movements north and south comes from
the seasonal appearance of films of the Antarctic diatom *Cocconeis*
ceticola on medium-sized and large males in south temperate waters
(Bannister 1969; Best 1969b), as well as distinct peaks in the abundance of different classes of sperm whales whose timing fits well
with a northward movement in autumn and a southward movement in spring
(Best 1969b; Gambell 1967).

If a longitudinal basis for stock separation is accepted, then
there also seems good reason for considering each oceanic region in
the southern hemisphere separately. The southern limit of female
sperm whale distribution varies between 45° and 50°S, at least in the
Indian and southwest Pacific Oceans (Ohsumi and Nasu 1970), so that
the land masses of South America, South Africa, and Australasia may
act as partial barriers to the movement of females from one ocean to
the next. Hence separate breeding stocks may be postulated for each
oceanic region. This approach has been followed by most workers in
the southern hemisphere in attempting to delineate their stocks. Bannister (1969) has taken the longitude of Tasmania (146°E) as the
western limit from the distribution of catches. Gambell (1972) has
for similar reasons taken the limits of the East African stock to be
between the longitude of Cape Agulhas (20°E) and about 70°E, while
Best (1970a) has taken the West African stock to be approximately
equivalent to Area IIE and Area IIIW in the Antarctic (that is, between about 30°W and 35°E).

Actual evidence of separate populations within the Antarctic is

scant, though of course this may be due to intermingling of males
from different breeding stocks in summer, much as in the North Pacific.
Two sperm whales marked by the Soviets on the eastern edge of Area IV
were killed one and two years later, respectively, the first being 220
and the second 350 mi northeast of the marking position. The latter
whale had moved into the western edge of Area V. A third whale,
marked by the Soviets in Area V, was killed five years later, 700 mi
east of the marking position, still in Area V (Ivashin 1970). These
returns suggest that sperm whales may enter the same area of the Antarctic each year. Cushing, Fujino, and Calaprice (1963) found some
indication that sperm whales between about 40°E and 55°E (in Area IIIE)
might differ in the frequency of their blood types from whales between
about 60°E and 90°E (on the eastern edge of Area IIIE and in Area IV),
although larger samples were needed to confirm this. There is also
some indication from morphometric analysis that sperm whales from
South Georgia to between 95°E and 150°E are a mixed stock containing
one component that does not differ in whales from Durban, the Bonin
Islands, Japan, and the southeast Pacific, and a second component that
has certain differences in the measurement of flippers and tail flukes
(Clarke and Paliza, in press). The latter component is strongest in
the west at South Georgia, suggesting that this represents an Atlantic stock.

The Bureau of International Whaling Statistics has made available
catch and effort figures for all sperm whales caught outside the Antarctic and south of 40°S by pelagic fleets before and after the baleen
whale season. These were provided by 10° squares of latitude and
longitude for the seasons 1961/62 to 1969/70. The total catch per
catcher day for all seasons and both sexes combined is illustrated in
Fig. 11-1 for each square. Certain regions of high density can be
detected. These are the areas between (a) 140°E and 160°W, (b) 150°W
and 100°W, (c) 60°W and 10°W, (d) 30°E and 60°E, and possibly (e) 70°E
and 90°E. Areas (a), (c), and (d) are represented by regions of high
density both north and south of 40°S, while for area (b) there is only
information south of 40°S. If each of these areas is really indicative
of high population density, then they could be designated as the fol-

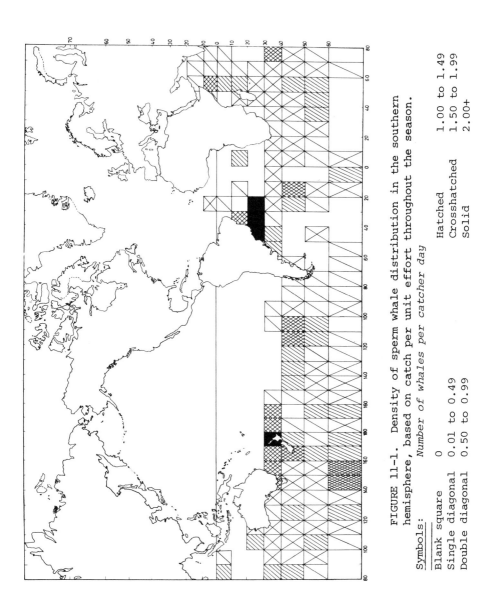

FIGURE 11-1. Density of sperm whale distribution in the southern hemisphere, based on catch per unit effort throughout the season.

Symbols:
Number of whales per catcher day

Blank square	0	Hatched	1.00 to 1.49
Single diagonal	0.01 to 0.49	Crosshatched	1.50 to 1.99
Double diagonal	0.50 to 0.99	Solid	2.00+

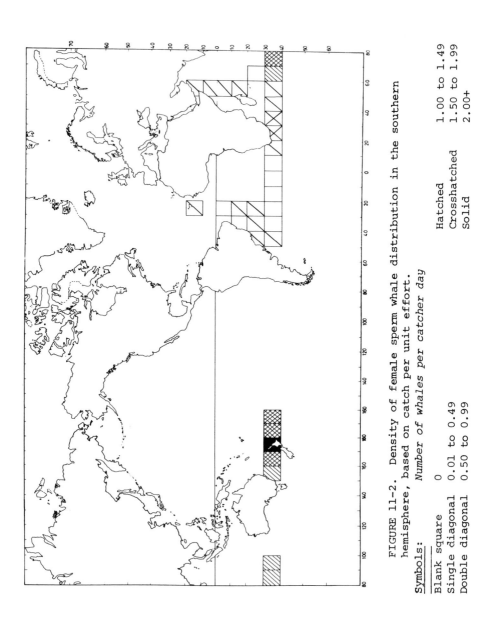

FIGURE 11-2. Density of female sperm whale distribution in the southern hemisphere, based on catch per unit effort.

Number of whales per catcher day

Symbols:

Blank square   0
Single diagonal   0.01 to 0.49
Double diagonal   0.50 to 0.99

Hatched   1.00 to 1.49
Crosshatched   1.50 to 1.99
Solid   2.00+

lowing stocks: (a) New Zealand and East Australian, (b) Central South Pacific, (c) Eastern South American, (d) East African, and (e) Indian.

In Fig. 11-2 the catch per unit effort has been plotted for females only; the returns of one nation were omitted, as north of 40°S they contained an unrealistically low proportion of females. Although the data coverage is poor, regions of high density appear in areas (a), (c), (d), and (e), suggesting that these may be separate breeding stocks.

Better coverage for the area north of 40°S is provided by the data in Slijper, van Utrecht, and Naaktgeboren (1964), based on sightings from Netherlands merchant ships. These were available as the number seen per 1,000 hours steaming per month per 10° square, and the data for all twelve months have been combined in Fig. 11-3 for the area south of 10°N. There seems to be a concentration of sperm whales on either side of the Atlantic, particularly off west Africa between 20°W and 20°E. In the Indian Ocean there is an area of high density off east Africa between 30°E and 50°E; an area of very high density from 90°E to 140°E, particularly in equatorial regions; and a possible third, central concentration south of India from 60°E to 90°E. Data from the Pacific are insufficient for any conclusions to be drawn.

Combining these two sets of data, the following stocks can be proposed.

(1)	Eastern South American	60°W to 20°W
(2)	West African	20°W to 20°E
(3)	East African	20°E to 60°E
(4)	Indian	60°E to 90°E
(5)	West Australian	90°E to 140°E
(6)	New Zealand and East Australian	140°E to 160°W
(7)	Central South Pacific	160°W to 100°W

There is very little information available here for the southeast Pacific, but it seems likely that there is a further stock (Western South American) from 100°W to 60°W.

These are basically the same units as those proposed at the 1963 Seattle meeting (Anon. 1963), with the addition of a central Pacific and a central Indian Ocean stock.

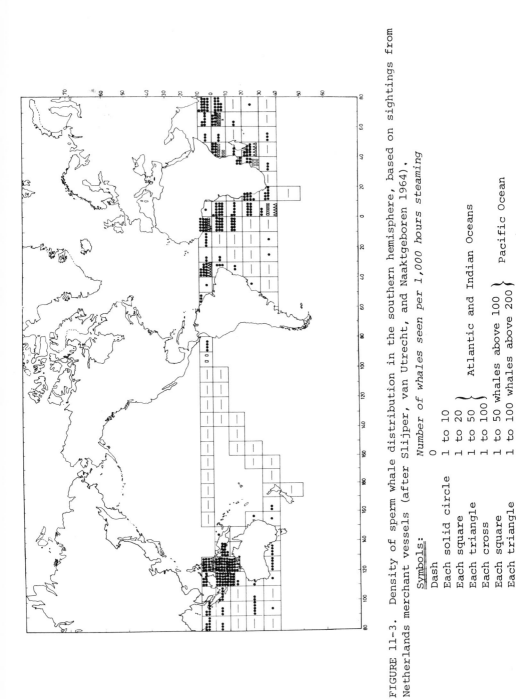

FIGURE 11-3. Density of sperm whale distribution in the southern hemisphere, based on sightings from Netherlands merchant vessels (after Slijper, van Utrecht, and Naaktgeboren 1964).

Symbols:

Number of whales seen per 1,000 hours steaming

Dash	0	
Each solid circle	1 to 10	
Each square	1 to 20	}
Each triangle	1 to 50	} Atlantic and Indian Oceans
Each cross	1 to 100	}
Each square	1 to 50 whales above 100	}
Each triangle	1 to 100 whales above 200	} Pacific Ocean

Segregation by Sex and Age

The exclusion in the southern hemisphere of female sperm whales from latitudes higher than 45°S to 50°S has already been mentioned. In the North Pacific, however, females are frequently found beyond 50°N, particularly in the northwest, their limit being about 57°N (Ohsumi and Nasu 1970). This is probably due to the more complex oceanographic structure of the North Pacific. The restricted distribution of female sperm whales has its significance in determining the geographic limits of breeding stocks, as described above. A temporary or permanent segregation of certain reproductive classes of females has also been suggested (Clarke 1956; Pike 1963), but there is no evidence of this on the west coast of South Africa, or on the east coast (Gambell 1972). If such a segregation did occur, it would have considerable importance in the calculation of the duration of one or more phases of the reproductive cycle (see below). In males, however, there is additional segregation by size and age.

The analysis of seasonal movements of sperm whales in several temperate coastal regions has revealed that males tend to move through the whaling grounds in different size groups, and that each group has a different pattern of migration (Bannister 1969; Best 1969b; Gambell 1967). Samples of the population taken over different periods of time are therefore likely to contain different proportions of these size groups, and this can affect both the apparent age at recruitment and the total mortality rate calculated from an age composition. The segregation of males above a certain age and size into higher latitudes also raises problems in the interpretation of the age composition of the male catch, for without knowledge of the segregation rate and its variation with age and time of year one cannot be sure of the validity of mortality estimates so calculated.

Ohsumi (1966) has constructed a mathematical model to describe changes in the male segregation rate with age, based on the difference between the age composition of males on the coast of Japan and in Aleutian Islands waters during summer and early autumn. Males apparently begin to segregate to higher latitudes at age 12, and

continue to do so at an increasing rate until age 25, after which they
disperse at a constant rate. The rate of increase in segregation from
ages 12 to 25 is given as 0.092. At about age 18, half of the males
have segregated to higher latitudes, and from age 25 on segregation
rates vary from 75% to 90%.

Certain biological evidence supports some of these conclusions.
The nature of cyamid infestation on male sperm whales off the west
coast of South Africa has been shown to change at a body length of
39 to 40 ft, or close to 12 m, when the whales apparently lose a species otherwise characteristic of females (*Neocyamus physeteris*) and
become infested with a second species (*Cyamus catodontis*): this is
considered to represent the stage at which 50% of males enter the
Antarctic for the first time (Best 1969a). This is equivalent to an
age of 19 years (Best 1970a), which agrees closely with Ohsumi's estimate for the age at 50% segregation.

In 1946, 1947, 1948, 1950, and 1951, whaling off Durban continued
throughout the year. The proportions of females and small (shorter
than 40 ft, or 12.2 m), medium-sized (40 to 45 ft, or 12.2 to 13.7 m),
and large (longer than 45 ft, or 13.7 m) males in the catch have been
calculated for each month and smoothed by taking a running average of
threes (Fig. 11-4). In summer (December to February) large males

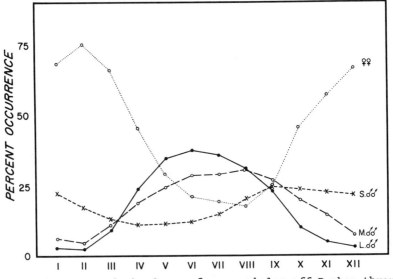

FIGURE 11-4. Seasonal abundance of sperm whales off Durban throughout the year (1946-1948, 1950-1951).

averaged 3.2% of the catch and were at their lowest abundance for the
year, whereas at their peak in winter (June to August) they comprised
36.1% of the catch. On these figures alone, the degree of segregation
of large males to higher latitudes in summer could be estimated as 91
to 92%. This, however, makes no allowance for seasonal variations in
sperm whale abundance. The catch per unit effort of sperm whales off
Durban in February and March is two or three times that in midwinter
(Gambell 1972). This means that the percentage of large males in mid-
summer should be multiplied by a factor of two or three, and under
these circumstances the proportion that segregate to higher latitudes
in summer can be calculated as 73 to 82%. These figures are all close
to Ohsumi's estimates.

As a consequence, the model proposed by Ohsumi for the North
Pacific also seems applicable to the Antarctic. The age at 50% entry
to this area can therefore be taken as 18 and the segregation rate of
males between ages 18 and 25 as 0.092, with no correction necessary
to estimates of Z (the total instantaneous mortality rate) above this
age. The proportion of large males absent from the Antarctic in summer is probably between 10% and 25%.

Age Determination

Three methods of assessing the age of individual sperm whales
have been tried, one of which (counting the number of laminae in the
bone of the lower jaw) has been shown to be of no use after an age
of about 13 (Nishiwaki, Ohsumi, and Kasuya 1961). The method in
almost universal use at present involves counting the number of growth
layers in the dentine of mandibular teeth. One growth layer is composed of two adjacent laminae, one translucent and the other opaque.
Although an accumulation rate of two growth layers per year was
originally proposed (Nishiwaki, Hibiya, and Ohsumi 1958), it is now
considered more likely that there is one layer per year (Ohsumi,
Kasuya, and Nishiwaki 1963; Best 1970a). The most convincing evidence in favor of this theory comes from the number of growth layers
in the teeth of marked sperm whales which had been at large for a

considerable time between marking and recapture. In 9 out of 16 such
whales, the number of layers divided by the number of years at large
was less than 2.0, and in four of them, less than 1.5. When it is considered that the age of the whale at marking was unknown, the accumulation rate of growth layers seems more likely to be one per year than
two (Best 1970a). Although Gambell and Grzegorzewska (1967) produced
evidence of a biennial accumulation rate from seasonal changes in the
nature of the most recently formed lamina, Gambell (1972) now considers it more reasonable to assume one layer per year.

There is some doubt, however, about the accumulation rate of
dentinal layers in very young sperm whales (Anon. 1969c). For this
reason a permit was granted by the South African Division of Sea Fisheries for the taking of fifteen calves in March 1971. Only nine were
taken. The number of growth layers in teeth from these animals has
been plotted against body length in Fig. 11-5, where similar data have
been included for a 15.5-ft (4.72 m) male calf from Durban (Best 1968),
six whales up to 23 ft (7 m) in length from whole schools shot under
special permit by the Japanese (Ohsumi 1970a), and three small whales
mentioned by Nishiwaki, Hibiya, and Ohsumi (1958). The mean length

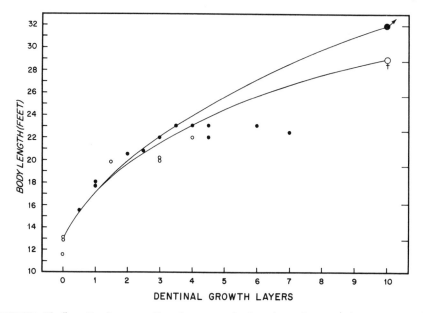

FIGURE 11-5. Early growth of sperm whales based on number of dentinal
growth layers.

at age 10 has been included from published growth curves for each sex
(Best 1970a). Approximate growth curves have been fitted to these
points by eye, making allowance for the fact that the sample is stratified by length (only animals below a certain size being included), so
that in older ages the lengths tend to fall below the growth curve.
The mean length at age one is about 17 ft (5.2 m); at age two, 19 to
20 ft (5.8 to 6.1 m); and at age three, about 22 ft (6.7 m). There
is reason to believe from the size composition of the population that
animals reach a length of 19 to 20 ft (5.8 to 6.1 m) one year after
birth (see below), and consequently two growth layers may be laid down
in the first year.

The third method used for age determination involves the accumulation of corpora albicantia within the ovaries, and so is only applicable to mature females. Its validity depends on the assumptions,
first, that the frequency of ovulation is fairly regular throughout
reproductive life, and second, that all corpora albicantia persist
in the ovaries as visible traces of both successful and unsuccessful
ovulations. The first of these assumptions seems reasonable from the
linear relationship between corpora counts and the number of growth
layers in the teeth over the main span of reproductive life (Best
1970a), even though the ovulation rate does seem to decrease in older
animals (Ohsumi 1965; Gambell 1972). The persistence of corpora
albicantia in the ovaries has been established by several authors
(Best 1967; Chuzhakina 1961; Gambell 1972; Ohsumi 1965).

Using a combination of data on the length of the reproductive
cycle and the frequency of ovulation (and its percentage success) at
various stages of the cycle, it has been possible to estimate the mean
ovulation rate per year. Best (1968) obtained an estimate of 0.59
for a four-year reproductive cycle, but this made no allowance for
the proportion of females falling pregnant during the cycle. If
these are taken into account, the total number of ovulations in the
four-year cycle becomes 2.131, or an average of 0.53 a year. This
is very close to the regression of corpora counts against dentinal
growth layers (0.52 ± 0.12) calculated by Bartlett's method (Best
1970a), indicating that the accumulation rate of dentinal growth layers
is probably one per year.

Gambell (1972) has made similar calculations for sperm whales off the east coast of South Africa. The annual ovulation rate estimated from reproductive data alone was 0.43 for an average four-year reproductive cycle, compared to a regression of corpora counts on dentinal growth layers (least-squares method) of 0.45.

Pike (1966) found distinct modes in the frequency distribution of corpora counts at intervals of 2.2 to 2.5 corpora, which, in a reproductive cycle of four years duration, suggested that the rate of accumulation was 0.5 to 0.6 per year.

Rice and Wolman (1970) calculate that the total number of ovulations per reproductive cycle varies from 1.15 to 1.36, but they estimate the typical reproductive cycle to last three years, giving an annual ovulation rate between 0.38 and 0.45. This compares with a least-squares regression of corpora counts on number of growth layers of 0.33.

This third method of age determination, besides being confined in use to mature females, has the disadvantage that individual variations in the ovulation rate reduce its suitability for the age determination of individuals.

Age at Recruitment into the Catchable Stock

This has usually been expressed as the first fully recruited year class in the catch, or in other words the year class that is most abundant in the catch. The figures in Table 11-1 are available for either sex and for different geographical areas; except where stated, they are based on age compositions from dentinal growth layer counts.

The majority of values for males in temperate waters fall in the range 15 to 18 years. The somewhat higher values recorded for Western Australia and North Pacific Area IV are probably due to size selection, although segregation by sex and age may also be a contributory factor. The much higher values for the Aleutian Islands and the Antarctic (27 to 28 years) are probably a valid reflection of the segregation of large males to high latitudes.

The first fully recruited age class of females in each area is

TABLE 11-1. Age at recruitment into the catchable stock, based on dentinal growth layer counts, except as noted.

	Age in years		
Area	Male	Female	Source
South Africa, west coast	18	20[a]	Best 1970a
South Africa, east coast	18	18	Gambell 1972
Western Australia	21-25	-	Bannister 1969
California coast	17[b]	-	Rice and Wolman 1970
Coast of Japan	15	-	⎫
North Pacific, Area III	18	⎫	⎬ Shimadzu 1970
North Pacific, Area IV	22	⎬ 22	⎪
North Pacific, Area V	18	⎭	⎭
Coast of Japan	12	13	⎫ Ohsumi 1966
Aleutian Islands	27	-	⎭
Antarctic, Areas II-V	28	-	Boerema 1970

[a] Based on ovarian corpora counts.
[b] Including special-permit whales.

similar to that of males caught in the same area.

These values for the age at recruitment assume knife-edge or instant selection at a particular age, whereas in fact recruitment to the catch takes place gradually over a number of years. In calculations of yield it would be more valid to use estimates of the mean age at recruitment. These have been made by comparing the frequency of new recruits in successive age classes in the left-hand limb of an age composition curve, making allowance for natural mortality in each previously recruited age class. For the west coast of South Africa the mean age at recruitment for males seems to be about 12 years, and for the Antarctic Areas II to V (from Boerema, 1970) about 23 years. These figures have been taken for the moment as representative of the mean ages at recruitment in temperate waters and in higher latitudes respectively.

Mortality Rates

Most calculations of mortality rates to date have utilized the age composition of the catch based on dentinal growth-layer counts. The slope of the right-hand limb of the age-composition curve has then been equated with Z, or the total mortality rate, assuming that one growth layer is laid down per year. The lowest values of Z recorded to date are, for males, 0.058 to 0.069 (from age 32 to 37+; Shimadzu 1970); 0.071 (age 24 to 66; Ohsumi 1966); and 0.074 (from age 21 to 25+; Bannister 1969). For females, the lowest Z values are 0.053 and 0.058 (age 13 to 50 and 20 to 48, respectively; Shimadzu 1970); 0.056 (age 13 to 47; Ohsumi, Kasuya, and Nishiwaki 1963); 0.056 to 0.074 (age range unknown; Whales Research Unit 1966—in Shimadzu 1970); 0.07 (age 16+; Rice and Wolman 1970); and 0.073 (age 12 to 52; Ohsumi 1966). These Z values for both sexes were derived from stocks considered to be only lightly exploited, and hence they have been equated with natural mortality, or M.

There are certain inaccuracies involved in calculating mortality rates from the slope of a single regression line fitted to an age composition, particularly in whales, because changes in the overall mortality rate become apparent only as new year classes enter the catch. Mortality rates derived in this way may be misleading when applied to current situations. An alternative method is to compare the abundance of a particular age class in one year (measured as the catch per unit effort) with its abundance as the next age class one year later. This method has been tried for sperm whales by Bannister (1969) and the author, but unsuccessfully, possibly because of faulty estimates of effort, deficiencies in age determination, or the effects of segregation.

A theoretical rate of natural mortality for females can be computed from the equation

$$(1 - e^{-M_2}) \, e^{mM_1} = \frac{p}{2} ,$$

where p = pregnancy rate, m = age at sexual maturity, M_1 = natural mortality rate in immature animals, and M_2 = natural mortality rate in mature animals (Anon. 1969a). In a stable or lightly exploited popula-

tion the best evidence (see below) is that the pregnancy rate is about
0.20 and the age at sexual maturity 9 years. If $M_1 = M_2$, the overall
natural mortality rate will be 0.060. This is very similar to the
calculations of Z given above, that were considered to be close to the
natural mortality rate.

This result, however, is based on the assumption that the natural
mortality rate of immature females is identical to that of mature females, whereas it is reasonable to assume that it should be higher.
Several of the published values for the total mortality rate of females
are actually lower than 0.060, the lowest being 0.053, as we have seen
above. These values refer to mature females. If a value of 0.055 is
adopted for M_2 and substituted in the above equation, M_1 (or the immature natural mortality rate) will be 0.070 if $p = 0.20$ and $m = 9$ —
that is, in an unexploited stock. These values for M_1 and M_2 are
therefore taken to be the best available estimates for the natural
mortality rate of immature and mature female sperm whales.

As part of the response of the population to exploitation, it is
to be expected that the natural mortality rate will decrease. The
exact amount of this decrease is unknown but will clearly be small,
as the stable mortality rate in an unexploited stock is so low. Ohsumi
(1970b) has assumed that M_2 will decrease in a linear fashion from
0.060 in an unexploited stock to 0.040 as the stock size approaches
zero. Assuming the same relationship between mortality rate and stock
size, an M_2 of 0.055 in an unexploited stock will decrease to 0.037
as the stock size approaches zero. Although Ohsumi has proposed a
more rapid decrease in M_1 with decreasing stock size than in M_2, it
is assumed for the present that M_1 and M_2 will decrease at parallel
rates.

The natural mortality rate of males is assumed to be similar to
that of females, and although male sperm whales become sexually mature
much later than females (age 19 as against age 9—see below), the
immature mortality rate is considered to apply only to males less than
10 years old.

Pregnancy Rate

The importance of the pregnancy rate in stock assessments has
been briefly indicated above. It is principally a measure of potential recruitment to the population. The simplest method for its
estimation is to determine the proportion of pregnant females in the
catch, adjusted to allow for the fact that the gestation period lasts
more than a year (14 to 16 months). Unfortunately this method assumes
that all reproductive classes of females are fully represented in the
catch, whereas there are a number of factors that may bias the catch
in this respect.

The chief of these is the protection afforded to lactating females (whales accompanied by calves or suckling whales) by the Schedule
of the International Whaling Commission. Furthermore, if size selection among schools of females is rigidly applied, it is also possible
that only the largest and oldest animals will be killed; in these the
pregnancy rate is below average (Best 1967; Gambell 1972; Ohsumi 1965;
Pike 1966). The segregation of reproductive classes in separate
schools or by different migration patterns has also been suggested,
but is not well substantiated; Ohsumi (1970a) has described the continuity of school structure in female sperm whales and the presence of
all reproductive classes within a school.

Some or all of these factors may account for the variations in
the pregnancy rates given for different areas (see Table 11-2). The
difference between a pregnancy rate calculated for the whole catch
and that for only a section considered to be the most fertile is
shown by the data for either coast of South Africa.

Estimates of the pregnancy rate given above vary from 0.19 to
0.29, all of which have been corrected in some way to allow for the
pregnancy overlap, and it is not possible to determine which is the
closest to reality. Some of the higher rates may reflect a response
on the part of the population to a high rate of exploitation, as
described for blue and fin whales (Laws 1961). Although pregnancy
rates in the North Pacific decline from east to west, in line with
the estimated degree of exploitation of the stocks (Masaki 1970b),

TABLE 11-2. Pregnancy rates in different areas; see text for discussion.

Area	Total	Maximum	Source
South Africa, west coast	0.20^a	0.22	Best 1968 (and unpublished data)
South Africa, east coast	0.19	0.23	Gambell 1972
Western Australia	0.28^b	-	Bannister 1970
California coast	0.29^c	-	Rice and Wolman 1970
Coast of Japan	$0.26-0.29^d$	-	Ohsumi 1965
North Pacific, Area I	0.28	-	
North Pacific, Area II	0.27	-	
North Pacific, Area III	0.22	-	Masaki 1970b
North Pacific, Area IV	0.19	-	
Azores	0.19	-	Clarke 1956
British Columbia	< 0.35	-	Anon. 1967

[a] May to October.   [b] June to October.

[c] May to December, using Ohsumi's (1965) exchange ratio 2.

[d] Includes any female with a corpus luteum in the ovaries.

the pregnancy rate on the California coast is higher than that estimated for any of the North Pacific areas. The relation between pregnancy rate and stock condition will be discussed in more detail below.

An alternative, more oblique method of estimating the pregnancy rate is to determine the average duration of each phase of the reproductive cycle and thus to obtain the proportion of the cycle occupied by pregnancy.

Of all three phases of the reproductive cycle (pregnancy, lactation, resting period), pregnancy is the one whose duration is known most accurately. From studying the rate of growth of the fetus seasonally, the length of gestation is estimated to be within the limits of 14 to 16 months (Bannister 1969; Best 1968; Gambell 1972; Ohsumi 1965; Rice and Wolman 1970).

Estimates of the duration of lactation are more controversial.

Because the size distributions of mammary gland thickness and of the
diameters of the largest corpora in lactating females are bimodal,
lactation is assumed to last 24 to 25 months (Best 1968; Gambell
1972). Chuzhakina (1961) also found large and small regressing corpora lutea in lactating females at the same time of year and deduced
that lactation possibly lasted 10 to 11 months. Off the California
coast, however, no such bimodality is apparent in the diameter of the
largest corpora, and lactation is taken to last about one year (Rice
and Wolman 1970). According to preliminary results of investigations
in the southeast Pacific, lactation is estimated to last either 5 to
6 months or 17 to 18 months (Clarke, Aguayo, and Paliza 1964). Apart
from the relative proportion of lactating females in the catch, the
main evidence given in support of this theory is the size of the fetuses
in three females simultaneously pregnant and lactating: all three animals are assumed to have conceived at a postpartum ovulation, but the
available evidence suggests otherwise (Best 1968). Other authors have
chiefly used the proportion of lactating animals in the catch to estimate the duration of lactation (Bannister 1969; Clarke 1956); this
procedure is open to the same objections as similar calculations of
the pregnancy rate.

An alternative approach to this problem has been from the examination of calves and animals taken to be close to the size at weaning.
Clarke (1956) has provided the only estimate to date of the length of
the sperm whale calf at weaning. Fifteen whales from 19 ft 4 in (5.9
m) to 25 ft 7 in (7.8 m) long were examined; milk was found in the
stomachs of only two, both 21 ft 7 in (6.58 m) long. The next largest
animal, 22 ft 4 in (6.8 m), apparently was already taking solid food
from the presence of nematodes in the stomach. The length at weaning
was deduced to be about 22 ft (6.7 m). The fetal growth curve was
then extrapolated to meet this point, which was plotted 13 months after
birth. As Clarke admitted, however, it could have been plotted two
years after birth. Ohsumi (1965) has drawn a growth curve for young
sperm whales up to 9 years of age, based on growth layer counts in an
undisclosed number of animals. A length of 22 ft (6.7 m) is reached
19 months after birth (in spring), but as weaning is thought to occur

in autumn, lactation is assumed to last 24 to 25 months.

In an attempt to obtain more data on this point, the South African
Division of Sea Fisheries issued a permit for the taking of 15 sperm
whale calves in March 1971. The sizes of the 9 animals taken have been
plotted against the month of capture in Fig. 11-6, and combined with

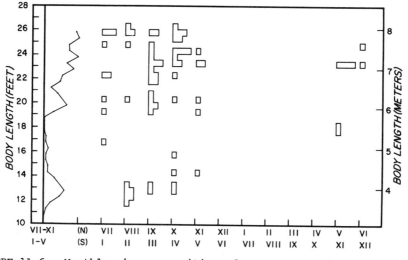

FIGURE 11-6. Monthly size composition of young sperm whales.

similar data from 64 calves and small sperm whales from other sources
(Bannister 1969—1; Best 1968—4; Clarke 1956—19; Ohsumi 1965—33;
Ohsumi 1970a—6; Rice and Wolman 1970—1). Records from the northern
hemisphere have been plotted on a scale shifted six calendar months
relative to those from the southern hemisphere. The majority of observations cover the period January to May (southern) and July to
November (northern), which is equivalent to the main calving season
(Best 1968; Gambell 1972; Ohsumi 1965). The records for these five
months have been summed and smoothed by threes.

The size composition of young sperm whales at this time of year
seems to consist of one group of animals 11.5 to 14.5 ft (3.5 to 4.4
m) long, quite distinct from a second group about 19 to 21 ft (5.8 to
6.4 m) long, with animals over 22 ft (6.7 m) in length forming a third
group in which no clear subdivisions can be made. The group of
smallest animals obviously represents newborn calves. The second

group about 19 to 21 ft (5.8 to 6.4 m) long might therefore be considered as animals one year old. The discontinuity in size between
newborn animals and this second group has been confirmed by field
observations made in the process of collecting the calves under
special permit.

On seven separate occasions schools of females were encountered,
five of which contained one or two newborn animals. In four of these
schools the smallest animals, apart from a newborn calf, were killed,
proving to be whales 19 ft 10 in (6.05 m), 19 ft 10.5 in (6.06 m),
20 ft 10 in (6.35 m), 22 ft (6.7 m), and 22 ft 5 in (6.84 m) in length.
Apart from a newborn calf, the smallest animal in each of the other
three schools encountered was estimated to be 20 to 22 ft (6.1 to 6.7
m), 25 ft (7.6 m), and 25 to 27 ft (7.6 to 8.2 m) long respectively.
On only one occasion was an animal seen that could have been intermediate in size between the two groups, this being an animal seen
briefly and estimated at 18 ft (5.5 m). In every case the discontinuity in size between the two groups was striking, and this seems
to confirm that the group of calves 19 to 21 ft (5.8 to 6.4 m) long
are one year old on the average, and therefore animals 22 ft (6.7 m)
or more in length are probably two years or more in age.

The stomach contents of all nine calves were examined. The
three newborn animals (11 ft 7.5 in or 3.54 m, 12 ft 8 in or 3.86 m,
and 13 ft or 3.96 m long) had all been feeding on milk and, judging
by the lack of nematode parasites, had probably not yet taken solid
food. All four animals from 19 ft 10 in (6.05 m) to 20 ft 10 in
(6.35 m) in length had been eating squid, but two of them had also
recently had a meal of milk, tests by chromatography for lactose and
triglycerides being positive. (I am indebted to Drs. M. N. Morris
and P. A. S. Canham of the Institute for Parasitology, Durban, for
designing and running these tests.) This information indicates that
sperm whale calves can suckle for at least one year, and that they
start to take solid food sometime in their first year. The two remaining calves, 22 ft (6.71 m) and 22 ft 5 in (6.84 m) long, had been
feeding on squid only. It is too early yet to determine how much
longer than one year lactation lasts, as more samples must be collected

in the spring (six months after the calving season). When these are available, it should be possible to follow the growth of the newborn group, to determine the incidence of milk in the stomachs of animals about 18 months old, and to investigate the possibility of the 6-month-old group starting to feed on squid. A permit for this purpose was issued by the South African Division of Sea Fisheries in August-September 1971.

The third phase of the reproductive cycle, the resting period, is also of doubtful duration. The majority of authors have assumed that it simply fills the period between the end of lactation and the next breeding season and consequently, under any scheme for the reproductive cycle, should not normally last longer than 12 months. The proportion of resting animals in the population should therefore never exceed the proportion of pregnant animals to any great extent at any time of the year.

The proportion of resting females in the commercial catch is shown in Table 11-3 relative to the proportion pregnant. The months

TABLE 11-3. Proportion of pregnant to resting females in the commercial catch in the southern and northern hemispheres.

Period	Number pregnant	Number resting	Ratio	Source
May to September (S)	63	113	1:1.79	Best (unpublished)
May to September (S)	80	193	1:2.41	Gambell 1972
June to September (S)	19	31	1:1.63	Bannister 1970
November (N)	8	12	1:1.50	Ohsumi 1965
November and December (N)	20	13	1:0.65	Rice and Wolman 1970
Total	190	362	1:1.91	

May to September in the southern hemisphere (S) were chosen because this is the period between the end of lactation and the breeding season, when resting animals should be most abundant and pregnant animals should mainly include those in midpregnancy. Very little information

is available for the corresponding period in the northern hemisphere
(N), which is November to March.

The ratio of pregnant to resting animals for the whole sample is
1:1.91, whereas on theoretical grounds it should be close to parity.
If subfertile young and old females are excluded from the sample, the
ratio still remains high (1:1.69—Best 1968; ca. 1:1.91—Gambell 1972).
Selection against lactating animals cannot be responsible for these
high ratios, as this would affect both pregnant and resting females
equally. The segregation of reproductive classes seems to be excluded
as a cause (see above), and the only remaining solution is that the
resting period must frequently last longer than a year. This implies
that either a significant number of animals fail to ovulate during
the first breeding season following the end of lactation, or that
they fail to conceive at such an ovulation.

Evidence on these points is difficult to obtain. Best (1968) has
produced some evidence to show that the majority of resting whales do
ovulate at the first estrus following the end of lactation and that
most are successful in conceiving then. Gambell (1972), however,
concludes that only about 69% of whales conceive in the main estrus,
and that the remainder undergo a further one to five years in a resting state before more than 99% have conceived. Nevertheless, these
"late conceivers" are balanced by about 21% of the population which
are estimated to conceive before or at the end of lactation; the mean
reproductive cycle is still calculated as four years, of which eight
to nine months are attributed to the resting period (Gambell 1972).

On the other hand, the prevalence of resting animals between the
peaks of the breeding and the weaning seasons in the northern hemisphere (April to August; Ohsumi 1965) strongly suggests that a considerable proportion of resting animals do not conceive at the main
estrus. At this time of year, resting animals should be at their
lowest abundance and pregnant animals at their highest, about twice
the normal rate, due to the simultaneous presence of animals in early
and late pregnancy. The ratio of pregnant to resting animals should
therefore be much less than 1:0.5, but in fact is about 1:0.79, as
shown in Table 11-4. This indicates that there must be a significant

TABLE 11-4. Proportion of pregnant to resting females between the
peaks of the breeding and weaning seasons in the northern hemisphere.

Period	Number pregnant	Number resting	Ratio	Source
May to August	87	31	1:0.36	Ohsumi 1965
April to August	30	31	1:1.03	Rice and Wolman 1970
June to August	29	53	1:1.83	Clarke 1956
Total	146	115	1:0.79	

conception failure rate at the first breeding season following the
end of lactation.

To summarize, the basic reproductive cycle would probably consist of 14 to 16 months pregnancy, 1 or 2 years lactation, and 8 to
10 months resting period, making a total of 3 or 4 years duration,
depending on the length of lactation assumed. The theoretical pregnancy rate would therefore be either 25% or 33%. The apparent frequency with which ovulations at the main estrus are unsuccessful, however, means that the resting period will average considerably longer
than 8 to 10 months, and the actual pregnancy rate will be somewhat
less than expected. Assuming a mean value of about 1:1.8 for the
ratio of pregnant to resting animals at maximum fertility during the
period between the weaning and breeding seasons, the actual pregnancy
rate can be calculated as about 21% if lactation lasts 2 years and
about 26% if lactation last 1 year.

The theoretical natural mortality rate of 0.06, based on a pregnancy rate of 20%, is close to the lowest observed rates of total
mortality (see previous section); as a consequence, an actual pregnancy rate of 21% calculated for a reproductive cycle including 2
years lactation seems more reasonable than a rate of 26%.

Changes in Pregnancy Rate as a Response to Exploitation

The concept of a sustainable yield implies a response on the
part of the population to exploitation. In the case of whales this

is usually assumed to take the form of an increase in recruitment (r)
and a reduction in natural mortality (M), so that the rate of net
recruitment ($r - M$) increases as exploitation intensifies. The major
component of recruitment is the number of calves born annually; so
that changes in the pregnancy rate with stock size have great significance in this respect.

There is no direct evidence yet of the relation between pregnancy rate and stock size in the sperm whale, but Ohsumi (1970b) has
produced a graphic model of fluctuations in the pregnancy rate with
changes in the population size of sexually mature females. In this
model the pregnancy rate is 0.25 when the stock is at its maximum
size, increasing to about 0.31 when the stock is half its original
size, and to a maximum rate of 0.33 as the stock size approaches zero.
However, these pregnancy rates may be too high. The actual pregnancy
rate has been estimated above to be about 21%: this figure refers
mainly to female populations that are considered to be lightly exploited (Bannister 1969; Best 1968; Gambell 1972). Some idea of the
maximum possible pregnancy rate can be obtained if the ovulations at
different stages of the reproductive cycle proposed by Gambell are
all assumed to be successful; this leads to a pregnancy rate of 31 to
32%. More realistic values for the pregnancy rate would therefore
seem to be 20% for an unexploited stock; using Ohsumi's curve, this
value rises to about 26% when the stock is half its original size and
to a maximum of 28% as the stock size approaches zero.

These figures assume a lactation period lasting two years. For
a one-year lactation period the corresponding values would be the
same as in Ohsumi's (1970b) original model.

Age at Sexual Maturity

There has generally been good agreement on the age of the female
sperm whale at sexual maturity, using dentinal growth layers for age
determination.

12-13 years	Bannister 1969
7 years (= growth layers)	Berzin 1964

9 years   Best 1970a

7 years   Gambell 1972

9 years (= growth layers)   Nishiwaki, Hibiya, and Ohsumi 1958

9.2 years   Ohsumi 1965

8 years   Pike 1966

9 years   Rice and Wolman 1970

As part of the response of the stock to exploitation, it is to
be expected that the mean age at sexual maturity will decrease, as
described for the fin whale (Anon. 1971a). Ohsumi (1970b) has suggested that the age of the female sperm whale at puberty in an unexploited stock, 9 years, will decrease in a linear fashion with stock
size to 7 years as the stock approaches zero. At present there is
no actual evidence to substantiate this prediction.

The age of male sperm whales at sexual maturity is not so well
established as that of females. Histological examinations of the
testis by different authors have given conflicting results, shown to
be caused by sampling different areas of the testis (Best 1969a). The
slow maturation of the testis also creates problems in the interpretation of the stage at which fertility is reached. The following
definitions and their diagnoses have now been proposed (Anon. 1971b).

Puberty: Stage at which 50% of whales are immature at the
center of the testis. Normally occurs at a body length of 30 to 32 ft
(about 9 to 10 m) and an age of 9 to 10 years.

Sexual maturity: Stage at which 50% of whales are immature and
50% are maturing or mature at the periphery of the testis. Normally
occurs at a body length of 39 ft (about 12 m) and an age of 19 years.

Social maturity: Stage at which equal proportions of males are
immature and mature at the periphery of the testis (maturing animals
not included). Normally occurs at a body length of 45 ft (about
13.7 m) and an age of 25 to 27 years.

The first stage corresponds with the onset of spermatogenic
activity, the second with the entry of males into higher latitudes
for the first time, and the third to a point at which testis growth
accelerates (Best 1969a).

The attainment of social maturity is also considered likely to

be equivalent to the advent of harem-master status. A point of considerable importance in calculating the potential yield of males, however, is whether under heavy exploitation the latter stage could be
acquired by any sexually mature (or even pubertal) male. When, in
fact, is the male sperm whale fully fertile for the first time? To
investigate this problem, samples approximating seminal fluid were
collected from 42 male sperm whales landed at Durban in 1970, only
the freshest animals available being chosen. Using an automatic
pipette, 1 or 2 ml of fluid were drawn from the convoluted part of
the vas deferens just anterior to its meeting point with the vas
deferens from the other side. This fluid was then diluted with 8 ml
of 10% formalin and stored in a screw-cap tube. The density of
spermatozoa in these samples was determined by the cytology laboratory, Department of Obstetrics and Gynaecology, Groote Schuur Hos-
pital, Cape Town, to whom I am indebted. The concentrations of
spermatozoa in the original seminal fluid are plotted against body
length in Fig. 11-7.

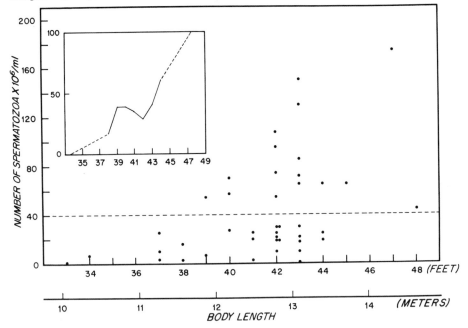

FIGURE 11-7. Variation with body length of spermatozoa density in
vas deferens of sperm whales.

A feature of this graph is an apparent discontinuity in spermatozoa counts between 30×10^6 and 50×10^6/ml, suggesting a rapid increase in spermatozoa production at this time. Chittleborough (1955) found very similar concentrations of spermatozoa in the vas deferens of humpback whales and has adopted a density of 50×10^6/cc as a criterion of fertility, based on values for human male ejaculate. Although it is inadvisable to compare spermatozoa densities in ejaculate with those in vas deferens fluid, as considerable dilution of the latter takes place in most mammals before ejaculation, Chittleborough (1955) has pointed out that this dilution may be somewhat less in cetacea because of the absence of seminal vesicles and Cowper's glands. For humpback whales, a value of 50×10^6 spermatozoa/cc gave reasonable results as a criterion of fertility, and from the observed discontinuity in densities in sperm whales, a figure of 40×10^6 spermatozoa/ml has been adopted.

Assuming that whales with more than 40×10^6 spermatozoa/ml are fertile, the percentage of fertile males at each body length has been calculated and smoothed by threes (Fig. 11-7), from which it is apparent that 50% of males are fertile at a body length of 43 to 44 ft. This is very close to the length at social maturity, and consequently it seems unlikely that sexually mature males are capable of maximum fertility before they reach the length and age corresponding to social maturity, or harem-master status. This is not to say that they are infertile, but rather subfertile. The situation seems to be analogous to that in the bottle-nosed dolphin *Tursiops truncatus*, where although 7-year-old animals are already producing spermatozoa, "the probabilities, based on sperm counts, are infinitely greater that a 20-year-old male will impregnate a female with one copulation than that a 10-year-old would in several copulations" (Caldwell and Caldwell 1968).

There is no evidence of the effect heavy exploitation might have on the age at social maturity in males. In order to maintain a high pregnancy rate, however, it seems essential that harem bulls be of high fertility; if the stock of mature males is so reduced that younger, subfertile males become extensively involved in reproductive

activity, the high pregnancy rates expected as a response on the part
of the stock to exploitation may not materialize.

Number of Females per Harem Bull

The yield of male sperm whales depends to some extent on the number of animals needed to maintain a proper ratio to mature females in
a harem school. The rate of exploitation is particularly sensitive to
the values chosen for a mean harem size (Anon. 1969a).

There are obvious difficulties in obtaining exact values for the
number of females per harem bull, due to the impracticability of accurately observing cetaceans in the field. Nevertheless, five separate
estimates of this parameter are available, derived from a study of
the composition of the catch from certain schools (Best 1969a, 1970b;
Ohsumi 1970a), a comparison of the age composition of males and females off the coast of Japan (Ohsumi 1966), and an analysis of aircraft sightings (Gambell 1972):

Harem size	Source
at least 10	Best 1969a
10-15	Best 1970b
10	Gambell 1972
16	Ohsumi 1966
14	Ohsumi 1970a

All these estimates refer to the number of mature females per
bull, rather than to the total number of females in a school. A
range of 10 to 16 females is obtained. How this figure will change
with exploitation depends largely on the degree of exploitation ac-
corded either sex. If there is a tendency for the male stock to be
reduced more rapidly than the female relative to their sustainable
yields, then the harem size is likely to increase rather than decrease.

Diving Behavior

Doi (1970) has provided a mathematical model for assessing whale
population size from sightings at sea, of which one of the most important parameters is the average duration of a whale's dive. Due to the

practical difficulties involved, there is very little authentic information on this subject available for sperm whales. With the introduction of asdic sets on whale catchers, however, it has become possible
to follow the diving behavior of a group of whales with some accuracy.

During March 1971 diving information was obtained during normal
whaling operations off Durban, by continuously recording the commentary
of the asdic operator on tape, adding comments where necessary. These
tapes were later analyzed on playback using a stopwatch, and the dive
times shown in Table 11-5 were recorded. More data are needed, especially from large and medium-sized males, and, if possible, some indication is required of how long the whales remain at the surface relative to each dive. The latter data might be best obtained from an
aircraft.

TABLE 11-5. Recorded dive durations off Durban, March 1971.

Duration of dive (mins)	Small males	Females	Duration of dive (mins)	Small males	Females
0-1	1	9	16-17	1	1
1-2	1	5	17-18	-	-
2-3	2	1	18-19	-	1
3-4	2	-	19-20	1	1
4-5	-	1	20-21	1	-
5-6	-	1	21-22	-	-
6-7	-	1	22-23	-	-
7-8	-	2	23-24	-	-
8-9	2	-	24-25	-	-
9-10	-	1	25-26	-	-
10-11	-	1	26-27	-	-
11-12	-	1	27-28	-	-
12-13	-	-	28-29	-	1
13-14	-	1	29-30	1	-
14-15	-	-	-------		
15-16	-	-	44-45	-	1

Calculations of Sustainable Yield

Using some of the results given above, it is now possible to make some estimates of the potential yield of sperm whales.

The stable catch of female sperm whales (C) per 10^4 mature females (N) can be calculated from the following function (Anon. 1969a):

$$C = N \frac{p}{2} e^{-tmM_1} - N(1 - e^{-M_2}).$$

For an unexploited stock, where the pregnancy rate (p) is 0.20, the immature natural mortality rate (M_1) is 0.070, the mature mortality rate (M_2) is 0.055, and the age at maturity (tm) is 9 years, the yield is naturally zero. As exploitation intensifies, the pregnancy rate is expected to increase, natural mortality to decrease, and the age at sexual maturity to become less. According to Ohsumi's (1970b) simulated model of how these parameters will change in response to changes in stock size, the maximum sustainable yield is reached when the mature female stock is about 60% of its original size. At this stage the age at sexual maturity has decreased to 8.2 and, from the figures for mortality and pregnancy rates reached in this paper applied to Ohsumi's models, the pregnancy rate will have increased to 0.257 and the natural mortality (M_2) will have decreased to 0.048. Natural mortality in immature animals (M_1) is taken to have decreased to 0.063. These values give a yield of 298 mature females per 10^4 mature females.

The fishing mortality required to produce this catch will depend on the age at recruitment into the catch. A function to calculate this is

$$R = \left[\frac{R}{(1 - e^{-M_2})} (1 - e^{-M_2(t_r - tm)}) + \frac{R e^{-M_2(t_r - tm)}}{1 - e^{-(M_2 + F)}} \right] \left(\frac{p}{2} \right) e^{-tmM_1}$$

where R = recruits to the female population under a stable level of exploitation, t_r = recruitment age, and F = fishing mortality applied to ages greater than t_r (Anon. 1969a). Because there is considerable variation in the age of females at recruitment, fishing mortalities have been calculated for three different values of t_r:

When t_r = 13, Z = 0.096, F = 0.048.

When t_r = 18, Z = 0.132, F = 0.084.

When t_r = 23, Z = 0.252, F = 0.204.

The sustainable yield of males is more complicated to calculate, as the age at recruitment is frequently below the age at which males are considered to begin to contribute to reproductive activity. Although sperm whales are polygynous, a certain number (depending on the average harem size) of mature males is necessary for breeding purposes. The potential yield of males is in turn dependent on the female population size which, although at present considered to be close to its unexploited level in most areas, would have to be significantly reduced to produce its maximum sustainable yield. Operations in high latitudes (especially in the Antarctic) also involve a higher age at recruitment for males than those in temperate regions.

In the first example, the mature female population is considered to be only lightly exploited, so that p = 0.20. Two values for harem size (n) are included, 10 and 16.

Values of the fishing mortality giving the stable yield of males are obtained from the function

$$\left(\frac{p}{2}\right) N \, e^{-M_1 t_r} \, e^{-(F+M_1)(tm-t_r)} = \frac{N}{n}(1 - e^{-(M_2+F)}),$$

and the actual yield per year from the function

$$\left(\frac{F}{F+M}\right)\left(\frac{p}{2}\right) N \, e^{-M_1 t_r}.$$

These equations are simplifications in that they ignore the seasonal effect of the fishery, but in extreme cases the actual yield will only be 5% higher or lower than the calculated yield (Anon. 1969a).

The theoretical yield of males has been calculated for an Antarctic fishery (t_r = 23) and a temperate fishery (t_r = 12), when N = 10^4 mature females and tm (age of male at social maturity) = 26. The value of M_1 after the age at recruitment and of M_2 in both instances have been taken as 0.055, but M_1 before recruitment is taken as 0.070 for the temperate fishery and 0.061 (the average of 9 years at 0.070 and 14 years at 0.055) for the Antarctic fishery.

harem size	$t_r = 12$	$t_r = 23$
10	203	163
	($F = 0.049$)	($F = 0.108$)
16	243	186
	($F = 0.071$)	($F = 0.169$)

If the female stock is now reduced to the level giving its maximum sustainable yield of mature females, then $p = 0.257$. The same two values for harem size and age at recruitment are adopted as in the previous example, but values of M_1 after the age of recruitment and of M_2 in both instances are taken as 0.048, and the value of M_1 before recruitment is taken as 0.063 in a temperate-zone fishery and 0.054 (calculated as above) in an Antarctic fishery. The yields per 10^4 mature females will be the following:

harem size	$t_r = 12$	$t_r = 23$
10	362	289
	($F = 0.072$)	($F = 0.168$)
16	399	309
	($F = 0.094$)	($F = 0.240$)

The maximum sustainable yield of both sexes of sperm whales per 10^4 mature females is therefore estimated to be between about 590 and 700, of which males comprise from 49 to 57%. This relatively low proportion of males in the catch, considering their polygynous behavior, is mostly a result of their high age at social maturity, which demands a steady degree of recruitment even when the mean age at recruitment into the catch is low.

Two further points should be stressed. Firstly, these estimated yields refer to stable stock conditions, and there is obviously a considerable surplus (especially of males) that can be removed in the process of reducing the stock to its size at maximum sustainable yield. This situation also applies to the fishing mortalities involved. Secondly, these yields simply refer to the number per 10^4 mature females, whereas it should be remembered that the size of the *original* stock may have been reduced considerably to reach the level giving maximum sustainable yield. According to Ohsumi's model (1970b), the stock of mature females is at 60% of its unexploited size when giving maximum sustainable yield, which means that in calculating the actual

yields per 10^4 mature females in the *original* stock, these figures
will have to be reduced by a similar amount. On this basis, 350 to
420 sperm whales would be the maximum sustainable yield from an
original stock of 10^4 mature females.

REFERENCES

Anonymous. 1963 (MS). Report of the meeting of the scientific sub-
    committee of the International Whaling Commission, Seattle,
    Washington, 18-22 November 1963.
_____ 1967. Sperm whale sub-committee meeting report, Honolulu,
    Hawaii, 10th-18th February, 1966. Rpt. IWC, 17:120-127.
_____ 1969a. Report of the IWC-FAO working group on sperm whale
    stock assessment. Rpt. IWC, 19:39-56, and 8 appendices to p. 83.
_____ 1969b. Report of the IWC-FAO working group on sperm whale
    stock assessment. Appendix 2. Summary of sperm whale marking
    investigations in the North Pacific. Rpt. IWC, 19:56-58, 69.
_____ 1969c. Report of the meeting on age determination in whales,
    Oslo, 26th February to 1st March, 1968. Rpt. IWC, 19:131-137.
_____ 1971a. Report of the special meeting on Antarctic fin whale
    stock assessment, Honolulu, Hawaii, 13th-25th March, 1970.
    Rpt. IWC, 21:34-39.
_____ 1971b. Report of the special meeting on sperm whale biology
    and stock assessments, Honolulu, Hawaii, 13th-24th March, 1970.
    Rpt. IWC, 21:40-50.
Bannister, J. L. 1969. The biology and status of the sperm whale
    off Western Australia—an extended summary of results of recent
    work. Rpt. IWC, 19:70-76.
_____ 1970 (MS). The biology and status of the sperm whale off
    Western Australia. (Paper submitted to the special meeting on
    sperm whale biology and stock assessments, Honolulu, 1970.)
Berzin, A. A. 1964. Growth of sperm whales in the North Pacific.
    Trudy vses. nauchno-issled. inst. morsk. ryb. khoz. okeanogr.,
    53:271-275.
Best, P. B. 1967. The sperm whale (*Physeter catodon*) off the west
    coast of South Africa. 1. Ovarian changes and their signifi-
    cance. Investl. Rpt. Div. Sea Fish. S. Afr., 61:1-27.
_____ 1968. The sperm whale (*Physeter catodon*) off the west coast
    of South Africa. 2. Reproduction in the female. Investl. Rpt.
    Div. Sea Fish. S. Afr., 66:1-32.
_____ 1969a. The sperm whale (*Physeter catodon*) off the west coast
    of South Africa. 3. Reproduction in the male. Investl. Rpt.
    Div. Sea Fish. S. Afr., 72:1-20.
_____ 1969b. The sperm whale (*Physeter catodon*) off the west coast
    of South Africa. 4. Distribution and movements. Investl. Rpt.
    Div. Sea Fish. S. Afr., 78:1-12.
_____ 1970a. The sperm whale (*Physeter catodon*) off the west coast
    of South Africa. 5. Age, growth and mortality. Investl. Rpt.
    Div. Sea Fish. S. Afr., 79:1-27.

_____ 1970b (MS). The sperm whale (*Physeter catodon*) off the west coast of South Africa. 6. Social groupings. (Paper submitted to the special meeting on sperm whale biology and stock assessments, Honolulu, 1970.)

_____ and Gambell, R. 1968. A comparison of the external characters of sperm whales off South Africa. Norsk Hvalf.-Tid., 57:146-164.

Boerema, L. K. 1970 (MS). Notes on sperm whale age-composition tables. (Paper submitted to the special meeting on sperm whale biology and stock assessments, Honolulu, 1970.)

Caldwell, D. K., and Caldwell, M. C. 1968. The dolphin observed. Nat. Hist. (N.Y.), 77:58-65.

Chittleborough, R. G. 1955. Aspects of reproduction in the male humpback whale, *Megaptera nodosa* (Bonnaterre). Austral. J. Mar. Freshwat. Res., 6:1-29.

Chuzhakina, E. S. 1961. Morfologicheskaya kharakteristika yaichnikov samok kashalota (*Physeter catodon*) v svyazi s opredeleniem vozraska. Trudy Inst. Morf. Zhivot., 34:33-53.

Clarke, R. 1956. Sperm whales of the Azores. Discovery Rpt. 28:237-298.

_____ and Paliza G., O. In press. Sperm whales of the southeast Pacific. Part III: Morphometry. Hvalråd. Skr.

_____ Aguayo L., A., and Paliza G., O. 1964. Progress report on sperm whale research in the southeast Pacific Ocean. Norsk Hvalf.-Tid., 53:297-302.

_____ _____ _____ 1968. Sperm whales of the southeast Pacific. Part I: Introduction. Part II: Size range, external characters and teeth. Hvalråd. Skr., 51:1-80.

Cushing, J. E., Fujino, K., and Calaprice, N. 1963. The Ju blood typing system of the sperm whales and specific soluble substances. Scient. Rpt. Whales Res. Inst., Tokyo, 17:67-77.

Doi, T. 1970. Re-evaluation of population studies by sighting observation of whale. Bull. Tokai Regional Fish. Res. Lab., 63:1-10.

Fujino, K. 1963. Identification of breeding subpopulations of the sperm whales in the waters adjacent to Japan and around Aleutian Islands by means of blood typing investigations. Bull. Japanese Soc. Scient. Fish., 29:1057-1063.

Gambell, R. 1967. Seasonal movements of sperm whales. Symp. Zool. Soc. Lond., 19:237-254.

_____ 1972. Sperm whales off Durban. Discovery Rpt. 35:199-358.

_____ and Grzegorzewska, C. 1967. The rate of lamina formation in sperm whale teeth. Norsk Hvalf.-Tid., 56:117-121.

Ivashin, M. 1970 (MS). Some results of marking sperm whales by the USSR in the southern hemisphere. (Paper submitted to the special meeting on sperm whale biology and stock assessments, Honolulu, 1970.)

Laws, R. M. 1961. Reproduction, growth and age of southern fin whales. Discovery Rpt. 31:327-486.

Masaki, Y. 1970a. Study on the stock units of sperm whale in the North Pacific. (Paper submitted to the special meeting on sperm whale biology and stock assessments, Honolulu, 1970.)

_____ 1970b (MS). Difference between pregnancy rates of sperm whales in the North Pacific by area. (Paper submitted to the special meeting on sperm whale biology and stock assessments, Honolulu, 1970.)

Nishiwaki, M., Hibiya, T., and Ohsumi, S. 1958. Age study of sperm
    whale based on reading of tooth laminations. Scient. Rpt. Whales
    Res. Inst., Tokyo, 13:135-154.
_____ Ohsumi, S., and Kasuya, T. 1961. Age characteristics in the
    sperm whale mandible. Norsk Hvalf.-Tid., 50:499-507.
Ohsumi, S. 1965. Reproduction of the sperm whale in the north-west
    Pacific. Scient. Rpt. Whales Res. Inst., Tokyo, 19:1-35.
_____ 1966. Sexual segregation of the sperm whale in the North
    Pacific. Scient. Rpt. Whales Res. Inst., Tokyo, 20:1-16.
_____ 1970a (MS). Some investigations on the school structure of
    sperm whale. (Paper submitted to the special meeting on sperm
    whale biology and stock assessments, Honolulu, 1970.)
_____ 1970b (MS). A trial to get mathematical models of population for
    sperm whale. (Paper submitted to the Scientific Committee, Inter-
    national Whaling Commission, 22nd meeting, 1970.)
_____ and Nasu, K. 1970 (MS). Range of habitat of the female sperm
    whale with reference to the oceanographic structure. (Paper sub-
    mitted to the special meeting on sperm whale biology and stock
    assessments, Honolulu, 1970.)
X_____ Kasuya, T., and Nishiwaki, M. 1963. Accumulation rate of
    dentinal growth layers in the maxillary tooth of the sperm whale.
    Scient. Rpt. Whales Res. Inst., Tokyo, 17:15-36.
Pike, G. C. 1963 (MS). British Columbia—sperm whales. In report
    of the meeting of the scientific sub-committee of the Inter-
    national Whaling Commission, Seattle, Washington, 18th-22nd
    November, 1963.
_____ 1966 (MS). Progress report on the study of sperm whales from
    British Columbia. (Paper submitted to the sperm whale sub-
    committee meeting, Honolulu, 1966.)
Rice, D. W., and Wolman, A. A. 1970 (MS). Sperm whales in the eastern
    North Pacific: progress report on research, 1959-1969. (Paper
    submitted to the special meeting on sperm whale biology and stock
    assessments, Honolulu, 1970.)
Shimadzu, Y. 1970 (MS). Natural mortality coefficient and rate of
    exploitation for sperm whales in the North Pacific. (Paper sub-
    mitted to the special meeting on sperm whale biology and stock
    assessments, Honolulu, 1970.)
Slijper, E. J., van Utrecht, W. L., and Naaktgeboren, C. 1964. Re-
    marks on the distribution and migration of whales, based on
    observations from Netherlands ships. Bijdr. tot de Dierk.,
    34:3-93.

CHAPTER 12

THE RELATION OF THE NATURAL HISTORY OF WHALES TO THEIR MANAGEMENT

G. A. Bartholomew

Effective management of heavily exploited wild species obviously requires that harvesting procedures be based on an accurate knowledge of their natural history. Although there is a large body of scientific data about the biology of whales, almost the only aspect of this knowledge that has been used by the whaling industry is information on the abundance of whales and when and where they can be found.

Since its inception a century and a half ago, commercial whaling has pursued a pattern of exploitation that has remained unfortunately aloof from biological knowledge. This sterile operational philosophy has produced an unbroken series of biological catastrophes. One by one the stocks of large whales have been depleted; population after population has been reduced to commercial and ecological insignificance. It can therefore do no harm to attempt to find new ways of thinking about patterns of management for the populations of the larger whales. Certainly the approaches that have been used in the past have been futile. One by one the whale populations have disappeared, and the entire commercial whaling enterprise is in a state of collapse.

Whales are, of course, extremely difficult to study, and biologists have had to be highly opportunistic in their research. Most of the detailed information we have has necessarily derived from animals obtained in the course of commercial whaling. Nevertheless, despite the formidable logistic and technical problems involved, a remarkable amount of information has been amassed. From these data it has been possible to produce progressively more refined population models.

Unfortunately, the predictions and recommendations based on these models have been largely ignored by the whaling industry. The dismal

consequence, as shown clearly by other chapters in this book, is that
the main result of the application of these models has been to document with melancholy precision the declines in whale populations of
the world and to quantify the extent of chronic overexploitation.

Since whaling is a marine enterprise, most of the patterns of
thought that have been devoted to the harvesting of whales have been
derived from fisheries biology, where it is customary to think in
terms of populations and aggregates rather than individual animals.
However, it may be profitable to look for some alternative to the current practice of basing our thinking about whale management on the
concepts and traditions which have been developed on the basis of our
knowledge of fish. After all, whales differ profoundly from fish in
their physiology, behavior, and ecology. Perhaps instead of thinking
about whales in terms of aggregates, we should think about them as
individuals operating in a social context that is maintained by complex individual social interactions. Certainly the larger whales are
hunted as individuals and have low reproductive rates.

The history of human thought is replete with unsuccessful attempts
to apply ideas based on data obtained from one frame of reference to
situations in which fundamentally different factors are operating. We
can perhaps avoid this pitfall if we use as our point of departure not
fishery biology, but the ecology and social behavior of large mammals
—which, to say the least, is what whales are. So let us use as our
frame of reference points of view which are appropriate for studying
large mammals rather than for developing statistical information about
schools of fish. If we think about whales as individual organisms,
we may be better off than if we rely on statistically derived abstrac-
tions based on adventitiously selected population samples. However
difficult this approach, we need only point out that the present methods,
despite the large amount of information they have yielded, have been
depressingly unsuccessful in leading to effective resource management.

If the patterns of harvesting can be based on the natural history
and the structure of the breeding populations, and can be supported
by analytical frameworks appropriate to each of the various species,
it may be possible to establish new and viable patterns of management.

Fortunately we do not need to know all about the biology of whales to
find scientifically sound bases for management. What is needed is
solid information on certain rather restricted features of whale biology under natural conditions. When these data are sufficient, effective management can proceed even when detailed knowledge on associated
aspects of their biology is scanty or indeed lacking completely.

There are apparently no exceptions to the statement that in all
species of mammals the sex ratio at birth is close to one-to-one. However, one male can fertilize many females. Consequently, among adult
mammals the sex ratio required for an optimal frequency of insemina-
tion of females may depart greatly from one-to-one. The extent to
which this ratio can be distorted without affecting the reproductive
performance of the population as a whole will depend, of course, on
the breeding behavior of the species involved. In monogamous forms,
any disturbance of the one-to-one sex ratio will have a deleterious
effect. In polygynous or promiscuous forms, the amount of disturbance
imposed by removal of males will vary with the degree of polygyny and
the completeness of promiscuity. In fact, from a strictly physiological point of view, there is almost no upper limit of the allowable
number of females per male, and even in the real world the effective
ratio of females to males in a breeding population of mammals during
a given year may be forty or more to one, as it is in some pinnipeds.

To guide us in our thinking about whale populations it will be
useful to examine the case of the northern fur seal, *Callorhinus
ursinus*. This species supplies an instructive analog not only because it is a large marine mammal, but also because it offers us the
most successful history of management found among marine mammals.
The reasons for the success of the management procedures applied to
this species by both the United States and the USSR are generally
familiar to students of marine mammals, but a brief review will assist
us in identifying the salient points. It should be emphasized that
the primary reasons for the successful manipulation of this species
are not to be found in remarkable insights by the persons who do the
managing, nor in the complexities of the operations associated with
their utilization. Rather, the reasons for success are functions of

the biology of this particular species (and indeed, of all members of
the family Otariidae).

The features of the life history of *Callorhinus* which preadapt
it to successful management are readily identified:

(1) It is the most highly polygynous of mammals. In any given
     year only a few of the available males participate signifi-
     cantly in the breeding performance. Consequently, from
     the standpoint of population dynamics, many of the males
     are in excess and therefore can be harvested without re-
     ducing the capacity of the population to reproduce.

(2) Its reproduction is both highly seasonal and highly local-
     ized geographically. Thus, during the breeding season,
     essentially all members of a given population can be found
     in the same place at the same time.

(3) In the rookeries, the different age classes segregate them-
     selves spatially, so that the young nonbreeding animals
     are found in areas that are physically separated from those
     occupied by the breeding females, the territorial males,
     and the pups. Thus, the animals presort themselves for
     harvesting, and this allows the reproductively idle young
     animals to be removed without significantly disturbing
     the breeding performance of the rest of the population.

(4) The females reach reproductive competence by the third
     year, but the males are not big enough and experienced
     enough to maintain territories until after they are 7, 8,
     or more years old. Thus, the reproductively idle popula-
     tion contains a preponderance of males, which of course
     facilitates harvesting.

(5) An additional factor which has not been taken advantage
     of in the management of fur seals, but which merits iden-
     tification because of its applicability to whales, should
     be mentioned. In *Callorhinus* (and most other otariids)
     the two sexes and the various age groups segregate them-
     selves geographically in the nonbreeding season. During
     fall and winter, the adult male *Callorhinus* remain in the

north in the vicinity of the Aleutian Islands and the Bering
Sea, while most of the immature animals of both sexes and
all of the females move south to the latitudes of California
and Japan. At least in theory, any animal taken by pelagic
hunting during the winter season in the north would be a
male and could be harvested without affecting the reproductive potential of the group as a whole as long as the harvest were not excessive.

Obviously, even in a species preadapted for sustained yield management, harvesting must be based on biological knowledge and adequate enforcement of regulations. The catastrophic results of pelagic sealing
during the summer in the early years of the harvest of *Callorhinus* are
a case in point. Pelagic sealers lay off the breeding grounds and
collected the females that went to sea to feed. Since the females do
not leave the rookery until they have been fertilized, and since estrus
is postparturitional, each female taken on the high seas left behind
on the rookery a pup that would starve to death, and she also carried
in her body a pup to be born the next season. Thus each female taken
at sea resulted in the deaths of three individuals: the female her-
self, the pup that remained on shore, and the embryo which she carried.
Obviously, even a modest pelagic harvest of the females during the
breeding season would have a profound effect on the population.

We can now ask ourselves what information we can usefully transfer from the biology and commercial history of the fur seal to the
management of other marine mammals, specifically large whales. It
is obvious that we are in a position to ask a number of biologically
meaningful questions. The key point at which each of these questions
is aimed is the possibility of uncoupling the harvest from the reproductive capacity of the population. Any device in the natural history
of an animal which separates males from females, adults from young, or
breeding males from nonbreeding males, offers a potential point at
which man may enter as a harvester without significantly affecting
the reproductive potential of the population as a whole.

(1) Are the large whales polygynous or promiscuous? If the
     breeding structure is such that sustained pair-bonds need

not be established, it should be possible to crop selectively the males without affecting the annual recruitment in the population as a whole.

(2) Are the large whales dimorphic, and is their dimorphism sufficiently great to allow the whaler to distinguish male from female? If males can be readily identified, obviously it might be possible to hunt them rather than females.

(3) Is there any time of year when it is possible to kill a female without also destroying her suckling young and embryo? This obviously also involves subquestions. Is estrus always postparturitional? How old are the young before they can sustain themselves without suckling?

(4) Does the distribution of males and females differ seasonally or geographically? If so, it should be possible to let these factors determine the time of hunting and thus allow the spatial and temporal pattern to determine whether males or females will be taken.

(5) What are the effects of age on fertility? What is the relation of natural mortality to age?

Some of these questions can be answered in part from data supplied elsewhere in this book, some can be answered from the personal knowledge of students of whales, others at present are unanswerable. From the data presented by Bannister and Best (see Chapters 10 and 11) there is reasonably convincing evidence that sperm whales are polygynous, and, of course, they are dimorphic. Thus it may be possible to harvest selectively a substantial number of males of this large and commercially important cetacean without reducing the reproductive capacity of the species as a whole. Moreover, sperm whales apparently segregate geographically by sex. Most of the individuals that occur at high latitudes are old males. Thus, as in the North Pacific sperm whale fishery, it may be possible to hunt those sperm whales that go to high latitudes without affecting reproduction in the species. In any event, the populations of sperm whales appear to have held up better than those of any other of the commercially important species.

The biology of sperm whales may well make them amenable to the maintenance of a sustained yield by a system of preferential harvesting of males, but the problem of uncoupling harvesting from reproductive potential is more complex in baleen whales. If these whales are sexually dimorphic, the dimorphism is too slight to be visually reliable to an observer on board ship.

Essentially no direct information is available on the breeding structure of populations of any of the baleen whales, but there appear to be differences in the seasonal movements of the various elements of the population of some species. For example, Jonsgård (see Chapter 4) has reported that the annual movements of the minke whale in the eastern North Atlantic show a pattern in which pregnant females segregate themselves somewhat during the northward migration and stay closer to the shore of Scandinavia than do other members of the population. By prohibiting the taking of this species near shore, the deleterious effects of harvesting on recruitment could perhaps be minimized. An analogous situation appears to exist among humpback whales in the southern oceans. Dawbin (1966) reports that the pregnant females have patterns of seasonal migration differing somewhat from those of other members of their populations. Similarly, Mitchell's report (Chapter 5) indicates that off Newfoundland and Labrador male finback whales travel through faster and go farther north than the females. Again, this situation might offer a key to sexually preferential harvesting.

Environmental Deterioration and Whale Populations

Environment and organism are inseparable. Whether it consists of bacteria or whales, a population cannot long exceed the carrying capacity of its habitat. In the terrestrial environment in most of the cases where man has been a primary factor, reductions in the size of mammal populations, or extinctions of mammal species, have been caused by destruction or alteration of habitat rather than by over-hunting. In the marine environment, however, the reductions of mammal populations and extinctions of mammal species have been caused

directly and specifically by overhunting. Nevertheless, at the present
stage in the history of the planet, man is more than just an overly
efficient hunter of marine mammals. He is an important factor in controlling the conditions which determine the environmental quality in
ever-increasing areas of the sea itself. The world ocean is large,
but it is not large enough to tolerate all of the activities of industrial man and still support the animals and plants which have hitherto
been an inseparable part of the marine ecosystem.

In the present context of whales and whaling, at the very least
we need to make an inventory of things we are doing to the marine habitat which can affect cetacean populations. A couple of examples will
suffice. Bays and lagoons are notoriously vulnerable to human activities, and those marine mammals which breed in such areas are as vulnerable as the habitat they occupy. Consider the case of the gray
whales which breed in the bays and lagoons of southwestern North America.
We are in a position to affect the numbers of this species, or even to
exterminate it completely, without hunting a single animal. San Diego
Bay in California was formerly an important breeding ground for gray
whales; it is now completely preempted by humans. In Baja California,
one of the species' principal calving areas, Scammon's Lagoon, is
undergoing the initial stages of exploitation by man, and enormous
Magdalena Bay is a prime target for commercial development.

Even relatively open waters have not escaped progressive pollution by the effluvia of industrial society. The quality of the southern parts of the North Sea is being steadily degraded. As an environment for the smaller toothed cetaceans, it has deteriorated to the
point where populations are noticeably diminished and restricted.

It is clear that man controls the destiny of whales not only as
a hunter, but as a casual and indifferent despoiler of the marine
habitat. The future of whales and whaling is not bright. On the one
hand, the deterioration of coastal marine habitats continues apace.
On the other, the commercial whaling interests appear to be completely
intransigent.

We are under no illusions about the reactions of the whaling
interests or of society in general to the points of view we have pre-

sented, but one has only to glance at the Twentieth Report (1970) of
the International Commission on Whaling to realize that the present
system of quotas and blue whale units is biologically absurd. Commercial whaling is an unguided missile that has already received its instructions to self-destruct. The only hope, however economically
unattractive in the short haul, is to convert to a biologically reasonable pattern of harvest.

SUGGESTED READINGS

Bartholomew, G. A.   1970.   A model for the evolution of pinniped
    polygyny. Evolution, 24:546-559.
    _____ and Hoel, P. G.   1953.   Reproductive behavior of the Alaska
    fur seal, *Callorhinus ursinus*. J. Mammal., 34:417-436.
Bonner, W. N.   1968.   The fur seal of South Georgia. British Antarc-
    tic Survey Rpts. No. 68, p. 81.
Chapman, D. G.   1964.   A critical study of Pribilof fur seal popula-
    tion estimates. U.S. F&W Serv. Fishery Bull., 63:657-669.
Craig, A. M.   1964.   Histology of reproduction and the estrous cycle
    in the female fur seal, *Callorhinus ursinus*. J. Fish. Res. Bd.
    Canada, 21:773-811.
Dawbin, W. H.   1966.   The seasonal migratory cycle of humpback whales.
    In Whales, dolphins, and porpoises, K. S. Norris, ed. Berkeley,
    University of California Press. Pp. 145-170.
Harrison, R. J., Hubbard, R. C., Peterson, R. S., Rice, C. E., and
    Schusterman, R. J., eds.   1968.   The behavior and physiology of
    pinnipeds. New York, Appleton-Century-Crofts.
Kenyon, K. W.   1960.   Territorial behavior and homing in the fur seal.
    Mammalia, 24:431-444.
    _____ and Rice, D. W.   1961.   Abundance and distribution of the Stel-
    ler sea lion. J. Mammal., 42:223-234.
Laws, R. M.   1956.   The elephant seal (*Mirounga leonina* Linn.).   II.
    General, social, and reproductive behavior. F.I.D.S. Sci. Rpt.
    13.
Nishiwaki, M., and Nagasaki, F.   1960.   Seals of the Japanese coastal
    waters. Mammalia, 24:459-467.
North Pacific Fur Seal Commission.   1969.   Report on investigations
    from 1964-66. Washington, D.C.
Peterson, R. S., and Bartholomew, G. A.   1967.   The natural history
    and behavior of the California sea lion. Amer. Soc. Mamm.,
    Spec. Publ. no. 1.
Roppel, A. Y., and Davey, S. P.   1965.   Evolution of fur seal manage-
    ment on the Pribilof Islands. J. Wildlife Management, 29:448-463.
Scheffer, V. B.   1958.   Seals, sea lions and walruses. A review of
    the Pinnipedia. Palo Alto, Calif., Stanford University Press.

PLATES

1   *Balaenoptera musculus*, blue whale, blowing. Queen Charlotte
Sound, British Columbia. (*Photograph by Raymond M. Gilmore.*)

2   *Balaenoptera musculus*, blue whale, wearing a Mather spaghetti
dart. (*Photograph by J. S. Leatherwood.*)

3   *Balaenoptera musculus*, blue whale, off southern California.
(*Photograph by J. S. Leatherwood.*)

4   *Balaenoptera physalus*, fin whale, off Newfoundland. (*Photograph* © *by Kenneth C. Balcomb.*)

5 *Megaptera novaeangliae*, humpback whale, breaching off Baja California. (*Photograph* © *by Kenneth C. Balcomb.*)

6 *Physeter catodon*, sperm whale and calf, off Pt. Conception,
California. (*Photograph* © *by Kenneth C. Balcomb.*)

7 *Eubalaena glacialis*, right whale, feeding at Cape Cod, Massachusetts. (*Photograph by W. A. Watkins.*)

PART FOUR

Management and Conservation

CHAPTER 13

THE ROLE AND HISTORY OF THE INTERNATIONAL WHALING COMMISSION

J. L. McHugh[1]

I restrict myself here to two subjects: the International Whaling
Commission and what it can and cannot do, and the history of the Commission and world whaling since 1946. I think all are aware that scientists have been concerned about the effect of whaling on stocks
since at least as early as 1910, when the subject was brought up at
an International Zoological Congress.

The years 1964 and 1965 were very important for the Whaling Commission. In those years events which began in the 1950's, or probably
much earlier, reached their culmination and threatened the very existence of the organization. In the end these tribulations strengthened
the Commission and forced it to begin to deal forthrightly with some
urgent problems which it had previously ignored.

From 1945 to 1965 various regulations were in effect on world
whaling. It was forbidden to kill some species, prohibitions were
in force on killing calves and lactating females, minimum size limits
had been set, substantial areas of the world ocean were closed to
pelagic whaling, open and closed seasons for whaling had been set,
and so on. The Antarctic was the major world whaling area and thus
the important regulations were those that applied to the Southern Ocean.
But the most important regulation of all in the Antarctic, the blue-
whale-unit quota, offered no protection at all to the resource. From
the time of the first meeting of the Commission established under the
International Convention for the Regulation of Whaling (1946) in 1949
to the disastrous meeting of 1964, almost all major actions or failures

[1]Manuscript completed during appointment as Fellow of the Woodrow
Wilson International Center for Scholars, Washington, D.C.

TABLE 13-1. World catch of the five major species of whales, 1920 to 1971.

Season[a]	Antarctic					Remainder of world ocean				
	Blue	Humpback	Fin	Sei	Sperm	Blue	Humpback	Fin	Sei	Sperm
1920	1,874	261	3,213	71	8	400	284	1,733	1,049	741
1921	2,617	260	5,491	36	31	370	343	1,413	651	720
1922	4,416	9	2,492	103	3	859	1,153	2,002	678	817
1923	5,683	517	3,677	10	23	1,186	1,462	3,046	888	576
1924	3,732	233	3,035	193	66	1,113	973	3,859	1,526	884
1925	5,703	359	4,366	1	59	1,845	2,983	4,755	1,092	1,380
1926	4,697	364	8,916	195	37	2,530	2,674	5,348	1,299	1,545
1927	6,545	189	5,102	778	39	2,170	2,359	3,506	1,219	1,275
1928	8,334	23	4,459	883	72	1,293	1,458	2,594	1,407	1,608
1929	12,847	59	6,690	808	62	916	256	2,443	741	1,699
1930	17,898	853	11,614	216	73	1,181	1,066	2,667	625	1,053
1931	29,410	576	10,017	145	51	239	206	571	60	48
1932	6,488	184	2,871	16	13	217	52	679	83	95
1933	18,891	159	5,168	2	107	171	65	915	29	433
1934	17,349	872	7,200	0	666	137	1,417	1,468	541	1,176
1935	16,500	1,965	12,500	266	577	334	2,123	1,578	696	1,661
1936	17,731	3,162	9,697	2	399	377	1,529	2,451	721	4,440
1937	14,304	4,477	14,381	490	926	332	2,135	3,305	746	6,126
1938	14,923	2,079	28,009	161	867	112	3,046	1,671	768	2,896
1939	14,081	883	20,784	22	2,585	71	510	1,838	793	2,926
1940	11,480	2	18,694	81	1,938	79	526	1,230	460	2,153
1941	4,943	2,675	7,831	110	804	86	258	1,319	704	4,778
1942	59	16	1,189	52	109	21	319	955	321	4,666
1943	125	0	776	73	24	26	277	1,018	433	5,350
1944	339	4	1,158	197	101	14	261	1,084	792	2,450
1945	1,042	60	1,666	78	45	69	243	987	140	1,616

TABLE 13-1 (cont'd.)

Season[a]	Antarctic					Remainder of world ocean				
	Blue	Humpback	Fin	Sei	Sperm	Blue	Humpback	Fin	Sei	Sperm
1946	3,606	238	9,185	85	273	69	255	1,837	662	3,181
1947	9,192	29	14,547	393	1,431	110	261	1,948	738	6,015
1948	6,908	26	21,141	621	2,622	249	489	2,887	952	7,228
1949	7,625	31	19,123	578	4,078	156	3,364	2,640	1,277	4,938
1950	6,182	2,143	20,060	1,284	2,727	131	2,920	2,842	1,187	5,456
1951	7,048	1,638	19,456	886	4,968	230	2,714	3,363	2,147	13,296
1952	5,130	1,556	22,527	530	5,485	306	2,467	3,078	2,593	6,072
1953	3,870	963	22,867	621	2,332	348	2,365	2,714	1,587	7,245
1954	2,697	605	27,659	1,029	2,879	312	2,550	3,676	1,462	10,664
1955	2,176	495	28,624	569	5,790	319	2,218	3,561	1,371	9,804
1956	1,614	1,432	27,958	560	6,974	373	2,448	3,538	1,516	11,616
1957	1,512	679	27,757	1,692	4,429	263	2,517	3,900	1,446	14,727
1958	1,690	396	27,473	3,309	6,535	305	2,527	4,207	2,361	15,311
1959	1,192	2,394	27,128	2,421	5,652	250	2,661	3,824	3,118	15,646
1960	1,239	1,338	27,575	4,309	4,227	226	2,238	3,489	2,726	16,117
1961	1,744	718	28,761	5,102	4,800	243	2,122	3,155	2,683	16,330
1962	1,118	309	27,099	5,196	4,829	137	2,127	3,056	3,608	18,487
1963	947	270	18,668	5,503	4,771	482	2,488	3,248	4,046	23,087
1964	112	2	14,422	8,695	6,711	260	316	4,731	4,995	22,544
1965	20	0	7,811	20,380	4,352	593	452	4,540	5,074	21,196
1966	1	1	2,536	17,587	4,555	242	58	4,156	5,480	22,823
1967	4	0	2,893	12,368	4,960	66	4	3,507	6,648	21,464
1968	0	0	2,155	10,357	2,568	0	2	2,930	6,742	21,512
1969	0	0	3,020	5,776	2,682	0	0	2,300	6,204	21,455
1970	0	0	3,002	5,857	3,090	0	0	2,055	5,338	22,752
1971	0	0	2,888	6,151	2,745	0	0	---	---	---

[a] 1920 means 1919/20 season in the Antarctic, calendar year 1920 elsewhere.

to act were governed by short-range economic considerations rather
than by the requirements of conservation.

The story is familiar to all of you and I do not need to go into
it in detail. It can be traced (see Table 13-1) from the reports of
the Bureau of International Whaling Statistics, published annually
since 1930 with a hiatus during World War II, the reports of the International Commission on Whaling, various reports of the League of
Nations, and the Norwegian Whaling Gazette (Norsk Hvalfangst-Tidende).
When the Convention entered into force in 1948 and the Commission held
its first meeting in 1949, the Antarctic region was the major world
whaling area, producing more than 70% of the world catch, and it remained so until the 1962/63 season (calendar year 1963 outside the
Antarctic) when, of a total world catch of 63,579 whales of all species, 33,420 or about 52% were taken outside the Antarctic. For the
purposes of the Commission, the Antarctic is considered to be that
area of the world ocean south of 40°S latitude. The reason for the
decline was that as the abundance of baleen whales in the Antarctic
declined the numbers of pelagic expeditions declined too, and some
were shifted to the North Pacific.

Despite the great reduction in total numbers of whales taken in
the Antarctic (from 41,127 in 1960/61 to 11,949 in 1969/70), this
part of the world ocean still yields more than half the world baleen
whale catch (over 78% in 1948/49, and still slightly over 50% in
1969/70). The Antarctic has taken second place to the rest of the
world ocean in total numbers of whales killed because the catch of
sperm whales has risen sharply, mainly in the North Pacific Ocean.
In 1948/49 about 20% of the world whale catch was sperm whales. In
1968/69 sperm whales made up more than half the total world catch
(58%).

One of the greatest weaknesses of the 1946 Convention was the
decision to use the blue whale unit (b.w.u.) as a basis for setting
catch limits in the Antarctic. In fairness to the negotiators of
the 1946 Convention, it must be pointed out that there was some
precedent for this strange decision. In the early 1930's the whaling
companies, at that time mostly British and Norwegian, were in economic

difficulties from overproduction of oil, probably aggravated by the
economic depression. Their problem was clear: overproduction would
cause the price of oil to drop, but underproduction would encourage
entry into the industry by others (Bock 1966). A production agreement
was reached, with only two companies abstaining, by which the various
species of whales were rated according to oil yield in the ratio 110
barrels per blue whale. The ratio was extended to certain other species by the formula: 1 blue whale = 3 humpbacks = 5 sei. The formula
used in the 1946 Convention is different: 1 blue whale = 2 fin whales
= 2½ humpbacks = 6 sei.

It should not be necessary to explain why the b.w.u. is illogical
as a management unit. It is sufficient to say that even if baleen
whale catches in the Antarctic had been balanced equitably between
species according to their individual equilibrium yields, the quotas
of 16,000 to 14,500 b.w.u. in force during the early period of the
history of the Commission were far too high. According to the best
present estimates of the Scientific Committee of the Commission, the
combined maximum sustainable yield (m.s.y.) for blue, humpback, fin,
and sei whales in the Antarctic is about 9,000 b.w.u. So the catch
limit should never have been higher than 9,000 b.w.u. at any time,
and because blue and humpback whale stocks almost certainly were already overharvested in the Antarctic when the Convention was ratified,
even 9,000 probably would have been too high.

As a practical matter, because the catches of individual species
were not equitably balanced, and could not be expected to be when
individuals of some species were much more valuable than others, it
is obvious that the quota should never have exceeded 3,000 b.w.u.,
the m.s.y. of Antarctic blue whales. Indeed, because the blue whale
stocks already were overfished in 1946, even 3,000 b.w.u. was too
high. But these things were not as clear in 1946 as they are today,
and at the end of the war economic considerations were given much
weight.

Table 13-2 shows the catch limits in the Antarctic since the first
postwar season (1945/46), the actual catch of the pelagic fleets, and
the total catch including land stations. The catch of land stations

TABLE 13-2. Blue-whale-unit catch limits established for the Antarctic
since the 1945/46 season compared with actual catches.

| Season | Catch limit | Catch (in b.w.u.) | |
		Pelagic	Total[a]
1945/46	16,000	7,391	8,308
1946/47	16,000	15,338	16,543
1947/48	16,000	16,364	17,592
1948/49	16,000	16,007	17,296
1949/50	16,000	16,062	17,283
1950/51	16,000	16,416	17,579
1951/52	16,000	16,008	17,104
1952/53	16,000	14,867	15,792
1953/54	15,500	15,456	16,940
1954/55	15,500	15,324	16,781
1955/56	15,000	14,874	16,259
1956/57	14,500	14,745	15,944
1957/58	14,500	14,851	15,136
1958/59	15,000	15,301	16,117
1959/60	none[b]	15,512	16,280
1960/61	none[c]	16,434	17,261
1961/62	none[c]	15,253	15,657
1962/63	15,000	11,306	11,306
1963/64	10,000	8,429	8,773
1964/65	none[d]	6,986	7,323
1965/66	4,500	4,091	4,201
1966/67	3,500	3,512	3,512
1967/68	3,200	2,804	2,804
1968/69	3,200	2,473	2,473
1969/70	2,700	2,477	2,477
1970/71	2,700	2,469	2,469

[a] Including land-station catches.

[b] Japan, Netherlands, Norway, and the United Kingdom
announced voluntary limits which totaled 14,500 b.w.u.
The USSR announced no limits.

[c] The five Antarctic whaling nations announced voluntary limits which totaled 17,780 b.w.u.

[d] The three remaining Antarctic whaling nations (Japan,
Norway, and the USSR) announced voluntary limits which
totaled 8,000 b.w.u.

in the Antarctic was not included in the b.w.u. quota until the season
1966/67. Some indication of the stormy history of the Commission can
be inferred from the table. After some success in reducing the Antarctic quota in the middle 1950's, the Commission raised it by 500 b.w.u.
at the 1958 meeting, then for the next three years was unable to agree
on a quota. When it did set quotas again in 1962 and 1963, it set
them far too high, and when in 1964 the Commission tried to resolve
the problem it again was unable to reach agreement. The sharp quota
reductions in 1965 and subsequently were a remarkable victory for the
Commission, as I will try to explain later. This was the first indication that the Whaling Commission might become an effective organization. Its critics have failed to recognize this important point or
have chosen to ignore it. The widespread recent publicity in the
press, in magazines, and on radio and television in the United States
has verged on the irresponsible and has led people generally to believe
that the International Whaling Commission has been completely ineffective. This was not strictly true even before 1965 (Gulland 1966), although Commission actions up to that time had merely delayed the decline
of the stocks, not stopped it. Now the fin whale catch in the Antarctic appears to have been stabilized for the past six seasons, and the
sei whale catch for the past three (see Fig. 13-1). I am encouraged
by this improved performance of the Commission, though far from satisfied, for stabilization at low stock levels is not enough. But my
concern about developments of the past year is that they may weaken
the Commission, or even sentence it to death, at a time when it is
beginning to come alive. If I could see viable and better alternatives, this would not bother me, but I do not. Certainly unilateral
action by the United States, or by any other nation, is not a viable
alternative, in my opinion. On the other hand, if events of the past
year can be used to strengthen the Commission and to help it to do
its job better, they may have been worthwhile.

There have been occasions in the last year when I have been
frustrated and irritated by the public furore about whales and whaling.
I have seen it as threatening the efforts, now beginning to be felt,
that several of us have been making to achieve rational management of

FIGURE 13-1. Catches of fin and sei whales in the Antarctic since
1920. The dates refer to the year in which the whaling season began
(1920 = 1920/21 season). Estimated levels of maximum sustainable
yield (m.s.y.) and present sustainable yield (p.s.y.) are indicated
by horizontal lines. For the sei whale, estimates of m.s.y. and p.s.y.
are the same and the scientists are agreed on 5,000 whales as the best
estimate. For the fin whale, the scientists are not agreed on point
estimates, and, therefore, the ranges of estimates are shown. The
range of p.s.y. given for fin whales is the average of the lowest
estimates (Allen—Canada) and of the highest (Japan).

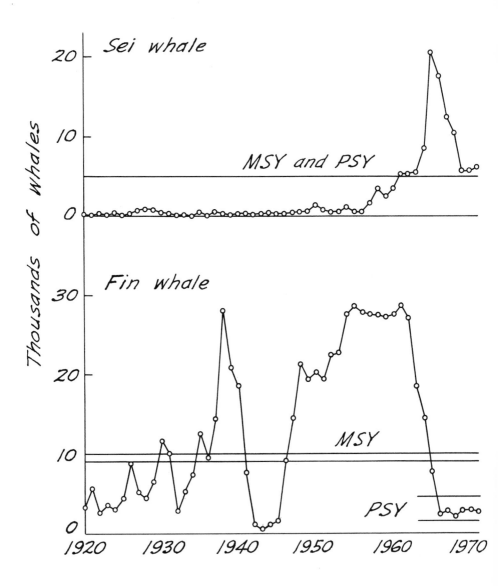

world whaling and restoration of the whale resources to full biological productivity for utilization by man. Yet in thinking about this
recent surge of public concern, and about some of the official actions
that have been stimulated by it, I recall events in 1965 at the special
meeting of the Whaling Commission and the annual meeting which followed.
Ambassador William C. Herrington was then active in whaling matters for
the United States government, and he concluded, and I agreed, that the
best strategy was to alert public opinion to the slow progress of the
Whaling Commission. We did what we could—and our magic incantations,
like those of the sorcerer's apprentice, appear to have been successful.
The problem now is to halt the forces we have set in motion before they
destroy the object of our efforts. The results of the forthcoming meeting of the Commission will show whether or not we have found the magic
words.

The International Convention for the Regulation of Whaling, 1946

The essential provisions of the International Convention for
the Regulation of Whaling are as follows:

A *preamble* which states the purpose of the agreement;

Article I, which defines the scope of the Convention, and the
activities and geographic areas to which it applies;

Article II, which defines terminology;

Article III, which describes the composition of the Commission,
its rules of procedure, staff, committees, expenses of members and
their advisers, and meetings;

Article IV, which states the functions of the Commission, as
follows:

> "a) encourage, recommend, or if necessary, organize
> studies and investigations relating to whales and whaling;
> b) collect and analyze statistical information concerning
> the current condition and need of the whale stocks and the
> effects of whaling activities thereon;
> c) study, appraise, and disseminate information concerning
> methods of maintaining and increasing the population of
> whale stocks."

This Article also describes arrangements for publication of reports.

    Article V, an important article which authorizes the Commission to amend the provisions of the Schedule as necessary to fix:

    a)   protected and unprotected species;

    b)   open and closed seasons;

    c)   open and closed waters;

    d)   size limits for each species;

    e)   time, methods, and intensity of whaling (including maximum catch of whales to be taken in any one season);

    f)   types and specifications of gear and apparatus and appliances which may be used;

    g)   methods of measurement;

    h)   catch returns and other statistical and biological records;

    i)   methods of inspection (this last authority was added by the Protocol of November 19, 1956).

Paragraph 2 of Article V provides that amendments of the Schedule:

    a)   shall be such as are necessary to carry out the objectives and purposes of the Convention and to provide for the conservation, development, and optimum utilization of the whale resources;

    b)   shall be based on scientific findings;

    c)   shall not involve restrictions on the number or nationality of factory ships or land stations, nor allocate specific quotas to any factory or ship or land station or to any group of factory ships or land stations;

    d)   shall take into consideration the interests of the consumers of whale products and the whaling industry.

Paragraph 3 of Article V is a rather involved section which says in effect that a member government may within 90 days of notification lodge an objection to any amendment of the Schedule, in which case the amendment shall not apply to that government. If one or more governments lodge objections, further time is allowed for other governments to do so, if they wish.

Article VI, which authorizes the Commission to make recommendations to any or all contracting governments on any matters that relate
to whales and whaling.

Article VII, which provides for prompt transmission by contracting governments to the Bureau of International Whaling Statistics, or
to any other body the Commission may designate, necessary statistical
and other information.

Article VIII, which authorizes contracting governments to issue
scientific permits for taking and treating whales, and provides for
appropriate reporting.

Article IX, which contains provisions for handling infractions
of the provisions of the Convention and for punishing violations.

Article X, which names the United States as depository government and provides for ratification and entry into force.

Article XI, which states procedures for withdrawal from the
Convention and lists the signatories.

The Schedule to the International Whaling Convention

The drafters of the Convention recognized that it might be
necessary to amend the whaling regulations from time to time and that
it would be unwieldy if this had to be done by formal agreements. For
this reason they placed all regulatory provisions in a Schedule to the
Convention, which, as already stated, can be amended by a three-quarters
majority vote of the Commission. The Schedule is a living document,
changing each year. It contains the following kinds of regulation, in
the order in which they appear in the Schedule:

1) provision for national inspectors on factory ships and
   at land stations;

2) bans on killing gray and right whales except by, or on
   behalf of, aborigines;

3) bans on killing calves or suckling whales or females
   accompanied by calves;

4) bans on killing blue whales in the North Atlantic and
   North Pacific Oceans;

5) bans on using factory ships or whale catchers attached
   thereto for taking or treating baleen whales except minke
   whales in most of the northern hemisphere except the North
   Pacific and in the southern hemisphere north of 40°S;

6) bans on killing humpback whales in the entire world ocean
   and blue whales in the southern hemisphere;

7) limitation of the baleen whaling season by pelagic fleets
   in the Antarctic to the period from 12 December to 7 April
   inclusive; limitation of the sperm whale season to a continuous period of 8 months out of any 12 months; and provision for one continuous open season of 6 months out of
   any 12 months for minke whales, this season not necessarily
   to coincide with the season for other baleen whales;

8) blue-whale-unit catch limit for the Antarctic, including
   the formula for determining b.w.u., notification of catches
   to the Bureau of International Whaling Statistics, and
   procedures for closing the Antarctic baleen whaling season.
   Since 1970 this paragraph has also included catch limits
   by species for baleen whales in the North Pacific Ocean;

9) minimum size limits for baleen and sperm whales, and
   prescribed methods for measuring whales;

10) limitations on seasons for catching baleen and sperm whales
    from land stations;

11) prohibitions on factory ships and land stations for treating
    protected species, or whales caught out of season or in
    closed areas or below minimum size limits; and provisions
    for full utilization of carcasses of whales taken legally;

12) maximum time limit in which a whale carcass shall remain
    in the sea, provisions for marking whales to identify
    catcher and order of catching, and provisions for keeping
    records and making reports;

13) provisions that gunners and crews of factory ships, land
    stations, and whale catchers shall not be remunerated
    according to the numbers of whales they take or for taking
    illegal whales;

14) provision for transmission to the Commission of national
laws and regulations;

15) details of the information that shall be transmitted to
the Commission and the Bureau of Internation Whaling Statistics under the provisions of Article VII of the Convention;

16) special provisions for factory ships which operate within
certain territorial waters;

17) definitions of terms.

The Commission has given standing instructions that the Schedule
shall be reprinted whenever any amendments made by the Commission come
into operation.

The International Whaling Commission

Much nonsense has been written and spoken recently about the
Commission and its powers and responsibilities. The International
Convention which created it is an imperfect document, as all treaties
are, because it represents a compromise between diverse interests. In
this respect it is not different from any other international fishery
convention to which the United States or any other nation adheres. The
basic conflict is very clear in the language of the Preamble to the
International Convention for the Regulation of Whaling, as Bock (1966)
has pointed out. The Preamble recognizes past overexploitation of
whales, states that continuation of such practices may endanger the
resource, and stresses the need to regulate the catch in various ways,
to restore some stocks to full biological productivity and to prevent
overfishing of others. Yet it also mentions the need to avoid economic and nutritional distress and to make possible the orderly development of the whaling industry. This conflict between rational management and economic development has been responsible for much of the
difficulty encountered by the Commission, as it has for all fisheries.

The basic conflict comes out even more clearly in the language
of Article V, which permits the Commission on the one hand to adopt
"regulations with respect to the conservation and utilization of whale
resources," but on the other hand to "take into consideration the

interests of the consumers of whale products and the whaling industry."
This often has placed the long-term objectives of "conservation, development, and optimum utilization of the whale resources" which "shall
be based on scientific findings" in conflict with short-term economic
gains. In the Whaling Commission, as in all international fishery
arrangements, an inverse correlation can be recognized between the
economic importance and political strength of that segment of the
national fishing industry and the ability of the national delegation
to give priority to scientific objectivity in reaching policy decisions.
Coming into existence shortly after the end of a major world war, when
the world faced a serious shortage of fats and oils, the Commission
was influenced from the start by short-term economic considerations.

Furthermore, it is not well understood that the Whaling Commission is far from autonomous. For example, the Schedule to the Convention, which contains all of the regulations adopted by the Commission
and which by definition "forms an integral part" of the Convention,
can be amended by the Commission at an annual meeting by a three-
fourths majority of those casting an affirmative or negative vote.
Such amendments are not necessarily binding on governments, however.
Article V of the Convention provides that member governments may,
within 90 days of notification of an amendment, enter an objection.
Unless that objection is withdrawn, the amendment "shall not become
effective with respect to" that government. This right to enter objections, although it is different specifically from the provisions
of other fishery conventions, is not different in principle. For
example, the Inter-American Tropical Tuna Convention and the Commission which it created have been held up by some authorities as
models for successful international fishery research and management.
Yet the original U.S. statute which implemented the Convention for
the United States contained no provisions for enforcement by the
United States. When research led to the conclusion that management
of the yellowfin tuna fishery in the eastern tropical Pacific was
necessary, the domestic statute had to be amended to make such management possible. The authority given by the amendment was not without
restrictions, however. The law states, in effect, that the regulations

will not be applicable to U.S. fishermen unless all other nations
which catch meaningful amounts of tuna in the regulatory area adopt
appropriate laws or regulations and are prepared to enforce them. The
question as to whether other nations have indeed met these requirements
has been a subject of bitter debate between the U.S. tuna industry and
government and in intergovernmental conferences since tuna regulation
began in the middle 1960's. The situation is complicated further because some nations which fish tuna in the area are not members of the
Inter-American Tropical Tuna Commission.

Many governments do not grant unconditional plenipotentiary
powers to the heads of their delegations to meetings of such commissions because they wish to reserve the right to disagree with majority
decisions of such bodies. In the United States, where the legislative
and judicial branches of government must be party to decisions reached
by the executive branch, such limitations on decision-making by international bodies are imperative.

It also is not understood by many that the International Whaling Commission, in addition to having limited powers to make regulations, has no power at all to enforce them. This is generally believed to be a great weakness of the Commission, and this may be true,
but again it is not unique. No international fishery commission has
power to enforce regulations, and none can make regulations contrary
to the policies or statutes of member governments. In the United
States, for example, even if the government does not exercise its
privilege to enter an objection to an amendment to the Schedule, the
amendment has no force against American whalers until a notice amend-
ing the domestic whaling regulations has been published and circulated
by the Secretary of Commerce. Some nations may automatically delegate
such authority to the head of their delegation, so that the regulation
is law when he agrees to it at a Commission meeting. I remember once
spending several hours in discussion with the representative of such
a country, who was completely unable to understand why a United States
Commissioner could not make an agreement which was binding on his
country. In this case the question was the amount of national contributions to the cost of operation of an international body. We

never were able to convince him that the democratic process in the
United States would not allow such delegation of authority to a presidential appointee. He finally dropped the subject and went away shak-
ing his head in wonderment.

The structure of the Whaling Commission is another source of
great confusion to most people. Many of the letters I receive illustrate this confusion:

> "There is great concern because the International Whaling
> Commission is not accountable for what it does."

> "To allot $1,000.00 for the International Whaling Commis-
> sion is surely a dead giveaway to the fact that the U.S.
> is not interested in its work."

> "Today all whales, however large and elusive, are fair game
> for the fleets of modern factory ships, and overfishing is
> the inevitable result."

Over and over again the inference is clear that people see the Commission as an irresponsible body, having no power, or not interested in
wielding its power, or simply endorsing the wishes of the whaling
industry. All this is not without cause, but it is also based on
misunderstanding of the nature and functions of international organizations. After all, such organizations have no magic powers of their
own. They are simply creatures of the governments that established
them, and they try to carry out the policies and wishes of the member
nations and their people. Their recommendations are subject to ap-
proval by the member nations and their people, and if their actions
are disapproved through misunderstanding or for good and sufficient
reasons, the member nations individually or together can do something
about it. This is what is happening in the United States now. The
American public is aroused, and this has had, and will have, profound
effects domestically and internationally.

The Bureau of International Whaling Statistics

The Bureau of International Whaling Statistics was established
by the government and whaling industry of Norway in the late 1920's

at the request of the International Council for the Exploration of the
Sea. The Bureau gathers, collates, and publishes annually reports on
the past whaling season, with details by species, areas of the ocean,
size of whales, numbers of factory ships, catchers, land stations, and
other information. It provides services to the International Whaling
Commission by keeping close records of the Antarctic catch as the
quota is approached, and notifies the Commission of the date on which
the Antarctic season must be closed.

This unique and valuable service to the whaling industry and
governments is still funded entirely by Norway, except for a small
contribution by the Commission in each of the last few years to cover
the extra cost of providing data for stock assessments. The Bureau
provided the basic data, going back to 1920, on which the Scientific
Committee of the Commission, the Committees of Three and Four, and the
FAO Stock Assessment Group drew for stock assessments and other studies.
Estimates of the condition of the whale stocks, sustainable yields,
and quotas needed to restore or maintain maximum sustainable yields
would have been impossible without the help given by the Bureau.

The History of Whaling since 1946

The great era of American whaling, when about 80 percent of
the world catch was taken by American whalers, ended before 1900. The
collapse came not from a scarcity of whales, as many people believe,
but from the discovery of petroleum in 1859 and destruction of much
of the American whaling fleet during the Civil War. The rebirth of
whaling in the twentieth century goes back to the invention of the
harpoon gun and explosive harpoon head by the Norwegian Svend Foyn
in the 1860's; but it was the development of the floating factory in
1903, and especially of the factory ship with stern ramp in 1925, both
also by Norwegians, which made expansion into all Antarctic seas possible.

The world whale catch did not exceed 10,000 animals until 1910.
The pre-World War I peak was reached in 1913 with a catch of over
25,000. This level was not reached again until 1926, but thereafter

catches climbed steadily, reaching nearly 55,000 in 1938. But the
total catch of all species does not tell the story. Already the blue
whale and the humpback had produced the greatest annual catches they
were ever to yield: nearly 30,000 blue whales in 1931 (over 99 percent
of them in the Antarctic) and over 6,600 humpbacks in 1937 (two-thirds
of them in the Antarctic). The record catch of 1938 was made up mostly
of fin whales—nearly 30,000 of them—by far the greatest catch of
this species ever made up to that time. It must be remembered also
that up to the Second World War there were no controls on whaling other
than a few local regulations by individual nations. All sizes and all
species were fair game. No method of whaling was prohibited, no area
of the world ocean was closed, and there were no limits on seasons.
The ink was hardly dry on the 1937 Convention, which would have ef-
fected such restrictions, but it never really had a chance to become
effective before the war intervened.

The Antarctic had become the major world whaling ground in the
1920's. At the end of the 1930's it was producing about 85 percent
of the world catch. In the rosy days following World War II, when
the Antarctic stocks had been given a respite of about six years,
everyone expected that the resource would have recovered. Concern
about the future of whaling led to an international conference in January 1944, at which a limit of 16,000 b.w.u. was set for the 1945/46
season in the Antarctic. The results were disappointing and a real
shock to the optimists—only about 8,300 b.w.u. were taken, as compared with the last two prewar seasons when catches had been 29,786
and 24,830 b.w.u. respectively!

The Commission held to the Antarctic 16,000 b.w.u. quota up to
and including the 1952/53 season (Table 13-2). The quota applied
only to pelagic whaling, not to land stations in the Antarctic. The
total catch, land stations included, exceeded the quota until the
1952/53 season, when the number of factory ships dropped from 20 the
previous season to 16, and the number of catcher boats was reduced
from 289 to 251. Before this season the Antarctic quota had been
caught only by a steadily increasing number of factory ships, from 9
in 1945/46 to 20 in 1951/52, and catcher boats from 93 in 1945/46 to

289 in 1952/52. The increase in total whaling power had been even
greater than these numbers would suggest, for the newer fleets were
more powerful and efficient and the skills of the whalers had been
increasing also. There had been a substantial drop in oil prices in
early 1952, and this probably led to the decision of Norway to with-
draw three expeditions and Japan one from the Antarctic in 1952/53.

   This reduction in Antarctic whaling effort was not to last for
long. In 1953/54 another factory ship was added, although the total
fleet of catcher boats was reduced still further. The reduction in
catchers was based on an agreement among the whaling companies, stimulated by the continued addition of more efficient catchers, as the
companies vied with each other to take the maximum share of the quota.
This cooperation, however, was not completely effective, and by
1961/62 the Antarctic whaling fleet had grown to 21 factory ships and
269 catcher boats. It is notable that, although the total Antarctic
catch including land stations again exceeded the pelagic catch quota
for several seasons following 1952/53, the pelagic catch did not
reach the quota again until 1956/57—and then only through reduction
in the quota and a further substantial increase in size and efficiency
of the fleet.

   By 1956 and 1957, after much bitter debate, the Commission had
reduced the Antarctic quota to 14,500 b.w.u., a token reduction at
best. At the 1958 meeting the Commission voted to retain the 14,500
b.w.u. quota, but the Commissioner of the Netherlands voted no. Subsequently his government lodged an objection, which triggered objections by the other Antarctic whaling nations. Consequently, the
quota for the 1958/59 season reverted to 15,000 b.w.u. This was the
beginning of a very difficult period for the Commission, which was
not to be resolved until 1965 or later.

   The authority of the Whaling Commission to recommend catch
limits is restricted to the total quota, or "global quota," as it is
usually called. The Commission has no authority to decide how the
quota should be allocated between nations, or even less between fleets.
This is a private matter between the nations concerned. Each year,
after the Commission agrees on the quota, the Antarctic whaling nations

meet and decide how that quota should be subdivided. Since 1966 a
similar procedure has been adopted by the four North Pacific whaling
nations. In fact, global quotas for the North Pacific were not ef-
fected by amendment of the Schedule until 1970.

For the 1959/60 season the Antarctic whaling nations were unable
to agree upon national quotas. Consequently, the Netherlands and
Norway withdrew from the Convention effective July 1, 1959, and no
global quota was in effect for the Antarctic in 1959/60. The individual governments, with the exception of the Netherlands and USSR,
stipulated maximum catches as follows: Japan, about 5,000 b.w.u.;
Norway, 5,800 b.w.u.; United Kingdom, 2,500 b.w.u. The Netherlands
company declared that it would observe a limit of 1,200 b.w.u. The
USSR made no commitment, but considering its catch in 1959/60 and
the previous agreement that the USSR would be allocated 20% of the
Antarctic quota, it can be assumed that the Soviet Union set for itself a quota of 3,000 b.w.u. (that is, 20% of 15,000 b.w.u.). Thus,
about 17,500 b.w.u. was the approximate goal for the 1959/60 Antarctic whaling season, a most alarming development. The catch fell considerably short of this, and this was one of the first in a series
of events which slowly convinced the whaling industry that the conclusions and warnings of the scientists were remarkably accurate,
although the whaling industry generally resisted these warnings until
1965, and to some extent still does today.

With two Antarctic whaling nations, the Netherlands and Norway,
out of the organization, the International Whaling Commission was
virtually ineffective. To bring the two countries back into the
Convention, the Commission decided at its 1960 meeting to suspend
the catch limit for the following two seasons, 1960/61 and 1961/62.
In lieu of the quota the Commission adopted the following resolution:

> "In view of the action taken to suspend the limit on the
> Antarctic pelagic whale catch until the season 1962/63,
> the Commission resolves it to be of extreme importance
> that each of the countries engaged in pelagic whaling
> should limit the size of its national catch to a level

in no event greater than that adopted for the season of 1959/60; and that the Secretary be instructed to so inform those Governments."

The five governments announced the following individual catch limits: Japan, 5,980; Netherlands, 1,200; Norway, 5,800; United Kingdom, 1,800; USSR, 3,000—a total of 17,780 b.w.u. It is interesting that Japan and the Soviet Union objected to suspension of the quota, which had the effect that for these two countries the old quota of 15,000 b.w.u. was in force. Since the other countries were not bound by a global quota, this action by Japan and the USSR merely expressed their interest in a principle.

The voluntary catch limits announced by the Antarctic whaling nations for 1961/62, the second season of the two-year suspension of the quota, were as follows: Japan, 6,680; Netherlands, 1,200; Norway, 5,100; United Kingdom, 1,800; USSR, 3,000—again a total limitation of 17,780 b.w.u. The increases in the Japanese stipulation from 1959/60 to 1960/61 and again in 1961/62 were related to the transfer of two British and one Norwegian factory ships to Japan, together with the shares of the quota that went with them (such transfers were sanctioned by the 1962 agreement between the Antarctic whaling nations).

Despite the massive whaling effort, which reached its all-time peak in 1961/62, the total of the voluntary quotas was not reached in either season. But the principal objective of the quota suspension was achieved: Norway reentered the Convention in September 1960 and the Netherlands in May 1962. The Netherlands had made her return conditional on agreement by the five nations on an apportionment of the Antarctic catch quota and other conditions. After the 1961/62 season the 15,000 b.w.u. quota was automatically reinstated and the Antarctic whaling nations met and agreed on the following apportionment: Japan, 41%; Netherlands, 6%; Norway, 28%; United Kingdom, 5%; and USSR, 20%. The scientists had warned that this quota was too high, and the 1962/63 season proved it. The catches of blue and fin whales declined sharply and the total catch was only 11,306 b.w.u. For the first time since the 1903/04 season the land stations on South Georgia did not operate.

Reduced catches and increasing costs of whaling in the Antarctic

forced a series of readjustments in the fleet. In 1962 two Norwegian
factory ships were sold for salvage, and a third did not operate in
1962/63. One British factory ship was sold to Japan. Japan used the
quota share that went with this vessel under the 1962 agreement, but
did not operate the vessel. Thus in 1962/63 there were 4 less factory
ships and 60 fewer catchers than in the previous season. Another
British factory ship was sold to Japan in 1963, eliminating the United
Kingdom from the whaling industry. Again Japan chose to use the quota
but not the vessel, so that in 1963/64 the fleet was reduced again by
1 factory ship and 11 catchers. Meanwhile the Commission had agreed
to a substantial reduction in the quota, an action unthinkable just
a few years ago, setting a catch limit of 10,000 b.w.u. for the 1963/64
season. The scientists once again warned that even this reduced quota
was far too high. The revised formula for subdivision was: Japan,
46%; Netherlands, 6%; Norway, 28%; USSR, 20%. At the close of the
1963/64 season the total catch was 8,429 b.w.u., with Netherlands and
Norway well below their assigned quotas. Two shore stations on South
Georgia reopened, this time under Japanese management, and took about
343 b.w.u. This disappointing season ended the Netherlands venture
into whaling. Her single fleet was sold and her quota share trans-
ferred to Japan.

Then followed the almost disastrous Commission meeting of 1964,
when it was obvious to all that the resource was in serious condition
and when the Commission had already prohibited killing of blue whales
in the Antarctic except for the area frequented by pygmy blue whales.
The 1964 meeting attempted to arrive at a quota consistent with the
scientific evidence, which as a result of the work of the Committee
of Three (later Committee of Four) and the FAO Stock Assessment Group
was undeniable. But the Antarctic whaling nations were unwilling to
accept the necessary drastic reduction in the quota and the meeting
ended without agreement on this point. Since the 1963 agreement had
been for the 1963/64 season only, this meant that there was no catch
limit for 1964/65. The three remaining pelagic whaling nations met
after the Commission adjourned and arrived at a voluntary quota of
8,000 b.w.u. distributed as follows: Japan, 4,160; Norway, 2,240;

USSR, 1,600. The argument did not end here, however. The Japanese
and Norwegian governments accepted these quotas, but the Soviet Union
called attention to the changed circumstances affecting distribution
of national quotas since the 1962 quota agreement was signed. These
changed circumstances were: elimination of the United Kingdom and
Netherlands and increase in the Japanese share through purchase of
factory ships. The USSR stated that revision of the quota arrangements
was a necessary condition to their acceptance of the voluntary catch
limit for 1964/65 and the International Observer Scheme. The problem
was not resolved before the 1964/65 season began, but the pelagic
catch was short by more than 1,000 b.w.u. of the 8,000 figure (Japan,
4,125; Norway, 1,273; USSR, 1,588). Again, the warnings of the scientists had been substantiated almost exactly.

Following the 1964 meeting of the Commission several member
countries, the United States included, had serious doubts about the
future effectiveness of the Commission and questioned the merit of
continuing to support it. It was obvious that the Antarctic whale
fishery was close to economic, if not biological, extinction and that
continued inaction by the Commission was unthinkable. Two possibilities were open: an appeal to the FAO for intervention, or a mass
withdrawal of most member countries. Recognizing that the Convention
and the Commission should be preserved if possible, it was decided to
try to convene a special meeting to deal with the subject of the Antarctic quota. Such a meeting of the Commission was held in May 1965.
The outcome was surprisingly successful and a great achievement by
the Commission; for it was agreed, over a three-year period, to get
the quota below the sustainable yield according to the best scientific estimates available in 1967. The quota for 1965/66 was set at
4,500 b.w.u. and at the 1966 and 1967 meetings it was reduced again
to 3,500 and 3,200 respectively, which in 1967 was just below the
estimated sustainable yield of 3,300 b.w.u.

The question of quotas by species was brought up, but it was not
discussed at length because the Antarctic whaling nations felt that
they would have sufficient problems adjusting to the economic strains
of this drastic reduction. Thus elimination of the blue whale unit

was not addressed seriously at this time.

The Antarctic quota remained at 3,200 b.w.u. for two seasons, 1967/68 and 1968/69, and commissioners, industry, and scientists alike believed that one of the major problems of Antarctic whaling had been solved. Even the problem of the blue whale unit became less urgent, at least temporarily, although this was more by accident than by plan. Six sei whales yield more meat than two fin whales, and Japanese whalers, to whom meat is especially important, had decided to concentrate on sei whales. This decision put most of the Japanese whaling fleet farther north, in somewhat more temperate waters, where sei whales are most abundant and fin whales scarce. It turned out that this allocated the catch between the two species in just about the right proportions. In fact, the catch of fin whales, which were over-fished, was held for a while slightly below the estimated sustainable yield, which was fortunate. Thus it was possible to set aside for the moment the difficult question of quotas by species in the Antarc-tic, so that the Commission could address itself to other important matters.

Meanwhile, the scientists had been concerned about their ability to read correctly the age of whales, by the indirect methods available. In 1968 a group of scientists from several member nations met for a thorough review of age determination in whales. They concluded that the method previously agreed upon was not the most accurate. The new index of age from ear plug laminations, and other adjustments affecting stock estimates, forced a downward revision of estimates of sustainable yields. In other words, the previous estimate of 3,300 b.w.u. in the Antarctic was too high; hence the Antarctic quota also was too high. This was a great shock to the Commission, for it had every reason to believe that the quota was about where it should be and that revisions, if necessary, would be relatively minor. It was a shock to the whaling companies also, for they had made the neces-sary economic adjustments and had no reason to expect that they would need to make more. Now, at the 1968 meeting, the Commission learned that it must reduce the Antarctic quota by at least 700 b.w.u. and perhaps more. The situation was indeed appalling, and it is not surprising that faith in the scientists was shaken.

If Norway had not decided to drop out of Antarctic whaling, the problem might not have been resolved. Norway did not wish to give up her quota entirely, for she had plans to reenter later, as she has now done. But she was willing to accept a smaller share of the Antarctic global quota, thus paving the way for a reduction in the quota without reducing the national quotas of Japan and the Soviet Union. The quota for 1968/69 was left at 3,200 b.w.u. on the understanding that this was in effect a quota of somewhat less than 2,500. The actual catch in the 1968/69 season was 2,473 b.w.u.

At the 1969 meeting of the Commission, the Antarctic quota was set at 2,700 b.w.u., with the understanding that Norway would be returning to the Antarctic, but would be catching mostly minke whales and few, if any, fin and sei whales. The catch in 1969/70 was just slightly higher than the previous season (2,477 b.w.u.). The quota was again set at 2,700 for 1970/71 with the assurance that it would not exceed 2,500 b.w.u. It did not. The 1970/71 catch was 2,469 b.w.u.

North Pacific Whaling

As Antarctic catches declined, attention turned to the North Pacific Ocean, the other remaining area of the world ocean in which pelagic whaling was permitted. Largely as a result of an increase in the USSR fleet from one expedition to four in the early 1960's, it became obvious that the North Pacific whale resources—which are much smaller than the Antarctic stocks—could not long withstand the heavy drain. The Commission established a special scientific committee on North Pacific whaling, and in 1966 there began a series of special meetings of the commissioners of the North Pacific whaling nations (Canada, Japan, USSR, USA). Before long, it was recognized that the stocks of blue and humpback whales were seriously reduced, and a worldwide ban on killing these two species was recommended by the Commission and accepted by the member governments.

As adequate stock assessments became available, quotas were placed on other species in the North Pacific. At present, the catches of fin, sei, and sperm whales are regulated by quota. From the begin-

ning, it was possible to avoid the blue whale unit in setting North
Pacific whaling quotas, although the price for this was an agreement
to allow the catch of fin whales to exceed the quota by 10% provided
that an equivalent reduction is made in the catch of sei whales and
vice versa. The quotas are perhaps not low enough, but with the exception of the sperm whale quota they are close enough to the best estimate of sustainable yield so that the resource will not be seriously
affected by one year's whaling. The sperm whale quota is voluntary
at present, but the Commission will be in a position to take action
at the 1971 meeting. This quota probably should be reduced substantially.

The International Observer Scheme

One great concern to the Commission and to conservationists around
the world has been the question of whether the quotas and other regulations are being observed strictly. There is indirect evidence that
infractions are sometimes substantial. The question will never be
settled without an adequate system of surveillance. The Schedule provides for national inspectors on factory ships and at land stations,
and it also mentions observers as follows: "and also such observers
as the member countries . . . may arrange to place on each other's
factory ships" and "at each other's land stations." Thus the observer
scheme is permissive rather than mandatory. This provision was authorized by the Protocol of 1959, which added the words "methods of inspection" at the end of paragraph 1 of Article V of the Convention.
At the 1959 meeting of the Commission it was agreed in principle that
factory ships in the Antarctic should carry observers of some other
nationality than the country operating the expedition, and a set of
rules governing funding and authority of the observers was drafted.

The International Observer Scheme was discussed again at the
Commission meetings of 1960, 61, 62, and 63, and in 1963 the language
was added to the Schedule to permit the scheme to be implemented.
Further discussions and special meetings were held each year thereafter, and in 1968 the Schedule was further amended to authorize a

global observer scheme which would include land stations as well as
pelagic expeditions. Despite all this discussion and planning, and
acceptance in principle by all member nations, the scheme still is not
in effect. During the 1970 meeting of the Commission it was agreed
that another effort should be made to implement an observer scheme. A
special meeting on the subject is to be convened in Washington on 16
June 1971. As one of the most important unsolved problems of the Commission, it is imperative that it be settled as soon as possible.

Conclusions

The International Whaling Commission has a history of acting too
little and too late, but it has been far from impotent. The turning
point came in 1965, when for the first time in its history the Commission agreed to establish a catch limit in the Antarctic lower than
the best scientific estimate of the sustainable yield. Despite some
gloomy predictions, the three-year agreement made in 1965 was concluded successfully in 1967, and the Antarctic baleen whale quota for
that year was set below the best estimate of the scientists. This
was an achievement of great significance when it is viewed against the
background of prior Commission deliberations.

The history of actions by the Commission and its member nations
in the North Pacific whaling area attests to the organization's intent
and capability to improve its performance. Faced with an increasing
rate of harvesting, stimulated by the decline of Antarctic whaling
resources and obvious overharvesting, the North Pacific whaling nations
moved quickly to avoid the mistakes made in the Antarctic—although
not enough to avoid those mistakes completely. It was necessary to
forbid further killing of blue and humpback whales, but quotas by
species were placed on the other baleen whales, not by blue whale units.
The North Pacific fin whale resource has been overharvested moderately,
but the catch quota is close to the best estimate of the present sustainable yield, and there is no reason to believe that the quota will
not be further reduced at the 1971 meeting to prevent further decline
of the stock. The North Pacific sei whale resource is being caught

at a rate greater than the maximum sustainable yield, but the stocks
are still larger than necessary to produce the m.s.y., and no difficulty is expected in reducing the quota to the desired level while the
resource is still at the point of maximum biological productivity.

The condition of the global sperm whale resource is less well
understood, and the scientific estimates are tentative at best. Despite this lack of precise estimates of the status of the stock, it
is encouraging that the North Pacific whaling nations agreed to a
voluntary catch limit for the present season. A catch quota should
be established by amendment of the Schedule at the forthcoming meeting
of the Commission, on the basis of improved scientific assessment.

In the light of these encouraging moves by the Commission since
1965, it is important to support and strengthen the International
Whaling Commission, not to destroy it. In my view, based on present
circumstances, the Commission has four important unsolved problems,
and I list them in order of priority:

1) implementation of an international observer scheme;

2) for whale stocks that are clearly overexploited, catch
   quotas that are sufficiently below the best estimates of
   present sustainable yield to ensure that the stocks will
   be rebuilt to maximum yields;

3) for whale stocks that are not overexploited, catch quotas
   no higher than the best estimates of maximum sustainable
   yields;

4) elimination of the blue whale unit as a management unit.

AUTHOR'S ADDENDUM—16 OCTOBER 1972

Historical reviews very quickly become out-of-date. This is
especially true of the recent history of the International Whaling
Commission. Two of the most important meetings of the Commission have
been held since this paper was written and presented. I appreciate
the opportunity to extend the record and describe the state of world
whaling today.

The 1971 Commission meeting, held in Washington, D.C., immedi-

ately following the Conference of which this paper was a part, was the
most important meeting in its history. The International Observer
Scheme was the subject of extensive discussions, and it was agreed
that all member nations should implement the scheme for the 1971/72
season. The catch limit for the 1971/72 season in the Antarctic was
reduced from 2,700 to 2,300 blue whale units (b.w.u.). Protocol made
it impossible to deal with the question of the b.w.u., but it was
agreed that for the 1972/73 season the b.w.u. would be eliminated and
quotas would be set by species. Catch limits for the North Pacific
in 1972 were reduced by 20% from 1971 levels, as follows: fin whales,
1,046; sei and Bryde's whales combined, 3,768; sperm whales, 10,841. It
was agreed also that the Scientific Committee would meet early in 1972
to review the status of sperm whale stocks and to seek practical methods
of limiting the catch of males. Modest increases in the financial
contributions of member nations were approved for 1971/72 and 1972/73.

Progress toward these objectives continued in the interim between
the 1971 and 1972 Annual Meetings. Especially encouraging progress
was made with the International Observer Scheme, following initial
disappointment when it was not possible to complete arrangements to
place observers on the Antarctic factory ship fleets before they sailed
for the 1971/72 season. But when the 1972 land-based whaling season
began, Australia and South Africa had placed observers at each other's
stations, as had Canada, Iceland, and Norway in the North Atlantic.
In the North Pacific, the United States placed observers at the Japanese
land stations, and Japan and the Soviet Union exchanged observers on
their high-seas whaling fleets. Meanwhile, Japan, Norway, and the USSR
had been holding discussions on an observer scheme for the 1972/73 Antarctic season, and I was privileged, in my capacity as Chairman of the
Commission, to attend the formal signing ceremony at the Japanese
Embassy in London during the 24th meeting of the Commission in June
1972.

At the 24th meeting all commitments made a year previously were
honored, if not already in effect. The b.w.u. was abandoned as a
means of setting catch limits in the Antarctic, and baleen whale quotas
for the 1972/73 season were set as follows: fin whales, 1,950; sei

and Bryde's whales combined, 5,000; minke whales, 5,000. For the North
Pacific in 1973 it was agreed to delete the provision that the catch
limit for fin or sei whales could be exceeded by 10% provided that
there was an equivalent reduction in the catch of the other. Baleen
whale quotas for the 1973 North Pacific whaling season were set as
follows: fin whales, 650; sei and Bryde's whales combined, 3,000.

The Scientific Committee met early in 1972, as planned, to review
the status of sperm whale stocks around the world. On the basis of
the scientific findings the Commission established sperm whale catch
limits as follows: North Pacific, 6,000 males and 4,000 females;
southern hemisphere, 8,000 males and 5,000 females. Since the purpose
of the new quotas on sperm whales is to provide a more equitable balance in catches of the sexes, it was agreed also to reduce the minimum
size limit to 30 ft, except in the North Atlantic Ocean, where the
minimum length will remain at 35 ft.

Other important accomplishments of the 24th meeting were: (1)
to establish a committee to recommend means of strengthening the
secretariat; (2) to make new approaches to nonmember whaling nations
to adhere to the Convention; (3) to establish a committee to develop
plans for a decade of intensified research on cetaceans; and (4) to
agree to extend indefinitely the present three-year ban on killing
blue and humpback whales in the North Atlantic, which was due to ex-
pire before the next Commission meeting.

Although the history of the International Whaling Commission up
to 1965 was in no way a credit to the Commission and its member nations,
and although it should not have taken seven years to achieve present
levels of control over whaling, the accomplishments of these seven
years have been substantial. Some problems remain, one of the most
important of which is to gain adherence by the citizens of those whaling nations not now subject to international controls. There is also
a great need to improve our scientific knowledge of all Cetacea, as
has been recognized by the Scientific Committee of the Commission for
a long time.

Another important need is improved public understanding of the
situation and the issues. There is a danger that overzealous and

uninformed people will continue to promote the notion that whaling continues unchecked and that a total moratorium is the only answer. Few people realize that a moratorium is in effect now on those species of Cetacea that can be considered to be "endangered," and that for some species this moratorium has been in force for more than a quarter of a century. Most people think that "whales are an endangered species," as I read in an article not too long ago. This view ignores the fact that, of the approximately 100 different kinds of Cetacea, less than 20 are being taken commercially, and less than 10 of these have been overharvested to the point at which they can be considered "endangered" according to any reasonable interpretation of the meaning of that term. Overharvesting of no marine resource can be condoned, but the fact remains that the Whaling Commission has declared a moratorium on certain whale species, has set catch limits on other overharvested stocks which should allow them to increase while still being taken commercially, and has set limits on the catch of others which should allow a catch to be taken indefinitely at levels of maximum sustainable yields. A total moratorium is not only irrational and unnecessary, but impossible to achieve. Continued pressure for total cessation of all whaling could well be counterproductive, by destroying the Commission just when it has finished resolving most of its major problems. It is to be hoped that reason will prevail, and that all efforts will be directed toward supporting the International Whaling Commission and strengthening it in every possible way.

REFERENCES

Bock, P. G. 1966. A study in international regulation: the case of
    whaling. Ph.D. thesis, New York University.
Gulland, J. A. 1966. The effect of regulation on Antarctic whale
    catches. J. Cons. Perm. Int. Explor. Mer, 30:308-315.
International Whaling Commission. 1950-1970. Annual Reports, London.
Jahn, G., Bergersen, B., and Vangstein, E. 1952-1970. International
    Whaling Statistics XXVII-LXIII. Sandefjord, Norway, Bureau of
    International Whaling Statistics (published in Oslo). Issued
    annually since 1930, under various names prior to 1952.

CHAPTER 14

ESTIMATION OF POPULATION PARAMETERS OF ANTARCTIC BALEEN WHALES

D. G. Chapman

Table 14-1 shows the catches of three important species of baleen
whales in the Antarctic (south of 40°S latitude) since 1924/25. These
are *Balaenoptera musculus, physalus,* and *borealis*—the blue, fin, and
sei whales, respectively. The data are taken from the International
Whaling Statistics, published by the Committee for Whaling Statistics,
Oslo. The decline in the catch of blue whales (obvious in the table)
and the decline in catch of fin whales per day's operation (not shown
in the table) led the International Whaling Commission to set up, in
1961, a special study group to provide an assessment of the stocks of
these species and to make recommendations to them on the levels of
sustainable catch.

During the period between the two world wars, research was initiated on estimating the parameters of wild populations and on developing
models to understand their dynamics and hence ways to manage them for
man's benefit. The importance of this was emphasized in fisheries,
where it began to be clear that without restrictions stocks might be
depleted and destroyed. Two names deserve to be mentioned from this
period—Baranov, whose pioneering work (1918) is now widely acclaimed,
though for a time his achievements were unrecognized, and Thompson.
Thompson and his coworkers (1930, 1934) developed methods of estimating
halibut population mortality rates and also developed models for management of that population.

These initial models were extended and developed in the postwar
period, particularly by Ricker (1948, 1958) and by Beverton and Holt
(1957). Thus in 1960 when the International Whaling Commission asked
for the study referred to, the study group that it appointed was able
to bring to bear on the problem a body of well-developed methodology
of population estimation and population dynamics.

TABLE 14-1. Baleen whale catches in the Antarctic (1924/25 to 1970/71).

Season[a]	Blue[b]	Fin	Sei	Season[a]	Blue[b]		Fin	Sei
1925	5,703	4,366	1	1948	6,908		21,141	621
1926	4,697	8,916	195	1949	7,625		19,123	578
1927	6,545	5,102	778	1950	6,182		20,060	1,284
1928	8,334	4,459	883	1951	7,048		19,456	886
1929	12,734	6,689	808	1952	5,130		22,527	530
1930	17,487	11,539	216	1953	3,870		22,867	621
1931	29,410	10,017	145	1954	2,697		27,659	1,029
1932	6,488	2,871	16	1955	2,176		28,624	569
1933	18,891	5,168	2	1956	1,614		27,958	560
1934	17,349	7,200	0	1957	1,512		27,757	1,692
1935	16,500	12,500	266	1958	1,690		27,473	3,309
1936	17,731	9,697	2	1959	1,192		27,128	2,421
1937	14,304	14,381	490	1960	1,239	(917)	27,575	4,309
1938	14,923	28,009	161	1961	1,744	(739)	28,761	5,102
1939	14,081	20,784	22	1962	1,118	(716)	27,099	5,196
1940	11,480	18,694	81	1963	947	(220)	18,668	5,503
1941	4,943	7,831	110	1964	112		14,422	8,695
1942	59	1,189	52	1965	20		7,811	20,380
1943	125	776	73	1966	1		2,536	17,587
1944	339	1,158	197	1967	4		2,893	12,368
1945	1,042	1,666	78	1968	–		2,155	10,357
1946	3,606	9,185	85	1969	–		3,020	5,776
1947	9,192	14,547	393	1970	–		3,002	5,857
				1971[c]	–		2,888	6,151

[a] 1932 refers to the 1931/32 season, etc.

[b] Including catches of pygmy blue whales; estimated catch of true blue whales shown in parentheses (1959/60 to 1962/63).

[c] Preliminary data.

The study group was also fortunate that at the urging of the International Council for the Exploration of the Sea in the 1920's,

Norway had established the Bureau of International Whaling Statistics.
Since 1930/31 this Bureau collected detailed statistics on the catch
of whales in the Antarctic and on the effort involved. In addition
to this catch and effort data and lengths of each captured whale,
several countries had, either through independent research expeditions
or by sending biologists with the whaling factory ships, accumulated
a great deal of biological information about whales. The most important of the early research expeditions were the Discovery Expeditions
of the United Kingdom in the 1930's. N. A. Mackintosh directed the
whale research of these expeditions. After the war the Norwegians
under the leadership of Johan Ruud made further studies, and subsequently Japanese biologists under the leadership of Omura and more
recently, Ohsumi, have provided a great deal of basic information.
The study committee of the Commission was able to obtain and compile
all the basic data from the different countries. Biological studies
of whales are obviously difficult, and most of the data utilized were
obtained through the catches. It was therefore valuable to assemble
and compare all of the available data. During the past decade scientists associated with the Scientific Committee of the Commission have
done further research on methods of population study for whales and
have developed new methods. Specific credits will be indicated when
these are discussed below.

The aim of this paper is to summarize the methods that have been
used, with primary emphasis being given to the baleen species (fin,
blue, and sei whales). The fin whales have been studied the most, so
more attention will be paid to these.

Methods of Analysis Based on Age and Catch/Effort Data

The basic assumption of fisheries management analysis is the
equation

$$C = qf\bar{N} \qquad (14\text{-}1)$$

where C = catch,

q = proportionality factor or catchability constant (assumed to
be constant if effort is measured correctly, regardless of
population size),

f = units of effort applied, and

$\bar{N}$ = mean population during the fishing season.

This simply says that if one standard unit of effort catches a fraction of the population, then f units of effort remove a fraction qf.

A second equation relates the size of a year class in year $t + 1$ to its size a year earlier in year t as follows:

$$N_{i,t+1} = N_{it}e^{-Z_t} \qquad (14\text{-}2)$$

where N_{it} = size of year class i at the beginning of season t,

$N_{i,t+1}$ = size of year class i at the beginning of season $t + 1$, and

Z_t = total mortality coefficient (instantaneous rate) in season t from fishing and from natural causes.

This is, of course, trivial but we further write

$$Z_t = M + F_t = M + qf_t$$

where M = natural mortality coefficient taken to be constant,

F_t = fishing mortality coefficient in season t, and

f_t = fishing effort in season t.

If we now consider only the number of animals in a year class, rather than the total population, then Eq. (14-1) is rewritten

$$(C/f)_{it} = q\bar{N}_{it} \qquad (14\text{-}3)$$

where $(C/f)_{it}$ = catch per unit of effort from year class i in season t. Finally we relate N_{it} and $\bar{N}_{it}$ as follows (using Paloheimo's approximation):

$$\bar{N}_{it} = N_{it}e^{-\frac{F_t + M\tau}{2}} \qquad (14\text{-}4)$$

where τ is the fraction of year that fishing operates. N_{it} and $\bar{N}_{it}$ are the initial and mean population sizes of year class i in season t respectively. From Eqs. (14-2) and (14-4),

$$\bar{N}_{i,t+1} = N_{it}e^{-Z_t}e^{-\frac{F_{t+1} + M\tau}{2}} \qquad (14\text{-}5)$$

and hence $\quad \ln \left[\dfrac{(C/f)_{i,t+1}}{(C/f)_{it}}\right] = -Z_t + \dfrac{F_t - F_{t+1}}{2}$.   (14-6)

It is frequently convenient to neglect the second term on the right, which is small unless fishing effort is changing rapidly.

In the case of fin whales for the early data, we do not have a breakdown of the catches by age, but we do have length data and age-length keys. Hence an age can be estimated for each whale for which length data are available. More usefully, the catch can be classified into age classes which are reasonably accurate for younger ages, but increasingly less so for older ages where mean length changes little with age. Thus in the earlier analysis it was necessary to work not with a single age class but with the total of all fully recruited age classes. In other words, the instantaneous mortality rate is estimated from the logarithm of the ratio of the catch per unit of effort of whales aged i and older in year t to the catch per unit of effort of whales aged $i + 1$ and older in year $t + 1$.

Such a ratio and hence the estimate of Z_t, the mortality coefficient, can be estimated by month (that is, comparing similar months one year apart) and by subarea. In this way there can be computed for each main area six estimates of Z which are then averaged to give a single estimate for the season and area.

These estimates tend to vary quite widely even after this averaging process; several causes may be noted. In the first place, migration of whales by time and place is variable. Secondly, the effort expended in any month and subarea fluctuates greatly. Thirdly, the catch of whales is greatly influenced by weather conditions, which also are highly variable. Another factor that contributed to variability of early analyses was the inadequacy of the age-length keys. Also because of inadequate data, it was necessary to combine data from several areas rather than using age-length keys for each individual stock.

To cope with the large variability of these mortality estimates, two procedures have been used. In the original studies by the Committee of Three, later Four (as the 1961 study group became known), mortality estimates were averaged over four seasons. These smoothed estimates were quite satisfactory. Subsequently another method of smoothing has been used by Doi and coworkers in Japan. After calculating Z_t by subarea and season, they calculate F_t (fishing mortality coefficient) as

$$F_t = Z_t - M.$$

Since $F_t = qf_t$, where f_t is the effort applied and q is the catchability coefficient, it follows that to each estimate of F_t there corresponds an estimate of q. The several estimates of q are combined in an average estimate, $\bar{q}$, and then a smoothed estimate of F_t can be calculated simply as $\bar{q}f_t$.

The method depends upon there being available an estimate of M, the natural mortality coefficient. Originally M was estimated from an analysis of the catch curve of the very earliest exploitation; in addition, an estimate was obtained by comparing the abundance of a year class in the 1939/40 and 1945/46 periods when there had been little exploitation in the intervening period. Both of these methods were somewhat suspect because they depended on age estimates made for whales caught in prewar years utilizing postwar age-length keys. It was conjectured that exploitation might have caused a change in the growth pattern and hence altered the age-length relationship.

However, another method is now available. In recent years the Japanese biologists have made age determinations of much larger samples of their catches than were available before. This makes unnecessary the use of age-length keys, which are reliable only for younger ages. The complete samples provide good information on the age composition to the very oldest ages. Consider now two year classes that were fully recruited prior to the onset of heavy exploitation, or perhaps even any exploitation, say the 1920 and 1921 year classes. These two year classes have experienced the same history of exploitation, and the ratio of the 1920 to the 1921 year class must reflect the survival rate of year classes in the preexploitation period, that is, the natural survival rate. More precisely, we plot the natural logarithm of the number caught against age, and the slope of the line so determined is the estimated natural mortality rate (sign disregarded).

Actually, all of these methods give estimates of M that are quite close and it is now agreed that the best estimate of M is 0.04. It is still an open question whether there may be a small difference in mortality rates between sexes. Estimates of M range from 0.034 (males) to 0.060 (females), but whether this difference is due to sampling error or due to difficulty of age determination of older females has not been resolved. While the differences indicated are small, they do have some bearing on estimates of present sustainable yield.

One other problem needs to be discussed here—the measurement of effort. What we need to measure is the area searched, and this will be proportional to days of operation if all other factors remain constant. As the catcher boats become more powerful and acquire more sophisticated gear, the area searched per day increases. While it might seem reasonable to measure searching power by the average speed of the boats, it has been found that tonnage rather than horsepower seems to reflect other changes in catcher operating efficiency and hence yield a better adjustment to provide a measure of area searched. In addition to adjusting raw effort data (catcher day's work) for changes in tonnage, it is also desirable to adjust for weather and a number of other items such as skill of the gunner. Corrections for weather are possible but have not been made, since weather variations over several seasons may be assumed to be a random factor. Information is not available to correct for differences between gunners.

The largest remaining complication in the measurement of effort is the fact that this is a multispecies fishery. At the present time, in evaluating a catcher day's work it is necessary to know whether the catcher was hunting for fin whales or for sei whales or for both. The same question could be asked for a catcher day's work in the early years, only at that time the primary alternatives were blue and fin whales. This alternation of interest between species has caused difficulty in estimation of population size in the 1960's. Fortunately during the period 1954/55 to 1962/63, by far the most important species was the fin whale. The blue whale density had dropped drastically, and in addition the season in which blue whales were permitted to be captured was shortened, while sei whales were not yet caught in any

appreciable numbers. It should be emphasized that the different species are to some extent found in different areas, so that emphasis
on one species to the exclusion of the others governs the areas of
operation and hence both the catches and the catch per day.

Least Squares Method

This method was developed by Allen (1966, 1968). It is based on
the approximate *recursion* formula

$$N_{t+1} = (N_t - C_t)e^{-M} + R_t \qquad (14\text{-}7)$$

where C_t = catch in season t,

$\quad N_t$ = population at the beginning of season t, and

$\quad R_t$ = recruitment in year t.

This is an approximation because it assumes that natural mortality
occurs after the fishing season; since the season is short, the approximation is quite good.

Now if estimates of the recruitment R_t are available, given N_0,
the initial population size, and M, the natural mortality coefficient,
we can calculate the successive estimates of N_t. Additionally, given
q, the catchability coefficient, from the calculated N_t, one can estimate the catches C_t. This, of course, uses Eq. (14-1). Denote these
expected catches as $\hat{C}_t$.

It is reasonable to determine the unknown parameters to minimize
the sum

$$S = \Sigma (C_t - \hat{C}_t)^2; \qquad (14\text{-}8)$$

hence the name least squares method. This can be done to estimate
N_0, M, and q or, slightly more easily, only N_0 and q, using the well-
established value of M. In any case it is desirable to program the
minimization procedure for a computer.

As stated, the input data include the estimated R_t, the annual
recruitment numbers. These must be estimated from the age composition data, basically obtained from the comparison of the ratio of
catches from a partially recruited year class to a fully recruited
year class with the same ratio of catches one season later. More
precisely, we have the equations

$$\frac{C_{it}}{C_{i-1,t}} = r_i \frac{N_{it}}{N_{i-1,t}} \tag{14-9}$$

where r_i = fraction of year class i that is recruited in season t and
where it is assumed that year class $i - 1$, one year older, is fully
recruited. For the catches of the next season

$$\frac{C_{i,t+1}}{C_{i-1,t+1}} = \frac{r_i N_{it} e^{-Z_t} + R_{i,t+1}}{N_{i-1,t} e^{-Z_t}}$$

where $R_{i,t+1}$ = recruits from year class i in season $t + 1$.
Note that

$$\frac{R_{i,t+1}}{r_i N_{it} e^{-Z_t} + R_{i,t+1}}$$

is the proportion of new recruits in the population $N_{i,t+1}$ (and hence
in the catch $C_{i,t+1}$ if this is now a fully recruited year class).
Denote this proportion by $\rho_{i,t+1}$. Then

$$\frac{C_{i,t+1}}{C_{i-1,t+1}} = \left(\frac{r_i N_{it}}{N_{i-1,t}} \right) \left(\frac{1}{1-\rho_{i,t+1}} \right) . \tag{14-10}$$

Let

$$B_{i,t+1} = \left(\frac{C_{it}}{C_{i-1,t}} \right) \left(\frac{C_{i-1,t+1}}{C_{i,t+1}} \right) .$$

Substituting Eqs. (14-9) and (14-10) and rearranging terms yields

$$\rho_{i,t+1} = 1 - B_{i,t+1} . \tag{14-11}$$

This analysis estimates the proportion of new recruits in the year
class fully recruited for the first time; by working backwards with
similar equations, it can be extended to estimate the proportions of
new recruits in all the catches. It should be emphasized that this computation of the proportion of new recruits for each year class and season
does not depend on estimates of other parameters.

If data on the age composition are not available, Allen's
method can be modified to find least squares estimates of N_0, q, and r,
the gross recruitment rate, assuming this to be constant. However, if
the population reacts to exploitation and resultant density changes by
changes in recruitment rate, this assumption is doubtful. Over short

periods of time it may be approximately correct and it has been used in this way by the present author. Allen has also developed other modifications of this procedure which are not discussed here. Some of the applications to Antarctic whale populations are found in Allen (1971).

Leslie-DeLury Methods

In the absence of age data, one is forced to rely on the effects of catch on changes in catch per unit of effort. If the population is closed, the procedure developed independently by Leslie and Davis (1939) and DeLury (1947) is straightforward and very satisfactory, provided measurements of effort are reasonable. A variety of modifications of the method have been applied to whale catch analysis to allow for the fact that the population is not closed; recruitment and natural mortality must be considered in addition to the catches. One simple method developed by Chapman (1970) uses two regressions as follows:

$$U_i = q[N_0(1 + (r - M)i) - K_i] \tag{14-12}$$

where U_i = catch per unit of effort in season i,

N_0 = population size at beginning of the period of study,

$r - M$ = net recruitment rate (gross recruitment—natural mortality),

and K_i = cumulative catch to midpoint of season i.

Further one can write another regression

$$K_i = di + f \tag{14-13}$$

in periods in which total catch per season is effectively constant. Equation (14-12) is an approximation, since recruitment is derived from the parent population while mortality applies to the actual population. After d and f are estimated, we have

$$U_i = q[N_0(1 + (r - M)i) - \hat{d}i - \hat{f}]. \tag{14-14}$$

Denote the regression of U on i as $U = a + bi$.

Then

$$N_0 = \frac{\hat{b}\hat{f} - \hat{a}\hat{d}}{\hat{b} - (r - M)\hat{a}} . \tag{14-15}$$

If $r - M$ is estimated, N_0 can be estimated. Other modifications of this method have been used, which will be referred to later.

Mark-Recapture Methods

Data from mark-recapture experiments have been widely used in
fisheries to estimate population sizes, exploitation rates, fishing
and natural mortality coefficients, etc. Such data have always to be
treated with caution and especially so for whales, where it is difficult to have a proper count of the whales actually marked, where the
number marked in any season has always been small, and where there is
a large nonrecovery of marks in the captured whales. Several experiments have demonstrated the latter phenomenon. For example, Doi,
Ohsumi, Nasu, and Shimadzu (1970) said that the reporting rates of
known marks (placed in whales experimentally) by Japanese expeditions
were as follows: during the flensing—70%; from the boilers—0.65%.
Previous comparisons have shown recovery rates from Japanese expeditions
to be higher than those of other countries.

Thus it is necessary to use methods that are not affected by this
problem or to make extensive adjustments for nonrecovery of marks.
The Seber-Jolly method (Seber 1962) falls into the first category.
This method requires that marks be put out at the beginning of two
successive seasons. Then the ratio of mark-recoveries of marking number 1 to marking number 2 in seasons 2 onward (adjusted for the different number of marks placed) can be used to give an estimate of the
total survival rate from season 1 to season 2. Since this is a ratio
of mark-recoveries under similar conditions, it is unaffected by mark
loss (on placement) or by nonrecovery of recaptured marks. It requires
only that the two experiments be conducted similarly, which is true
for whale marking experiments. Since we have a firm estimate of M,
from estimates of total mortality from mark-recaptures using this
method, it is possible to estimate fishing mortality and hence population size.

Recoveries of prewar marks have also been used (Anon. 1967, pp.
35-39) to estimate the efficiency of mark reporting with respect to
postwar marks. Though the estimates are quite variable, the smoothed
curve is consistent with known changes in the behavior of the whaling
fleet, for example, the increasing proportion of the catch taken by
the Japanese fleets, which have the highest mark-recovery efficiency.

Also, as the industry has made increasing use of whales for meat rather than for oil, the efficiency of mark-recovery has increased. This second method has a variety of problems which are discussed in the reference cited (pp. 38-39); despite these problems the two methods give extremely close estimates of the average total mortality rate for the period 1953/54 to 1959/60.

Methods for Analysis of Antarctic Blue Whale Stocks

The same methods can be used to estimate the population parameters of the blue whale stocks, though some modifications are necessary. The most important of these results from the paucity of age data. One age-length key was determined on the basis of counts of ovarian corpora for mature females, but no estimate of natural mortality rate is available for blue whales. However, it is reasonable to assume that M is close to that for fin whales. With this assumption, the blue whale stock has been reconstructed from 1933/34 through 1957/58 (Table 9-7 of this book).

There is good agreement between the limited analysis based on catch and age data and that based on modified Leslie-DeLury methods. It should be pointed out that these methods are more reliable for blue whales than for fin whales since the decline was more rapid; also during the early seasons blue whales were the preferred species, and hence the measurement of effort as applied to this species may be quite accurate.

While direct analysis of marking data is rendered almost pointless by the small number of recoveries, it is possible to compare the recoveries of blue whale marks with those of fins. This was done to a limited degree in the original study (Chapman, Allen, and Holt 1964, pp. 43-45). Further comparisons have been made since that time but not published. These confirm that the estimates, both prewar and in the 1950's, are reasonably close. Still another confirmation has recently been made. Doi, Ohsumi, and Shimadzu (1971) have analyzed the sighting data of their factory ships with regard to prohibited species. It is difficult to evaluate the data directly but by comparison with the fin whale population, they gave estimates of 4,160, 3,180, 5,840, 10,940, and 8,000 for the blue whale stock for the seasons 1965/66 to 1969/70.   There is a clear tendency for increase, though too much weight cannot be put on this since the areas of operation vary considerably from season to season. Neglecting any such tendency, the five-season average is 6,424. Allowing for the commercial kill from 1958/59 to 1964/65 and estimating the net rate of increase at 4% per year if the population was 3,800 at the beginning of 1958/59, it should be about 5,000 now. The difference between this estimate and the adjusted sighting estimate might be simply sampling error. If not, either the 1958/59 population estimate was about 1,000 low or the rate of increase has been greater than 4%.

Methods of Analysis of Antarctic Sei Whale Stocks

Prior to 1957/58 the pelagic whaling operations had no interest in sei whales: in no season were as many as 1,000 caught. The catch increased gradually from 1957/58 to 1963/64, but no great attention was paid to the species in the statistics or by the biologists. Then in 1964/65 the expeditions altered their pattern of operations to exert primary effort on sei whales, and the catch in this season went up to 20,330. Since that time more sei whales have been caught than fin whales in every season, but it is difficult to partition the effort between the two species. Furthermore, the determination of age in sei whales is more difficult than in fin whales—the layers in the ear plug are more variable and indistinct. This may be associated with their less clear-cut migration pattern, particularly at younger ages. Thus where the age of median recruitment of fin whales is between 5 and 6, it is asserted to be 17 for sei whales (Doi and Ohsumi 1970, p. 88). Also, because there was little early interest in sei whales, relatively small numbers of this species have been marked.

Despite these difficulties a great deal of information has been assembled on sei whale population parameters and essentially all methods outlined above have been used, though to date no attempt has been made to estimate recruitment by age using Allen's method.

The primary methods used have been the modified Leslie-DeLury
analysis and a reconstruction of the population based on Eq. (14-7)
with a theoretical reproduction model (Doi and Ohsumi 1970). For
four seasons the catch totaled 60,000, and this heavy kill resulted
in marked changes in catch per unit of effort in subareas. Thus it
was possible to obtain population estimates.

In 1966/67 the Japanese government operated two research expeditions in the Antarctic and invited scientist observers from other
countries to participate. The sighting data from these expeditions
have been useful as a check on the order of magnitude of the population estimates calculated by other means. However, comparison of
sighting and other estimates for fin and blue whales suggests that
sighting estimates are too low.

Finally there are the analyses based on comparisons between
indices for fin and sei whales. These are difficult to make because
even the smallest statistical unit used in the reports (10° squares)
is sufficiently large that the whale distribution within them may be
quite uneven. Thus if the expeditions seek one species and have less
interest in the other, the relative catches may not reflect the relative abundances. With the large statistical subdivisions which are
often used in the reported data (an area zone is $60^{\circ} \times 10^{\circ}$), this problem is even more acute. Additionally, there are month-to-month changes
and it is not known whether the time pattern of movement of the two
species is the same within or between years. All of these raise questions about the comparisons, so that at best they can provide some
positive support if estimates agree but limited negative evidence if
estimates disagree.

Conclusion

A review of the data collected and the analyses made will show
that the management of Antarctic baleen whales has had the benefit of
as much scientific information as any large managed natural animal
population in the world. In fact, in few fisheries is it possible to
find a comparable body of data and of analyses of these data, while
one does not need to look far to find resources being managed with

much less adequate data. I mention only two disparate examples of
many—the Peruvian anchoveta and the Arctic polar bear.

While all this is true, anyone who has worked with whale data
could cite pressing needs for more information—for example, short-
term needs like better age determination methods for sei whales, and
intermediate-range data such as more information on whale migration.
Finally there are long-term questions: how do we determine what are
the population control mechanisms in whales, and how are they affected
by exploitation? Some of these and other problems are discussed in
other chapters in this book.

REFERENCES

Most of the analyses on Antarctic whale populations are included
in reports of the International Whaling Commission in recent years.
These may be included as annexes to the Scientific Committee report
or as appendices to the main report. The reports of the Scientific
Committee, the special reports, and the analyses by the FAO assessment
group should be consulted in the reports from 1965 (Fourteenth Report)
to 1971 (Twenty-First Report).

Allen, K. R. 1966. Some methods for estimating exploited populations.
    J. Fish. Res. Bd. Canada, 23:1553-1574.
_____ 1968. Simplification of a method of computing recruitment
    rates. J. Fish. Res. Bd. Canada, 25:2701-2702.
_____ 1971. Notes on the assessment of Antarctic fin whale stocks.
    Rpt. IWC, 21:58-63.
Anonymous. 1967. Report of the I.W.C./F.A.O. joint working party on
    whale stock assessment held from 26th January to 2nd February
    1966 in Seattle. Rpt. IWC, 17:27-47.
Baranov, F. I. 1918. On the question of the biological basis of
    fisheries. Izvestiia, 1:81-128.
Beverton, R. J. H., and Holt, S. J. 1957. On the dynamics of ex-
    ploited fish populations. London, United Kingdom Min. Agr. and
    Fish., Fish. Invest. Ser. 2(19).
Chapman, D. G. 1970. Reanalysis of Antarctic fin whale population
    data. Rpt. IWC, 20:54-59.
_____ Allen, K. R., and Holt, S. J. 1964. Reports of the Committee
    of Three Scientists on the special scientific investigation of
    the Antarctic whale stocks. Rpt. IWC, 14:32-106.
DeLury, D. B. 1947. On the estimates of biological populations.
    Biometrics, 3:145-167.

Doi, T., and Ohsumi, S. 1970. On the maximum sustainable yield of sei whales in the Antarctic. Rpt. IWC, 20:88-96.

_____ _____ Nasu, K., and Shimadzu, Y. 1970. Advanced assessment of the fin whale stock in the Antarctic. Rpt. IWC, 20:60-87.

_____ _____ and Shimadzu, Y. 1971. Status of stocks of baleen whales in the Antarctic 1970/71. Rpt. IWC, 21:90-99.

Leslie, P. H., and Davis, D. H. S. 1939. An attempt to determine the absolute number of rats in a given area. J. Animal Ecology, 8:94-113.

Ricker, W. E. 1948. Methods of estimating vital statistics of fish populations. Indiana Univ., Publ., Science Series 15.

_____ 1958. Handbook of computation for biological statistics of fish populations. Ottawa, Fish. Res. Bd. Canada Bulletin 119.

Seber, G. A. F. 1962. The multisample single recapture census. Biometrika, 49:339-350.

Thompson, W. F., and Bell, F. H. 1934. Biological statistics of the Pacific halibut fishery. (2) Effect of changes in intensity upon total yield and yield per unit of gear. Seattle, International Pacific Halibut Commission, Rpt. No. 8.

_____ and Herrington, W. C. 1930. Life history of the Pacific halibut. (1) Marking experiments. Seattle, International Pacific Halibut Commission, Rpt. No. 2.

CHAPTER 15

RECRUITMENT TO WHALE STOCKS

K. Radway Allen

I want to draw attention to one of the most important problems
that underlies the effective management of the whale stocks, whether
we wish to ensure the continued existence of the species or to obtain
the best possible continuing yield from it. This is the problem of
recruitment; that is, the quantity naturally added to the stock each
year to build it up to to replace what has been lost through natural
deaths or catching. The aspect of recruitment which is particularly
important concerns the way in which the rate of recruitment changes
in response to changes in the size of the population.

Upon occasion Dr. D. G. Chapman has remarked that there is a very
good level of agreement between different workers using a variety of
methods in their estimates of the Antarctic fin whale stocks for the
later 1950's and early 1960's, but that there was considerably more
variation in the estimates of current stock size. In addition, there
is proportionately even greater variation in the estimates of current
sustainable yield. Sustainable yield is, in general terms, the catch
which can be taken without affecting the level of the stock, but it
can be defined precisely in several rather different ways. These lead,
in turn, to different estimates of the exact magnitude of the sustainable yield under current conditions. The relatively large differences
between the estimates of different workers remain, whichever definition
of sustainable yield is adopted. This brings out very clearly the
significance of the recruitment problem in assessing whale stocks and
devising suitable management practices.

Analysis of the causes of this situation of good agreement about
earlier stock levels, some variation in estimates of current population
size, and large variations in estimates of current sustainable yield

shows that the variations arise essentially from the differences in
the approach to the recruitment problem.

For the period of the late 1950's and early 1960's Antarctic
whaling concentrated heavily on fin whales, and the catch per unit
effort provides a good index of the abundance for this species. As
Chapman has shown in Chapter 14, this index can be used in various
ways to obtain population estimates, but since the basic data are essentially the same, the results are naturally in quite good agreement and
are relatively little affected by any assumptions or calculations which
are made regarding the recruitment rate during this period.

From about 1964, however, the whaling fleets devoted considerable
attention also to sei whales, and it is therefore no longer satisfactory to use catch per unit effort as an index of abundance for fin
whales. The best method of achieving estimates of population size
for this period, through to the present time, is therefore to extrapolate from year to year, starting with an estimate obtained from other
sources, subtracting the catch, which is known exactly, and then
adjusting for estimated natural mortality and recruitment. In this
way successive annual estimates of population size are obtained. There
is general agreement that the natural mortality rate has an instantaneous value (M) of about 0.04 and thus this factor contributes little
to the differences in current population estimates. It is obvious,
however, that estimates obtained in this way will be very sensitive
to differences in the recruitment rates used, and that the population
estimates will tend to diverge more and more widely as the length of
the extrapolation series increases. The higher the value of the recruitment rate used, the higher will be the current estimate of population size.

At any population size the sustainable yield is roughly the population multiplied by the net recruitment rate, the latter being the
difference between gross recruitment and natural mortality rates. As
a consequence, use of a higher recruitment rate will not only produce
a higher estimate of current population size, but also will imply that
the sustainable yield constitutes a greater proportion of this higher
estimated population. Therefore, when populations are estimated by

extrapolation, any differences in recruitment rates used in the calculations must lead to proportionately greater differences in estimated
sustainable yields than in population estimates.

Since extrapolation from the relatively good population estimates
of about 1960 is the best method we have at present for estimating
Antarctic fin whale stocks and sustainable yields, it is obviously
extremely important that the values we use for the recruitment rates
in these calculations be as accurate as possible.

Theoretical considerations tell us two things regarding the relationship between recruitment rate and population size. The first is
that over most of the range of possible population sizes there must
be some sort of tendency for recruitment rate to decrease as population size increases. This is essential to ensure that under natural
conditions the species will have a reasonably stable population. This
truism only applies to the net recruitment rate (that is, the difference between total recruitment and natural mortality) and not to the
total recruitment rate. Theoretically, at least, it is possible for
the total recruitment rate to be constant or even positively correlated
with population size if the natural mortality rate also has a sufficiently strong positive correlation.

The second general principle is that at small population sizes
the first principle breaks down, so that if the population continues
to decrease, the recruitment rate no longer increases or may, in fact,
decrease. It is this aspect of the relation which may cause a species to have a critical population size; if it is reduced to this
level, further decline and biological extinction will inevitably occur.

While general theory leads us to these two principles, it does
not lead to any particular conclusions as to the shape of the curves,
either for recruitment rate or for number of recruits against population size. It is often convenient to assume that these are simple,
and that they may be represented by smooth symmetrical curves for
rate and by straight lines for numbers of recruits. The postulation
of such relationships greatly simplifies our mathematical models, but
does not in itself give evidence of their validity. It is true that
for some animals, and particularly for some kinds of fish, quite good

evidence of simple curves does exist, but I believe that we should be cautious in applying them too confidently to species for which they cannot be supported by direct evidence.

Recruitment to either the total or the mature or the exploited population is influenced by a number of components, all of which could be density-controlled. These include the age at which females become mature, the pregnancy rate, the rate of successful births per pregnancy, the rate of survival from birth to recruitment, and the age at recruitment. For the total population, recruitment occurs, of course, at birth. For recruitment to the mature population, the age at recruitment becomes the age at maturity. For recruitment to the exploited population, age at recruitment may change at any time with changes in the practices of the fishery or, if size at recruitment remains more or less constant, it will change with any density-induced changes in growth rate.

All these components contribute to the overall recruitment process and all may be subject to density-dependent changes. Where such changes are of significant magnitude, it seems reasonable to assume that they will follow the general principles outlined above. Again, however, we should be cautious in assuming that the density-dependent changes in individual components will, in fact, follow conveniently simple or symmetrical models. I believe that it is perfectly possible that some components could have fairly constant rates over much of the possible population size range and then respond quite abruptly as the population passes a critical size near either the upper or lower end of the range. Such critical points could also occur at different population levels for different components. Provided that this caution is observed, the construction of models incorporating the principal components provides a useful method of examining the problem of overall recruitment.

The Scientific Committee of the International Whaling Commission has noted that there is evidence that the age at maturity in female Antarctic fin whales is probably over 10 years in unexploited stocks and may decrease to 6 or 7 years in heavily exploited stocks (Anon. 1971, p. 36). It has also found evidence in the same stocks that the

pregnancy rate has increased from about 0.3 in prewar years to about
0.4 in the 1950's and early 1960's. Since the size of the Antarctic
fin whale population has declined by about 75 to 80% since the war,
these changes are in the direction which is required to produce an increased recruitment rate with a decreasing population.

Another approach to the recruitment problem is to determine the
limiting values of the recruitment rate for an unexploited stock and
for a stock on the verge of extinction, and to examine the effects on
the population and sustainable yield estimates of various possible
curves lying within these limits and embodying the principles referred
to earlier. Doi, Ohsumi, Nasu, and Shimadzu (1970) have undertaken a
study of this kind for the Antarctic fin whale stocks. They estimated
the recruitment rate to the exploited stock in a previously unexploited
population to be 0.0461 and the maximum recruitment rate (that occurring when the stock is on the verge of extinction) as somewhat less
than 0.1870. They constructed a number of curves lying within these
limits and selected for further study those which gave, for the years
near 1960, population estimates consistent with those obtained by
other methods. Using these selected curves, they obtained, by extrapolation, estimates of 89,200 to 94,000 for the 1970 stock and 5,200 to
6,000 for the 1970 sustainable yield. The method, in itself, provides
no means of selecting between various possible curves.

A third approach is to estimate the total recruitment to the exploited population directly from the age structure of the catches
(Allen 1966, 1968). Application of this technique to the Antarctic
fin whale population has given no evidence of any major change in the
overall rate of recruitment to the exploited population. This appears
to have remained quite close to 6% of the size of the parent stock
throughout the 1950's and 1960's (Allen 1971). Thus the overall recruitment rate, as estimated in this way, has apparently remained
constant during a period of population decline in which two of the
components of recruitment seem to have shown appropriate responses.
This apparent discrepancy emphasizes the need for thorough study of
all aspects of the recruitment problem. Since these estimates of
recruitment rate are less than the values given for corresponding

population sizes by the curves of Doi, Ohsumi, Nasu, and Shimadzu, the
resulting estimates of 1970 population size and sustainable yield are
also lower. They are about 56,300 and 600 to 1,200 respectively (Allen
1971).

It is probably obvious that we have also to be cautious in applying results and relationships discovered for one species to other
whales. As an example, one may compare briefly the history of the
gray whale of the eastern North Pacific with that of the right whales
in the higher latitudes of both hemispheres. The right whales, being
the earliest target of heavy exploitation, were in most areas reduced
to such a low level that little further exploitation has been possible
for at least 100 years, and they have been formally protected for over
30 years. Despite this long period of virtual protection, there is
very little evidence of any appreciable recovery in the sizes of the
populations and clearly none of them are approaching anywhere near
the original levels. The gray whale, on the other hand, was also reduced to low levels, first by heavy catching between 1850 and 1890,
and then by a further fishery from 1924 to 1946. It has been completely
protected since 1946, and, because it is a whale that is relatively easy
to observe, a number of studies have been made on the changes in the
size of the population. Although there are some discrepancies, there
seems good evidence that it has recovered since 1946 at quite a rapid
rate, which may have been as high as 10% per annum. The population is
now probably about 11,000 (Rice and Wolman 1971), but there are no
data for the original population to provide a basis for comparing the
original and present stock size.

A number of factors may have contributed to this difference in the
response to protection between the right whales and the gray whale.
The geographical concentration of the gray whale breeding populations
in a few lagoons may well be one of these. This behavior could perhaps assist the species in maintaining its reproductive efficiency,
since it would improve the chances of encounters between breeding animals at very low population levels. Thus, the critical population
size below which reproductive failure occurs might be a smaller proportion of the unexploited stock size than it would be in more scat-

tered species such as the right whales. It is, of course, also possible
that the right whales were reduced to a relatively lower level than the
gray whales, so that even if the critical population levels are similar,
the right whales were brought closer to the population level beyond
which recovery is impossible. So far as I am aware, however, we have
no indication that any major right whale stocks have actually been
brought to extinction by hunting.

REFERENCES

Allen, K. 1966. Some methods for estimating exploited populations.
     J. Fisheries Res. Bd. Canada, 23:1553-1574.
_____ 1968. Simplification of a method of computing recruitment
     rates. J. Fisheries Res. Bd. Canada, 25:2701-2702.
_____ 1971. Notes on the assessment of Antarctic fin whale stocks.
     Rpt. IWC, 21:58-63.
Anonymous. 1971. Report of the special meeting on Antarctic fin whale
     stock assessment, Honolulu, Hawaii, 13th-25th March 1970. Rpt.
     IWC, 21:34-39.
Doi, T., Ohsumi, S., Nasu, K., and Shimadzu, Y. 1970. Advanced assess-
     ment of the fin whale stock in the Antarctic. Rpt. IWC, 20:60-87.
Rice, D. W., and Wolman, A. A. 1971. Life history and ecology of the
     gray whale. Amer. Soc. Mamm., Spec. Publ. 3.

CHAPTER 16

FURTHER DEVELOPMENT OF WHALE SIGHTING THEORY

Takeyuki Doi

Whale sightings not only are valuable for the evaluation of stocks, but also supply much information concerning the distribution and migration of whales. Sighting data, however, have been difficult to use to estimate the absolute number of whales. In previous papers I established a basic sighting theory and tried to show how to analyze sighting data (Doi 1970, 1971a). This work was continued, and special sighting data were collected during the 1969/70 Antarctic season. With this information the theory has reached the stage of practical use in estimating the absolute number of whales in a given area of the sea (Doi 1971b; this chapter is an English condensation of the original Japanese text).

Basic Theory

The basic equations described in the latter paper are as follows:

$$f(x) = \begin{cases} 1 & z \geqslant z_0 \\ \dfrac{z}{z_0} & z < z_0 \end{cases} \qquad (16\text{-}1)$$

and $\bar{P} = \dfrac{x_0 + \dfrac{1}{2VT}[(x_1 y_1 - x_0 y_0) + \ell^2(\theta_1 - \theta_0) - \cot\theta_1(x_1^2 - x_0^2)]}{x_1}$ (16-2)

where

$f(x)$ = probability of sighting a whale present in the scouting zone;

$\bar{P}$ = average sighting rate, that is, the proportion of whales sighted in the rectangle traced out by the boat's path, of width $2(\ell)\sin\theta_1 = 2(4.90) = 9.80$ mi;

V = velocity of the boat;

T = whale's dive duration;

ℓ = limit of visual range = 6.4 mi;

θ_1 = angle of whole visual range = 50°;

x = distance measured in a direction at right angles to the movement of the boat;

y = distance measured along the direction of the boat's movement;

z = distance along a line parallel to the y-axis (direction of boat movement) from the arc which is the limit of the visual range to the intersection of the ray at an angle θ_1 (50°) from the vertical axis. Thus z represents the length of the line parallel to the y-axis at a distance x from it, visible to the observer at θ who is scanning an angle of 50° on each side of the forward direction;

x_1 = $\ell \sin \theta_1$, the half-width of the rectangle in which whales may be sighted (with ℓ = 6.4 mi, $\theta_1 = 50^\circ$, x_1 = 4.90 mi);

y_1 = $\ell \cos \theta_1$ = 4.11 mi;

z_0 = VT;

x_0 = x-value at $z = z_0$;

y_0 = y-value at $z = z_0$;

θ_0 = θ-value at z_0 = $\sin^{-1} \dfrac{x_0}{\ell}$.

These symbols can be understood with the help of Fig. 16-1. $\bar{P}$ is the sighting rate, assuming that an observer can find every whale within his sight ($2\theta_1$). However, since he constantly moves his field of vision, the observer may miss some whales. Thus a correction factor K is needed:

$$K = 1 - (1 - \frac{\theta_p}{2\theta_1})^s, \qquad (16\text{-}3)$$

where

θ_p = visual angle of an observer;

s = number of observers engaged.

Therefore, a real sighting rate P is:

$$P = K\bar{P}. \qquad (16\text{-}4)$$

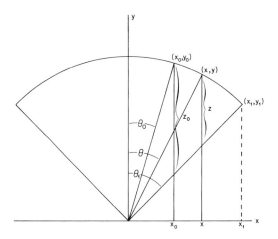

FIGURE 16-1. The fan-shaped area of scouting. The symbols are explained in the text.

In this technique such elements as the boat's velocity, the
ecology of the whale, the varying capability of observers, etc., can
be introduced.

Dive Duration and the Average Rate of Sighting

In order to obtain a numerical value for the average or real
sighting rate it is necessary to determine the duration of diving (T).
From the special program of sighting carried out by Japanese scouting
boats during the 1969/70 Antarctic season, I have been able to obtain
the mean values of T by species and to calculate $\bar{P}$ by use of Eqs.
(16-1) and (16-2). Both T and $\bar{P}$ are shown in Table 16-1, with the
appropriate values of V, ℓ, and θ_1.

Test and Modification of the Average Sighting Rate

Theoretical results should be tested by the data obtained from
the special sighting program. The best test is to compare the theoretical $f(x)$ of Eq. (16-1) with actual values. From such observations
it is found that the interval on which $f(x) = 1$ is much less than that
(x_0) suggested by the theoretical model. This smaller interval is
denoted by m. The difference between the theoretical and the actual

TABLE 16-1. Duration of whale dives in minutes (T) and average rate of sighting ($\bar{P}$) during the 1969/70 Antarctic season.

$$V = 12 \text{ knots}$$
$$\ell = 6.4 \text{ mi}$$
$$\theta_1 = 50° \ (0.8727 \text{ radian})$$

	From special sighting data		Calculated from Eqs. (16-1) and (16-2)				
Species	Number of whales sighted	Mean duration of diving, T (min)	z_0 (mi)	x_0 (mi)	y_0 (mi)	θ_0 (radians)	$\bar{P}$
Sei	2,252	4.21	0.842	4.4631	4.5870	0.7717	0.9565
Sperm	630	8.56	1.712	3.9556	5.0312	0.6663	0.9072
Blue	172	5.67	1.134	4.2989	4.7413	0.7365	0.9403
Fin	1,066	4.87	0.974	4.3896	4.6574	0.7557	0.9496
Right	65	3.70	0.740	4.5190	4.5320	0.7840	0.9618
Humpback	20	8.13	1.626	4.0083	4.9893	0.6768	0.9122
Minke	99	1.30	0.260	4.7723	4.2644	0.8416	0.9871
Pygmy blue	15	9.70	1.940	3.8134	5.1398	0.6387	0.8929

circumstance seems attributable to (1) the difficulty of finding whales at long distances, (2) the inattention of the observers to the direction of boat movement, and (3) a lack of consideration of weather and sea conditions. Thus an experimental modifying function $g(x)$ must be introduced:

$$g(x) = \begin{cases} 1 & x \leqslant m \\ e^{-a(x-m)} & x > m. \end{cases}$$ (16-5)

From curve-fitting, m and a are estimated by species. From Eqs. (16-1) and (16-5) a modified average rate of sighting, $\tilde{P}$, can be calculated:

$$\tilde{P} = \frac{\int_0^{x_1} f(x)g(x)\,dx}{\int_0^{x_1} dx}.$$ (16-6)

Table 16-2 shows the values of $\tilde{P}$, m, and a.

TABLE 16-2. Modified average rate of sighting.

Species	Parameters of modifying function $g(x)$ (Special sighting data of 1969/70 Antarctic season)		Modified average rate of sighting $\tilde{P}$
	m (mi)	a	
Sei	1.25	1.3166	0.4083
Fin	1.00	1.0410	0.3282
Sperm	0.75	1.1516	0.3908
Blue	0.50	1.0240	0.2480
Minke	0.25	1.4067	0.2467
Right	1.50	1.2642	0.4646

Observer Failure and the Breathing (Blowing) of Whales

Observers are unable to scout all the visual angle of $2\theta_1$ perfectly, since their actual visual angle is θ_p. These relationships

can be treated mathematically in the simplified model, using the cor-
recting factor K of Eq. (16-3). However, when the whale's breathing
time at the surface (τ) is taken into account, Eq. (16-3) must be
amended. Introducing an angular velocity of eye-scanning, ω, an angle
of $\omega\tau$ should be added to θ_p as the visual angle. But within the angle
($\theta_p + \omega\tau$) the observer cannot find all whales, because some whales will
have blown before or after θ_p passes. The rate q of finding whales
within the angle ($\theta_p + \omega\tau$) is defined as follows:

$$q = \frac{1}{\tau}\int_0^\tau \frac{\theta_p + \omega\tau - \omega t}{\theta_p + \omega\tau} \, dt = \frac{\theta_p + \frac{\omega}{2}\tau}{\theta_p + \omega\tau} \, ,$$

where t is time and changes from 0 to τ.

Therefore the effective visual angle amended for time of blowing is

$$(\theta_p + \omega\tau)q = \theta_p + \frac{\omega}{2}\tau.$$

The amended correction factor, K_ω, can be described as in Eq. (16-7):

$$K_\omega = 1 - (1 - \frac{\theta_p + \frac{1}{2}\omega\tau}{2\theta_1})^s. \tag{16-7}$$

The calculated results are shown in Table 16-3, using values of τ
by species from the special sighting investigation carried out during
the 1969/70 Antarctic season.

Real Rate of Sighting

The real rate of sighting, P, can be obtained from the following
formula:

$$P = K_\omega \tilde{P}. \tag{16-8}$$

In calculation of the real rate of sighting, it is necessary to
clarify the values of s and ω. Ordinarily two observers are up the
mast and several observers are on the upper bridge. In the special
sighting investigation the total ability of observers on the upper
bridge was considered to be equivalent to 0.5 ~ 1.0 person up the
mast, judged from the number of occasions of sighting. The angular
velocity of the eye, ω, is more difficult to determine than the adjusted

TABLE 16-3. Amended correction factor. $\theta_p = 7^0$ is determined from property of binoculars.

Species	From special sighting data in the 1969/70 Antarctic season Breathing time at surface, τ (sec)	Observers engaged s (persons)	Angular velocity ω (degrees/sec) 5	10
Sei	3.66	2	0.297	0.442
		4	0.506	0.689
Fin	4.57	2	0.335	0.508
		4	0.557	0.758
Sperm	2.72	2	0.257	0.370
		4	0.448	0.603
Blue	5.08	2	0.355	0.543
		4	0.584	0.791
Minke	3.23	2	0.279	0.409
		4	0.480	0.651
Right	3.56	2	0.293	0.434
		4	0.500	0.670

number of observers. When ω is very large, observers cannot sight whales for psychophysical reasons. As observers told me, ω will be about 5 to 10^0 per second. Accordingly, I regard the intermediate values of Table 16-3 as the most likely estimates of the real rate of sighting. Table 16-4 shows the real rate of sighting thus obtained.

TABLE 16-4. Real rate of sighting (see text).

Species	Real rate of sighting P
Sei	0.197
Fin	0.211
Sperm	0.138
Blue	0.141
Minke	0.112
Right	0.221

Abundance of Whales during the 1969/70 Antarctic Season

When we know the real rate of sighting, it is easy to estimate
abundance, which is equal to n/P, where n is the number of whales
sighted. This abundance, however, is only the number of whales within
the range of the sighting zone ($2x_1 \times L$), where L is the boat's distance of travel. Therefore, in order to obtain the total number of
whales in the sea, extrapolation is necessary as described below:

$$N = \frac{n}{P} \cdot \frac{A}{2x_1 L} \, , \tag{16-9}$$

where A is an area under consideration. I considered "squares" 5° of
longitude by 5° of latitude in the Antarctic. In each square I calculated abundance separately by use of Eq. (16-9), then summed the individual abundances to obtain a total number of whales. Another method
of estimating total numbers of whales is to use contour lines of the
same density of whales. The results are shown in Table 16-5. In Antarctic Areas I, V, and VI, where scouting boats did not survey, abundances cannot be calculated because of lack of sighting data.

TABLE 16-5. Estimated abundance of whales during the 1969/70 season
in the Antarctic (in thousands).

Species	Antarctic area				Total
	I	II	III	IV	
Sei		28.06	16.18	42.50	86.74
Fin		3.30	11.62	8.61	23.53
Sperm		11.88	30.33	56.31	98.52
Minke	0.08	5.06	8.20	17.92	31.26
Blue			5.77	0.78	6.55
Right		0.45	0.19	0.57	1.21

Comparison of the figures in Table 16-5 with population sizes
estimated by other indirect methods shows the application of sighting
theory to be reasonable and practical in the estimation of approximate
absolute numbers of animals. I have already explained that scouting

boats can identify whales close to the boat and avoid multiple sightings. The same distant whales may be reported repeatedly by the same
scouting boats. But this problem does not impair the validity of Eq.
(16-9) because multiple sightings of distant whales do not much change
the values of n/L, which indicates the relative index of abundance.

Future Problems and Uses of Sighting Theory

Whale sighting can be used directly to estimate the population
sizes of whales. Generally, population size is estimated indirectly
through analysis of information collected in the fishery. If direct
estimates by sighting are possible, we can eliminate many sampling
biases. Especially, such sighting data will be the only source of
information on the stocks of the protected species of whales. Such
direct estimates have particular value for the correct evaluation of
population size, since they can be compared with the results of indirect analysis based upon catch statistics, biological information,
and the results of marking and fishing effort statistics.

All the whale sighting data used here were taken by Japanese
whale scouting boats. Since the Japanese carefully avoid counting a
whale twice by recording estimated length, body color, distribution
of diatom films, and the direction of movements of the whale, the
possible error from double counting is slight.

Thus, whale sighting data can be logically treated. The velocity
of the boat, ecology of whales, number of observers, and the mechanism
of scanning have been taken into consideration. In order to make
sighting theory more accurate and more useful for stock assessment,
the following parameters should be obtained exactly from systematic
sighting observations and from mathematical models, including simula-
tion analysis:

(1) Whale biology

    (a) Duration of diving of a whale by species

    (b) Duration of respiration (blowing) of a whale by species

    (c) Swimming patterns of whales by species

    (d) Direction of movement of schools of whales

(2) Theoretical treatment of modifying functions

(3) Scanning

   (a) Visual angle of an observer

   (b) Angular velocity of an observer

   (c) Psychophysiology of the eye

REFERENCES

Doi, T. 1970. Re-evaluation of population studies by sighting observation of whale. Bull. Tokai Regional Fisheries Res. Lab.,
   63:1-10.
_____ 1971a. Diagnosis methods of sperm whale population. Bull.
   Tokai Regional Fisheries Res. Lab., 66:89-143.
_____ 1971b. Further development of sighting theory on whale.
   Bull. Tokai Regional Fisheries Res. Lab., 68:1-22.

CHAPTER 17

REFLECTIONS ON THE MANAGEMENT OF WHALING

Scott McVay

Can he who has discovered only some of the values of whalebone and whale oil be said to have discovered the true use
of the whale? Can he who slays the elephant for his ivory
be said to have "seen the elephant"? These are petty and
accidental uses; just as if a stronger race were to kill us
in order to make buttons and flageolets of our bones.

—Henry David Thoreau (1853)

Can he who has discovered only some of the values of whalebone
and whale oil be said to have discovered *the true use of the whale*?
The true use of the whale to date has only been glimpsed or hinted
at, and our knowledge of the whale is still a shadow of a shadow. The
inadequate state of our current knowledge should be a strong impetus
to bring the business of killing whales under rational management. Up
to now, it has not been—quite the contrary.

Today we are continuing witnesses to the destruction of the larger
marine mammals, whose patterns of life from birth to death, whose
social structures, and whose curious adaptations to the sea are per-
ceived only dimly. The exploitation of the whale tribe is going forward at the time when man is entering the sea in increasing numbers
and for diverse purposes unforeseen a few years ago. Our very lack
of information concerning the larger inhabitants of the watery seven-
tenths of the earth should stimulate us to assure the persistence of
these life forms in sufficient numbers to the end that, as our humanity
and technology evolve, the chance to understand whales better will remain. Their physiological adaptations to the sea, encompassing such
phenomena as their ability to "see" by sound and to dive to great
depths, obviously contain unlearned lessons for man—not to mention

lessons not yet imagined, and aesthetic and ecological considerations. So far, the lack of fundamental knowledge about whales—even acknowledging a rudimentary sense of what happens to populations that are overexploited—has not deterred the killing.

The convictions voiced by leading whale scientists have a direct bearing on the management of whaling. If they minimize what has happened to the great whale species as a result of the mechanized slaughter of the past fifty years, it is not likely that persons with the capacity to curtail the industry will take the necessary steps to do so. For if scientists speak with an uncertain voice, the whaling industry will surely continue to capitalize on that uncertainty for short-term gain.

The continuing destruction of many of the larger forms of mammalian life on the earth has raised but few voices in alarm from the scientific community. We intuit that the quality of human life is diminishing at the same time that the larger forms of life diminish, but the process is so pervasive, so persistent, that our capacity to check it often remains incipient. What happens to the whale, for example, will affect the quality of human life and impair the human spirit. Because of the lateness of the hour for the great marine mammals, a touch of futility haunts us. We are beginning to suspect, reluctantly, that the kind of life left on earth—if we do survive— is being forged from such brutal passages in human history as the whaling debacle, where fortunes have been made at the cost of a priceless legacy to the future.

Because the voices of scientists who study whales are still largely unheeded regarding the proper relation of man to whales, it will help to recall voices from the past, voices little heeded in their own day but which have gathered resonance with the years and which ring out in our time.

To present a conservationist's view of the whale problem, a passage from Thoreau's journal (1856), which does not happen to mention whales, illuminates our task:

> When I think what were the various sounds and notes, the
> migrations and works, and changes of fur and plumage which

ushered in the spring and marked the other seasons of the year,
I am reminded that this my life in nature, this particular
round of natural phenomena which I call a year, is lamentably
incomplete. I listen to [a] concert in which so many parts
are wanting. The whole civilized country is to some extent
turned into a city, and I am that citizen whom I pity. Many
of those animal migrations and other phenomena by which the
Indians marked the season are no longer to be observed. I
seek acquaintance with Nature—to know her moods and manners.
Primitive Nature is the most interesting to me. I take in-
finite pains to know all the phenomena of the spring, for
instance, thinking that I have here the entire poem, and then,
to my chagrin, I hear that it is but an imperfect copy that I
possess and have read, that my ancestors have torn out many
of the first leaves and grandest passages, and mutilated it
in many places. I should not like to think that some demigod
had come before me and picked out some of the best of the
stars. I wish to know an entire heaven and an entire earth.
All the great trees and beasts, fishes and fowl are gone.

If Thoreau listened to a concert in which so many parts were wanting 115 years ago, how much more so is that the case today! It is true
that many of the first leaves and grandest passages of nature's great
poem have been torn out and mutilated in many places. Some of us wish
to know an entire heaven and an entire earth, and an ocean with whales
in number and diversity as it was before the slaughter of the bowhead
in the seventeenth century. But that cannot be again for many generations, if at all. The outcome of current discussions will help to
determine whether the international community of cetologists insists
on the rational management of whaling.

From a conservation standpoint, we must look for a while longer
to the International Whaling Commission as the only existing mechanism
with the strength to regulate the industry. Now, for the second time
in its history, the Commission faces a situation that may threaten its
survival. The first crisis occurred in the middle 1960's.

Concurrently, we must take a longer view of the whaling business
and, at the risk of repeating what is already well known, I shall review briefly the recent history of regulation. *All* of the great whale
species should be kept in mind, not just the few still commercially
hunted today.

The highly mechanized whale "fishery" dates from Svend Foyn's
introduction of the harpoon gun in the 1860's. In 1905, the first
factory ship was sent to Antarctica. Another development, which appeared in 1925, was the slipway in the stern of the factory ship for
hauling whales on deck in the open sea. Otherwise, the industry has
remained essentially unchanged except for the introduction of sonar,
helicopters, and catchers with increased horsepower. More than two
million whales have been reported taken worldwide over the past forty
years. The world catch of *Balaenoptera* whales over this period is
shown in Fig. 17-1.

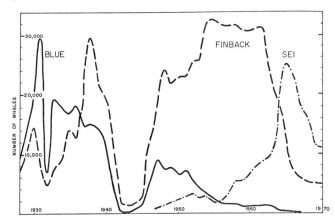

FIGURE 17-1. World catch of *Balaenoptera* whales.

In its early years the quota set by the IWC for Antarctic whaling
was so high, 16,000 blue whale units year after year, that the whaling
powers had to scramble to achieve that target. At the end of the 1950's
E. J. Slijper (1962, p. 415) contended "that all the arguments [for
substantially reduced catches] are still based on very little evidence,
and that whalers can and will restrict their activities only on the
most incontrovertible of arguments." The late Dr. Slijper was for
years a defender of the industry.

Slijper wrote (p. 393): "Now while this danger [of man eradicat-
ing whales] is by no means imaginary, it is far smaller than is generally believed." And, in another place (p. 392), he looks far ahead
and voices the view that "the writing for Cetaceans . . . is clearly
on the wall, and they would become extinct in the near (geologically
speaking) future, even if man left them severely alone." While his
evolutionary forecast is neither well founded nor widely accepted, it
is worth remembering because his views on the "harvesting" of whales
prevailed for two decades.

At a conference on the biology of whales in Washington in 1963,
Dr. Slijper reaffirmed that he saw no problem in conserving the whale.
Even if we recognize that the conference was not concerned with management problems, it may nevertheless be noted that at no point was the
status of whale populations seriously raised and discussed. One can
only conjecture that if the conservation issue had been aired and
acted upon at that time, the subsequent history of exploitation might
have been altered.

Throughout the 1950's, however, Johan Ruud, the Norwegian biologist, and Remington Kellogg, vertebrate paleontologist and U.S. Commissioner, kept reminding the Commission of one of its basic charges,
to provide for the proper conservation of whale stocks. Spokesmen
for the industry answered that the scientific data were not good
enough and that as long as this information remained inconclusive,
they were not inclined to curb their intensive whaling. Part of the
problem lies in the basic conflict in the original guidelines established for the Commission, which on the one hand provide for the conservation of whales based on scientific findings and on the other
"shall take into consideration the interests of . . . the whaling
industry."

Finally, in 1960, the Commission accepted the appointment of a
committee of three scientists to assess the problem and agreed to
abide by its recommendations. Then there did not exist scientists
who had been specifically trained in the biomathematics of exploited
marine mammal populations, but the Commission was fortunate to recruit
Sidney J. Holt (then with the FAO), Douglas G. Chapman (of the United

States), and K. Radway Allen (then of New Zealand), all able fish population biologists. After making interim reports, the Committee submitted its final report in 1963. The Commission failed to act on the
scientific recommendations, as it had previously pledged. When additional information was submitted in 1964, the Commission again balked
at the recommended quotas and pleaded for more time. Johan Ruud
threatened to resign as chairman of the Scientific Committee if the
Commission did not set rational quotas and provide protection for blue
and humpback whales; because the Commission did not respond, Ruud
resigned. The Commission was teetering on the verge of collapse,
seemingly unable to act on the firm set of recommendations made by
the Committee of Three Scientists.

At a special meeting of the Commission, convened in May of 1965
for the express purpose of setting reasonable quotas, a beginning was
made in regulating the industry. The quotas, still unfortunately in
blue whale units (a scheme that has been the bane of efforts at regulation), were lowered, and blue and humpback whales were finally protected.

Why does the blue whale unit seem irrational today, and how does
it work? One blue whale unit used to equal one blue whale, or two
finbacks, or two and one-half humpbacks, or six sei whales; today only
fin and sei whales are so tallied. First, by talking in "units"
rather than "whales," you make it arithmetic, not biology. And bad
arithmetic. Furthermore, such arithmetic removes the killing of whales
from our concern for the viability of each species. The blue whale unit
repudiates rational management, since what is not taken from one species
can be taken from another irrespective of what the latter can sustain.
As if this were not enough, the blue whale unit quota has been usually
set above scientific recommendations, thereby further reducing the remaining whale stocks.

The decade of the 1960's began with five major whaling powers
engaged in the highly mechanized "fishery" in the Antarctic. The three
European countries ceased whaling—England in 1963, the Netherlands in
1964, and Norway in 1968—because whale stocks had become so depleted
that it no longer paid to outfit expeditions. Their quotas and ships

were sold to the Japanese and Russians; the pattern of predation continues. Today, 85% of the world catch is taken by Japan and the Soviet
Union. The third largest whaling nation, taking 5% of the world catch
and still declining to become a member of the IWC, is Peru. The
Peruvian shore station at Paita is now under lease to a large Japanese
concern.

Some of the factual aspects of the present problem have been reported elsewhere in some detail (McVay 1966, 1971a, b, and c). Furthermore, the status of individual whale populations is beginning to
get a little more attention, although the Commission's response to
scientific recommendations is still marked by lethargy. In one sense
we are "beyond data," since a resolution of the whale problem is now
more in the domain of politics and economics than science. Yet, in
another sense, we urgently need reassessments of the data on fin and
sei whale populations in the Antarctic and the North Pacific; only
three parties are currently scrutinizing these data on a part-time
basis, and their interpretations vary widely.

It may be instructive to sample a few facts from whaling history:

(1) The bowhead, right, and gray whale populations were devastated by whaling in the nineteenth century and earlier. Only the
gray whale clearly has begun to come back from its low point of the
mid-1930's of an estimated few hundred to ten or twelve thousand today. This is only the well-publicized California population; its
Asian counterpart, which once used to migrate in the western Pacific
past the islands of Japan and along the coast of Korea, has been
sighted rarely in recent years.

(2) The blue and humpback whales have been brought to danger-
ously low numbers in our own time. In "The Blue Whale" (1971) George
Small keeps our attention riveted on "the largest animal known to have
lived on land or sea since the beginning of time." All too often a
species of whale is written off by the whalers and then forgotten by
others. By an impressive array of graphs and specific information
culled from a variety of sources, including the tedious verbatim
records of the International Whaling Commission, Small has succeeded
in not letting us fail to remember the magnitude of the tragedy.

The tragedy of the blue whale on a worldwide scale has been repeated in one whale stock after another. To provide a "local" example
for the humpback whale, a California whaling company is said to have
all but wiped out a "homesteading" population of humpbacks in the late
1950's and early 1960's, that was seen regularly between Monterey and
San Francisco, before the IWC banned the killing of this species in
1965. The scientist responsible for monitoring this station in the
middle 1950's, Raymond M. Gilmore (personal communication), asked the
company to impose a voluntary quota before the whaling commenced.
The company refused. He then made a formal request through the Bureau
of Commercial Fisheries to have a quota set on the catch of humpback
whales, but to no avail. The catch statistics on humpbacks taken by
California whalers, year by year from 1956 to 1965, reflect a familiar
pattern: 133, 199, 115, 140, 67, 62, 39, 55, 27, and 4.

(3) Despite intensifying efforts of the whaling industry, its
yield has fallen off markedly. For example, the catch of fin whales
in the Antarctic dropped in the 1960's from 27,000 annually to below
3,000. If the whalers had shown more restraint, if the industry had
been genuinely interested in the long-term yield, the Antarctic could
provide indefinitely 12,000 fin whales per annum.

The lady from Boston, as the story goes, who took up an ancient
if not honorable profession "rather than dip into capital" understood
the concept of principal and interest. Simply stated, a bank account
or investment portfolio can bear resemblance to a whale population.
If you consistently take capital as well as income, both eventually
will be wholly depleted.

In any discussion about the management of whaling, the notion of
"maximum sustainable yield (m.s.y.)" is iterated with some reverence.
This is, by definition, the largest number of animals that may be
"cropped" year after year without further depleting the population.
Current theory holds that maximum sustainable yield for a whale population occurs when that population reaches a level of 50% of its
original or unexploited population; the thinking is similar to that
for fish populations, the dynamics of which are better understood.
With whales, however, we do not yet know with assurance if the maximum

yield that can be sustained year after year occurs at 50, 60, 70, 80, or 90% of the original population.

Even today all "responsible" discussion of the management of whaling turns on this magic phrase, "maximum sustainable yield." Yet if our rationale for conserving whales derives solely from a market-place mentality—which reaches full flower in this concept—then the future will be a tragic repetition of the past. For, unfortunately, while the words "maximum sustainable yield" have been on the lips of everyone associated with the industry for the past ten or twelve years, the principle has been ignored every time the quotas were set. In many contexts m.s.y. may be the only viable argument, but so far it has had a strictly marginal effect on the whale fishery, which is still run largely on the basis of maximum return over the short haul. The maximum yield is not the maximum sustainable yield.

One hears the familiar refrain, "we have reached an approximation of the sustained yield in the whale fishery (where it is regulated). Decimation is not apparently going on for the regulated species. The whales are *already decimated*, true, but the process has stopped . . ." This is poppycock. We can hardly contend that the Antarctic situation is under control when quotas are still set in blue whale units, when the median estimate of the sustainable yield for fin whales in the Antarctic by Chapman and Allen is 2,200 and the whalers may well take 3,000 (representing a further incursion into the stocks which have already been depleted to 20% of their original numbers), when there is still no quota for sperm whales in the Antarctic, when there is no quota for minke whales after that fishery has been underway for four seasons. Also, a detailed biometric study has reached us, which states that the sei whale, supposedly being taken at a level of "maxi-mum sustainable yield" in the Antarctic—that is, about 5,000—may currently be under exploitation at more than double its sustainable yield (Jones, MS). If this is an accurate assessment, the sei whale population is not being "harvested" at a level of m.s.y. (that is, 50% of its original population), but rather it is approaching the serious condition of the fin whale population. And, to top it off, the ob-server plan has never been put into operation so that one cannot know

the extent of illegal catches. The process of decimation has not
stopped. Those who make such assertions have once again been taken
in by the industry and its spokesmen.

Another measure of what has happened to the Antarctic rorqual
fishery recently is that the catches must be taken in lower and lower
latitudes. This means smaller whales. Twelve years ago 89% of the
catch was taken south of 60^0 South; last year the figure was only 11%.
The catch per unit of effort has fallen to 25% of what it was a decade
ago.

In every year from 1958 to 1968, peak whaling catches worldwide
were more than 60,000 whales annually. In each successive year,
smaller and smaller whales and smaller and smaller species comprised
the catch. Fewer and fewer blue whales were taken as the years
passed, a decline that began in 1931. In addition to the fin whale,
the whalers began to hunt the long-ignored sei whale, which supposedly
yields only one-sixth the oil of the blue and which is preferred for
human consumption in Japan, for example. The pursuit of the smaller
whales has, in effect, subsidized the potential extermination of the
larger species. Now even the ranks of the smaller whales are so depleted that the catch has begun to decline sharply. Yet, in 1970,
42,266 whales were reported taken, or about one every 12 minutes.

These data give an indication of the scope of the dismal problem
and emphasize the difficulty in trying to turn the situation around.
Since the summer of 1970 more attention by the American public and
within the government has been directed at the whale problem than in
any previous period. In November 1970 the Department of the Interior
placed eight species of great whales on the endangered list, which
means that none of these whale species or their products may be im-
ported into the United States. This action followed an intensive
review by the Department of information from several sources, submitted over a six-month period, which supported the conclusion that
the now "protected" species—the bowhead, the right, the gray, the
blue, and the humpback—as well as the three remaining species that
still are commercially hunted—the fin, the sei, and the sperm whale
—belong on the endangered list. The decision was based on many

factors; one of them was the definition of what "endangered" really
means.

When is a species endangered? A species is endangered when the
exploitation rate is so high that, on the average over time, each adult
less than replaces itself in the next generation. When a policy of
exploitation prevails such that as the population decreases, the effort
per unit of catch increases at lower and lower population densities,
then the species in question is endangered. Furthermore, the exploitation trend must be viewed over years and not simply from one season to
the next, as often happens at international commissions.

In announcing the decision, Secretary of the Interior Walter J.
Hickel said:

> The decision will be controversial because certain interests
> have urged that we hold off listing the fin, sei and sperm
> whales until it is actually proven that they are on the verge
> of extinction . . .
>
> It is the clear intent of the Endangered Species Conservation
> Act, which took effect June 3, 1970, that we *prevent* condi-
> tions that lead to extinction. It is also clear that if the
> present rate of commercial exploitation continues unchecked,
> these three species will become as rare as the other five . . .
>
> We are not going to wait until all these species are on the
> brink of extinction before we take positive action.—U.S.
> Department of the Interior (1970)

I am not invoking some gloomy inevitability, nor am I suggesting
that commercial extinction is the necessary destiny of the fin, sei,
and sperm whales. Rather, I am suggesting that when the stocks of
whales have reached a level of commercial insignificance as they have
for five major species (the bowhead, the right, the gray, the blue,
and the humpback), they are no longer a significant food source and
may not be for a very long time to come. Further, I am suggesting
that they no longer play a significant role in the ecological scheme
of the oceans, whatever that role may have been, and that learning
about these plundered species is already exceedingly difficult.

It seems to me that the International Whaling Commission has a final opportunity to reinvigorate and strengthen its stewardship through two major actions:

(1) Implementation of the International Observer Plan, which will place an observer on every factory ship and at every land station to make sure that the regulations of the IWC are honored. Perhaps no objective is more important than this, because without observers one cannot know whether "protected" species or undersized whales are killed or whether other rules currently on the books are observed.

(2) Assignment of quotas to each whale species by region on the basis of the best available scientific evidence. In the past the Commission has often set quotas substantially above the recommendations of the Scientific Committee. Now it has become urgent to set quotas below the sustainable yield, especially for the fin whale, in order to allow the stocks to rebuild. Also, we should insist on abolition of the destructive blue whale unit and the 10% "slippage" in the North Pacific quotas, reminiscent of the blue whale unit.

Even these measures may not be enough. A moratorium of ten years on whaling has been advocated. This proposal will receive serious attention in the near future, because anything short of a moratorium will decrease the chances of recovery for certain whale stocks as well as slow down the rate of that recovery.

What is it in our nature that propels us to continue a hunt initiated in earlier times? Are we like some lethal mechanical toy that will not wind down until the last bomb explodes in the last whale's side? What is it that makes so small a thing of eliminating in our lifetime the oceanic role of the largest creature that has lived on our planet? What is it that kills the goose that lays the golden egg? Is it already too late? Is our own obituary scrawled in the fates of the bowhead and right whale, the blue and the humpback—all species that no longer contribute to the biological systems of which they were a part for millions of years? What is the true use of whales beyond bone, beef, and blubber? Of what value will whales be to men over the next thousand years if their stocks are allowed to rebuild?

Cetology has for a long time been a "dead" science. By that I mean that man's first glimpses of the whole leviathan were through occasional strandings, and our knowledge today of the order of Cetacea still consists of a number of glimpses that have yet to be woven into a coherent pattern. Very little of what we know of whales has been gleaned from studies that seek to find out about the life pattern of individual species over long spans of time at sea. Rather, the bulk of the scientific reports are based on data taken from dead whales and these data consequently are industry dependent. This means that, wittingly or unwittingly, the whale scientist may often be in a parasitic relationship to the whaling industry. He is often dependent upon the leavings of industry to do the kind of work he has learned to do—whether he develops estimates of population and sustainable yield; studies of distribution, migration, and morphology of a stock or species; studies of ear plugs, ovaries, teeth, vertebrate fusion, inventories of stomach contents, parasites, deformities, etc. This is all useful, yet one would hope that the continuing work on freshly killed whales would pay more attention to the brain and nervous system, building on the work of such men as Ogawa (1935). Furthermore, what has been missing from the equation has been any systematic study of the whole organism and its relation to group and environment.

But a fresh breeze is blowing. While attention to the natural history of cetaceans is not new, the beginnings of a stronger orientation toward living cetaceans are found in such work as the phonograph record and booklet produced by Schevill and Watkins in 1962. Scientists now are determined to know the whale in its natural habitat of the sea, still a strange place for us. I hope that current discussion produces a Magna Carta for Whales, and that we shall be capable of moving beyond the death-dealing relationship that has haunted almost all human transactions with whales for eight centuries. I hope that the men who have felt a special calling to study whales will bend their efforts to do more work on live whales and encourage young researchers to go to sea.

If cetology becomes truly a "life science" rather than a "death science," the great whales are more likely to survive. Indeed, work

over the next decade and more may provide sufficient information to
assure not only survival but eventual restoration of the great whales
to their former role in the oceans of the world. One day, then, we
may come to know "the true use of the whale."

REFERENCES

Gilmore, R. M. Personal communication.
Jones, R. (MS). Population assessments of Antarctic fin and sei
     whales.
McVay, S. 1966. The last of the great whales. Scientific American,
     215:13-21.
_____ 1971a. Can Leviathan long endure so wide a chase? Natural
     History, 80:36-41, 68-72.
_____ 1971b. Does the whale's magnitude diminish?—Will he perish?
     Bull. Atomic Scientists, 27:38-41.
_____ 1971c. Whales: a skirmish won, but what about the war?
     National Parks & Conservation Magazine, 45:1, 3, 14-19.
Ogawa, T. 1935. Beiträge zur vergleichenden Anatomie des Zentral-
     nervensystems der Wassersäugetiere: Ueber die Kleinhirnkerne
     der Pinnipedien und Zetazeen. Arbeiten anat. Inst. kaiserl.—
     Japan. Univ. zu Sendai, 17:63-136.
Schevill, W. E., and Watkins, W. A. 1962. Whale and porpoise voices.
     Woods Hole, Mass., Woods Hole Oceanographic Institution (24 pp.
     and phonograph record).
Slijper, E. J. 1962. Whales. A. J. Pomerans, trans. New York,
     Basic Books.
Small, G. L. 1971. The blue whale. New York and London, Columbia
     University Press.
Thoreau, H. D. 1853/1972. The Maine woods. Princeton, Princeton
     University Press.
_____ 1856. Original manuscript of journal dated March 23, 1856.
U.S. Department of the Interior. News release, 24 November 1970.
     Secretary Hickel bans imports of products from eight endangered
     species of whales.

PART FIVE

Field Methods

CHAPTER 18

RADIO-TELEMETRIC STUDIES OF

TWO SPECIES OF SMALL ODONTOCETE CETACEANS

W. E. Evans

Early attempts at tracking radio-tagged cetaceans (Schevill and
Watkins 1966) indicated transmitter antenna deployment and short
signal acquisition times could be significant among the many vari-
ables affecting success. Taking advantage of the lessons learned
from these initial observations, a receiver system that determines
signal azimuth electronically rather than manually was developed, and
species for which a live-capture technique existed (delphinids) were
chosen as the experimental subjects.

Since July 1968, eight saddle-back porpoises (common dolphin)
Delphinus delphis Linné 1758 and one pilot whale *Globicephala scammoni*
Cope 1869 have been captured, tagged with radio beacons, tracked for
up to 72 hours, and relocated by aircraft up to two weeks later (Evans
1970). For three of the *Delphinus* and the one *Globicephala*, the radio
beacons also transmitted data relating to maximum depth of dive. Each
radio beacon and data package was attached to the animal's dorsal fin
by means of a corrodible bolt designed to release the package from
the animal after 30 to 60 days (Fig. 18-1).

Instrumentation

The radio beacons used during these preliminary studies were
crystal-controlled to operate in the 11-m r.f. band. The maximum
transmission output power was 0.1 watt on seven of the beacons and
1.0 watt on two of the beacons. Maximum signal reception range from
a surface vessel is 20 to 30 km with an antenna height of 20 m above
the sea surface. Reception range is significantly affected by sea

FIGURE 18-1. Maximum-depth-of-dive data package attached to the dorsal fin of a Pacific pilot whale (*Globicephala scammoni* Cope).

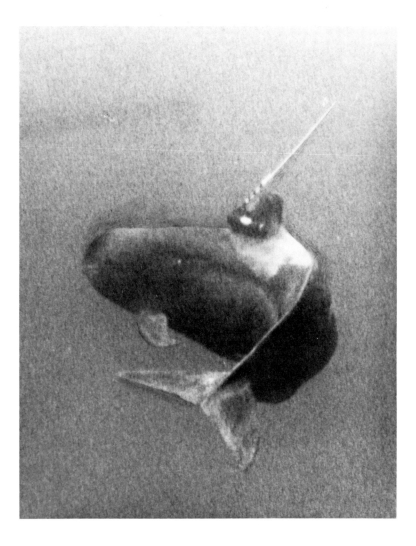

state, becoming intermittent but still readable with wave heights in excess of 3 m. Radio signal reception range is 50 km from an altitude of 400 m with the 0.1 watt radio beacons.

The mechanical configuration and actuation of the porpoise and whale radio beacons used are similar to those of the Model ST 202 submerged radio beacon manufactured by Ocean Applied Research, Inc., of San Diego, California (Martin, Evans, and Bowers 1971). A self-contained alkaline battery-pack provides power for 100 hours, with actuation via a seawater connection between the antenna tip and battery cap. The total transmission time available, assuming that the antenna is exposed 10% of the time, is approximately 40 days. Actual exposure time will vary with different species. The beacons which transmitted data on depth of dive generated a signal consisting of an amplitude-modulated r.f. carrier with an audio signal (continuous tone) which varies as a function of the maximum depth of dive (800 to 1,500 Hz). The tone's frequency was determined by a capacitor-controlled, voltage-controlled oscillator (VCO) in the transmitter. Voltage level on the VCO varies with maximum pressure sensed by a transducer externally mounted on the beacon case. Each time the animal surfaced, the transmitter turned on and continually emitted the signal indicating the maximum depth of the last dive. A 0.75-second time constant is used in the circuit to hold the maximum depth signal in case surface transmission is cut off due to the antenna tip being slapped by a wave without the animal actually initiating another dive.

The signals from the beacons were received and processed by two separate radio receivers, an OAR Model 210 ADF (automatic direction finder) and an OAR Model PDAS 810 (data receiving system). The ADF presents the signal both in audio form (800 to 1,500 kHz tone) and as a visual display on a cathode ray tube, which gives an instant bearing relative to the heading of the tracking vehicle with as little as 50 milliseconds of signal. This was especially valuable since some surface times for *Delphinus* were as short as 2 seconds. The antenna system consists of two loops, one parallel to and one at right angles to the tracking vehicle's heading, and an omnidirectional sense antenna which provides a reference to resolve the 180° ambiguity.

Details of receiver design are presented in Martin, Evans, and Bowers 1971. The PDAS 810 consists of a crystal-controlled radio receiver with a band-pass filter and a discriminator circuit. The audio output of the PDAS 810 receiver (800 to 1,500 Hz tone) is monitored on a panel speaker, and an external audio recorder can be connected at an output tap provided. A d.c. analog of depth ± 2% is derived by the discriminator circuit, and this signal is presented at a strip-chart recorder output tap. Recorder output taps are provided for other kinds of data (temperature, electrical conductivity, animal velocity, water turbidity) to be collected during future studies. This entire system is portable and can operate from a variety of surface vessels and aircraft. Data have been collected from a 29-m sailboat, a 30-m converted coastal minesweeper, a 14-m motor launch, single-engine and twin-engine fixed-wing aircraft, and U.S. Navy helicopters.

Results

The data collected from instrumented *Delphinus* and *Globicephala* have been indicative of significant diurnal differences in both respiration pattern and diving behavior. Although different in terms of time base, the respiration rate and depth of dive of *Globicephala* as a function of time of day follow patterns similar to those observed for *Delphinus* by Evans (1971). Thirty-minute samples of the diving pattern

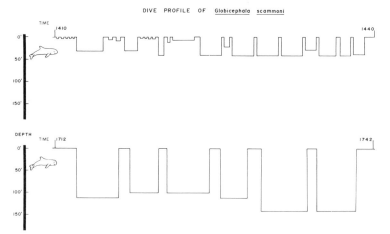

FIGURE 18-2. Midafternoon and sunset diving pattern (down time versus maximum depth of dive) for a 3-m male Pacific pilot whale.

for the one pilot whale tracked during midafternoon and at sunset are
shown in Fig. 18-2.

Early observations of the diving pattern of *Delphinus* indicated
a close relationship between the onset of dives greater than 10 fathoms
and the migration of animals associated with the acoustic deep scat-
tering layer (DSL) to the surface after sunset. Subsequent observations of *Delphinus* have further verified this relationship between the
onset of diving behavior and the migration of the DSL to depths above
20 to 50 fathoms. On days where the ambient light level is reduced by
heavy overcast or fog and the DSL does not completely descend, diving
patterns similar to those normally observed only after sunset continue
intermittently throughout the day (Fig. 18-3). In addition midwater
trawls at maximum dive depths while tracking a radio-tagged *Delphinus*
caught 40% bathylagids, 30% myctophids, 30% engraulids, unidentified
deep-sea squid, and unidentified crustaceans (mysids and euphausids).
All fish species collected have been observed in the stomach contents
of *Delphinus* (Fitch and Brownell 1968). In the case of the *Globicephala,*
large concentrations of the squid *Loligo opalescens*, known to represent
a high percentage of this species' diet (Norris and Prescott 1961),
were present from the surface to the maximum depth of dive.

Although all of the observed diving behavior was in close proximity
to prominent submarine topography (escarpments, seamounts, and canyons),
none of the dives were to depths close to the bottom, which was at a
minimum of 100 fathoms and an average of 200 to 500 fathoms. The maximum dive depth recorded from *Delphinus* was 141 fathoms. Mean diving
depth observed for the three individuals studied was 40 fathoms. The
maximum dive depth for the single *Globicephala* observed to date was
60 fathoms, although a captive *Globicephala* has been trained to dive
consistently to depths of 200 fathoms on command (Martin, Evans, and
Bowers 1971).

Discussion

Data on *Delphinus* herd movements from individual tracks, although
informative, do not provide much insight toward answering basic questions about total population size, home range, and other factors vital

FIGURE 18-3. Diving depth as a function of time of day. Note that
when the light level is reduced by overcast or fog, the DSL does not
migrate as deep as on days with high light level; under these conditions relatively deep dives continue throughout the daylight hours.

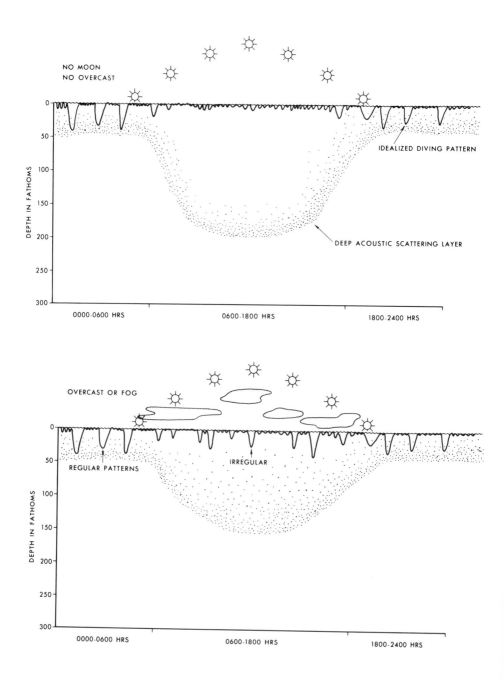

to a comprehensive program of management and conservation. When the data accumulated from tracking several individuals and their associated herds over a four-year period are combined with data from weekly aerial surveys of location and movement, definitive patterns do begin to emerge. In the specific case of *Delphinus* in the eastern Pacific Ocean some investigators, for example, Banks and Brownell (1969), using museum specimens, have separated two species of *Delphinus*: *D. delphis*, limited to a range essentially off the coast of California, and *D. bairdii* Dall 1873, ranging along the coast of Baja California and south along the mainland coast of Mexico. The patterns of movement and distribution derived from tracking instrumented individuals offers a different picture. Rather than forming specific groups limited in distribution to a finite part of the eastern Pacific, herds of *Delphinus* may move as far as 120 km in a 24-hour period. The routes of these movements are not random, but well defined (Fig. 18-4). In many cases herds marked in the coastal waters of California may move hundreds of miles south along the coast of Baja California, Mexico, following coastal escarpments, and then work their way north again. The reverse of this case has also been observed with animals identifiable as *D. bairdi* not being restricted to Mexican waters as stated by Banks and Brownell, but coming north of the southern California Channel Islands (this taxon is probably better called *Delphinus capensis* Gray 1828, but see van Bree and Purves 1972). These observations certainly do not deny the possible existence of more than one species of *Delphinus* in the eastern Pacific. They do, however, tend to illustrate that defining the distribution of any species of cetacean on limited observations can be misleading. It is quite premature to interpret the present data in terms of defining a population of *Delphinus* per se, but the specific patterns that have evolved provide us with valuable insight into the complexities of the concept of population, especially when dealing with delphinids.

The equipment and procedures used in these initial studies proved to be quite reliable, with data successfully collected on nine of the twelve animals tagged with beacons. The three failures were due primarily to minor equipment malfunctions associated with attempts to use power sources on board the tracking vehicles. With the data pack

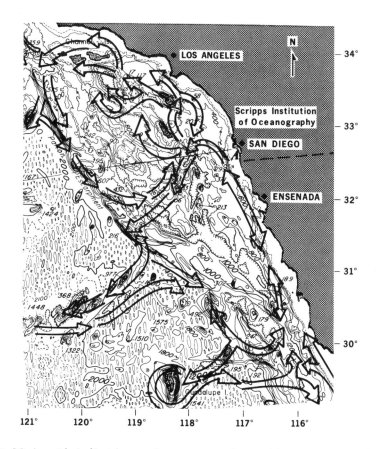

FIGURE 18-4. Distribution and movement of *Delphinus delphis* Linné in
the coastal waters of southern California and Mexico. Note the close
relationship between herd location, movement, and prominent features
of the bottom topography.

and radio beacon reliably attached, useful data should be obtainable
from any species of cetacean. Most small delphinids which can be
captured alive and reasonably uninjured are potential subjects for
study. Larger species which are difficult to catch alive and handle
at sea present a problem, but one that could be overcome if efficient
and reliable methods of remotely attaching packages can be developed.
The instrumentation must be so attached as to allow the greatest time
of antenna exposure while the animal is at the surface and also so as
to cause the least hindrance to the animal's mobility. Recent availability of more efficient power sources and advances in transmitter

circuit design have resulted in smaller packages, with greater output
power, and a transmission life in excess of nine months. The development of this type of commercially available system and an associated
decrease in the cost of individual instrumentation packages has provided a reliable and field-tested tool potentially invaluable for the
assessment and long-term management of whale stocks.

Acknowledgment

The success of any field study of cetaceans is highly dependent
on the assistance and support of many people representing a broad
range of expertise. The results of this and past radio-telemetric
studies of delphinids has been aided greatly by the assistance of Mr.
B. C. Parks, Naval Undersea Research and Development Center (NUC),
who captured and helped instrument all of the *Delphinus* discussed.
In addition, Mr. Parks was invaluable in providing navigational data
during the active tracking stages. Mr. Morris Wintermantel, also of
NUC, captured the pilot whale shown in Fig. 18-1. Many hours of
standing radio watches during the longer radio tracks were endured
by Messrs. J. D. Hall, J. S. Leatherwood, and L. McKinley. Mr. McKinley
also provided all the technical advice on the engineering aspects of
this research. Finally, none of this research effort would have been
possible without the support of all of the Navy personnel operating
the tracking vessels used: Captain C. B. Bishop, Commanding Officer
of NUC; Dr. W. B. McLean, Technical Director; Mr. George Anderson,
Head, Ocean Sciences Department; and Dr. C. S. Johnson, Head, Marine
Life Sciences Laboratory.

REFERENCES

Banks, R. C., and Brownell, R. L., Jr. 1969. Taxonomy of the common
     dolphins of the eastern Pacific ocean. J. Mammal., 50:262-271.
van Bree, P. J. H., and Purves, P. E. 1972. Remarks on the validity
     of *Delphinus bairdii* (Cetacea, Delphinidae). J. Mammal., 53:372-374.
Evans, W. E. 1970. Uses of advanced space technology and upgrading the
     future study of oceanology. American Inst. of Aeronautics and
     Astronautics Paper No. 70-1273. Pp. 1-3.
_____ 1971. Orientation behavior of delphinids: radio telemetric
     studies. Annals N.Y. Acad. Sci., 188:142-160.

Fitch, J. E., and Brownell, R. L., Jr. 1968. Fish otoliths in cetacean stomachs and their importance in interpreting feeding habits.
J. Fish. Res. Bd. Canada, 25:2561-2574.

Martin, H. B., Evans, W. E., and Bowers, C. A. 1971. Methods for
radio tracking marine mammals in the open sea. Proc. IEEE, Conf.
on Eng. Ocean Environ., Sept. Pp. 1-6.

Norris, K. S., and Prescott, J. H. 1961. Observations on Pacific
cetaceans of California and Mexican waters. Univ. Calif. Publ.
Zool., 63:291-402.

Schevill, W. E., and Watkins, W. A. 1966. Radio-tagging of whales.
Woods Hole Oceanographic Inst. Ref. 66-17. Pp. 1-15.

CHAPTER 19

NEW TAGGING AND TRACKING METHODS FOR THE STUDY OF

MARINE MAMMAL BIOLOGY AND MIGRATION

Kenneth S. Norris, William E. Evans, and G. Carleton Ray

Many aspects of our knowledge of marine mammal biology, including
those important to management, rest upon shaky ground. This is not at
all surprising to those of us who attempt to gather such data, particularly upon the open sea. It simply isn't easy to work with these elu-
sive, often bulky creatures from the unstable platform of a ship's
deck. In fact, it is remarkable how much we have been able to learn
about the large whales, for instance, as most data have been drawn
from the nonrandom sampling of industry. Crucial data, especially on
population identity and movements, remain to be gathered. This is as
true for marine mammal populations of precarious status as for exploited species. This paper concerns some new approaches that we feel
will soon allow a fresh look at the biology of marine mammals and that
will lead to better information on which to base management practices.

One of the most needed developments is to design better instrumentation for tracking and telemetry. A vital part of any population
study is information on movement and overall range. Without such information it is impossible to assess the discreteness of various populations. Such data are basic for management on a sustainable yield
basis. Most of our knowledge of large whale migrations has come
through the use of Discovery marks—slim, numbered, metal cylinders
shot into the whale (Brown 1962). Data from these marks are limited;
they are most useful for age and growth, but not so useful in determining movements, since only two points on the whale's track can be
recorded, the point of marking and the point of death. Further, these
two points are often in the same geographic area, since most marking
has been done on the feeding grounds where the catch is also made.

Only rather recently have many marks been placed in whales far from
the high latitudes (Rice 1963; Clarke 1962). Even so, there is doubt
about whether rorquals (Balaenopteridae) regularly cross the equator
(Slijper, van Utrecht, and Naaktgeboren 1964; Norris 1966). It is
also uncertain what routes most species travel in their movements
toward the equator—and whether they feed in temperate or tropical
waters. Moreover, for most species there are few data which adequately
show the degree of mixing of the various migratory populations. Dawbin
(1966) describes the mixing of various stocks of Antarctic humpback
whales, *Megaptera*. Next to nothing, however, is known of commercial
species of whales in the warm-water grounds of the open sea. To
clarify these important questions, we need to trace the paths of
whales with repeated fixes on marked individuals.

Scarcely anything is known of the movements of small porpoise
and whale schools, except for those species involved in minor fisheries, such as the pilot whale *Globicephala* (Sergeant 1962) or for species currently being radio-tracked (Evans 1971 and Chapter 18 of this
volume). The increasing importance of porpoises in commercial catches
or as adventitious kills in the tuna fishery (Perrin 1970) makes the
need for such data very great. It is reasonably certain that most of
the smaller odontocetes do not migrate such hemispheric distances as
large baleen whales, but it is also certain that in many places their
local occurrence is seasonal; such data exist only for a handful of
local populations. At this point it is fruitless to ask how big the
population of a given porpoise species might be, what its recruitment
is, whether or not its schools are units having temporal continuity,
or what the natural fecundity and mortality levels are. It is most
important to learn to track these animals too—not on the basis of
two points along their route, but in such a way that school inter-
mixture, migration paths and speeds, and population numbers begin to
become evident.

Finally, many other aspects of natural history are significant
for informed management. One needs to know about growth and metabolic
relationships in conjunction with knowledge of food sources and feeding behavior. We need information about discontinuous distribution

as related to hydrographic features and primary and secondary productivity. Such information will help to determine whether localized
exploitation decimates a population, or whether wider areas of the
ocean must be managed. In addition, one needs to know when and where
mating and calving take place, and how crucial environmental health
is to whale population health. Must humpbacks come into shallow
water to breed? Where do young gray whales (*Eschrichtius robustus*)
go after birth (Rice and Wolman 1971)? What happens to breeding
structure when familial or school relationships are disrupted by
human exploitation? What factors influence infant mortality? A
start on gathering behavioral data was made on captive porpoises
over two decades ago (McBride and Hebb 1948) and much has been learned
since, but precious little has been gathered from animals in their
natural habitats. Hence the technology of data-pack design and attachment to marine mammals is of much importance and promises to produce diverse and valuable information.

We will discuss three separate departures that are either operational or in design stages at present. Common to all three is the
equipping of marine mammals with data packs and/or radio beacons,
and this is our subject here. Methods range from the use of radios
on small cetaceans for subsequent tracking by boat or aircraft to
equipping marine mammals with recoverable packages for data from
shorter time periods and to instrumenting large whales for satellite
tracking. There are two aspects which we will not discuss here;
these are the r.f. link, i.e. transmission of signal, and the receiving system itself. Both of these deserve special consideration, and their exclusion here is not to indicate their lesser importance. However, they are problems more electronic than strictly
biological, hence their lesser emphasis here.

Radio-Telemetric Studies of Small Odontocete Cetaceans

The possibility of accomplishing the tasks previously discussed
would have been speculation ten years ago. The feasibility of applying some of these techniques to the study of marine mammals in the

field has now been demonstrated on a few species of small odontocetes.
Techniques for the capture of these animals, developed principally by
several oceanaria, have made possible the attachment of instruments
to them for subsequent radio or acoustic monitoring.

Radio-beacon circuit designs, successfully used for several years
in the recovery of untethered instrument packages at sea (Martin and
Kenny 1971), provided the basis for radio-tracking and data-monitoring
of delphinid cetaceans (Fig. 19-1). Within the past three years
several common dolphins (*Delphinus delphis*), two pilot whales (*Globicephala scammoni*), one killer whale (*Orcinus orca*), and one Hawaiian
spinner porpoise (*Stenella* cf. *longirostris*) have been successfully
tracked at sea. The *Delphinus* (Martin, Evans, and Bowers 1971) and
the *Stenella* (Norris and Dohl, in preparation) have been followed for
as long as 72 hours. The *Stenella* tracking was accomplished with the
equipment described by Martin, Evans, and Bowers (1971).

In addition to determining the movements of these individuals,
it has been possible to obtain the maximum depths achieved during each

FIGURE 19-1. A dorsal fin radio-beacon pack carried by *Delphinus delphis*. (*Photograph by the U.S. Naval Undersea Research and Development
Center*.)

dive by telemetry (Evans 1971 and Chapter 18 of this volume). As an
illustration of the potential of these techniques, especially on the
large whales, a summary of Evans's results follows:

(1) Schools of *D. delphis* follow and dive over prominent
     features of the ocean bottom (seamounts, escarpments,
     canyons).

(2) The depth of most *D. delphis* dives is closely corre-
     lated with the depth of the vertically migrating, deep
     scattering layer (DSL).

(3) Herd movements of *D. delphis* appear to correlate well
     with seasonal shifts in populations of certain fish
     such as the northern anchovy (*Engraulis mordax*) and
     the Pacific hake (*Merluccius productus*).

(4) Although herds of *D. delphis* may spend several days
     in a restricted area, movements of up to 150 to 200 mi
     in a 48-hour period occur.

(5) Movement and diving behavior of *D. delphis* varies
     significantly as a function of light-dark cycles.

Data such as these exemplify how information on niche and habitat
may be determined by using telemetric techniques.

Natural History Data Packs for Marine Mammals

In addition to tracking animals over long periods of time, there
are specific shorter-term needs. Information on behavior and physiology is needed for single dives. The carrying capacity of the environment for any animal population is determined to a great extent
by feeding behavior and trophoenergetics. The optimum way to gather
such data is to record over shorter periods and in more detail than
the aforementioned technique of radio-tracking allows.

Some work with data packs has been done by De Vries and Wohlschlag (1964) and by Kooyman (1966), who recorded dives of unrestrained
Weddell seals (*Leptonychotes*) under Antarctic sea ice. Progress in
telemetry (Mackay 1970; Fryer 1970) indicates that a recording instrument package could be developed to record several data channels for

later analysis. Formidable problems of miniaturization are partially
obviated by the large size of most marine mammals.

Accordingly, in cooperation with Ocean Applied Research Corpora-
tion of San Diego, California, we have designed and built a data pack
(Fig. 19-2A) the purpose of which is to make hour-long recordings
from free-swimming animals. A small tape recorder comprises the bulk
of the data pack and determines the diameter of the tubular plexiglass
housing, which we have specified for 600 m water depth. A clock in
the data pack switches it on at a preset time, whereupon the data are
recorded on tape. After the data pack is retrieved and the tape removed, demodulation of the data is by a deck unit and strip-chart
recorder. Package retrieval implies the inclusion of a radio beacon
as a feature of design.

The tape recorder largely determines the length of recording time.
Such data as temperature may be recorded digitally, and tape speed may
be slow. However, we also wish to record *in situ* sounds made by the
animal, and this must be done as analog, for which tape speed must be
much faster. A major design problem is miniaturization of the tape
recorder for longer operation in the analog mode. Our present data
channels include three for external temperature (two for animal skin
surface and subsurface, one for ambient water or air temperature),
and a fourth for calibration. In addition, a radio pill (Fig. 19-2B)
transmits internal (stomach) temperature, which tells two things,
core temperature and feeding (by a sudden drop in the temperature
indicated).

External temperatures are recorded by means of thermistors on
wire leads. A difficult problem concerns design of "pill" temperature transmitters with sufficient stability against variations in
both voltage and the temperature of their surroundings. Some previous studies present data which appear suspect, but in very few cases
are sufficient data presented on the electronics of circuitry. Pill
stability is discussed by Fryer (1970), and Wartzok, Ray, and Martin
(in press) discuss electronic design problems of this sort.

Attachment of the data pack is by a harness which has been
developed for pinnipeds (Figs. 19-2A and 19-3). Both otariid and

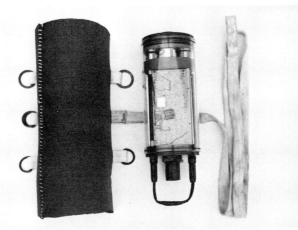

FIGURE 19-2A. A recording instrument package containing a tape recorder for recording both internal and external temperatures from an unrestrained pinniped. External leads to thermistors (not shown) are plugged into the bottom of the pack. A receiver picks up transmission from an internal pill. The entire pack is placed in the neoprene sleeve for protection. (G. C. Ray photograph.)

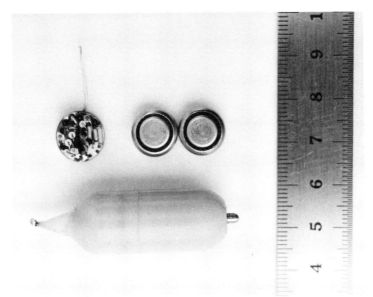

FIGURE 19-2B. A radio pill for transmission of internal temperature. The pill is enclosed in steel and coated with silastic. The antenna is a dipole. Shown above the pill are the transmitter (which is stable under changing conditons of temperature and voltage) and the batteries. (G. C. Ray photograph.)

FIGURE 19-3. A Steller sea lion, *Eumetopias jubata*, wearing the
recording instrument package shown in Fig. 19-2A. The animal is a
male of about 200 kg. (*G. C. Ray photograph.*)

phocid seals can carry relatively large packages by this means. Attachment is a function of body size, shape, and especially flexibility.
The package illustrated here is relatively simple electronically. The
major problem is one of design and it was believed wise to solve problems of attachment, durability, and transmission from the internal
pill before adding other channels such as for heart rate, swimming
speed, diving depth, and the like. Many of these data are obtainable
with present technology and with only a moderate increase in size of
the data pack. Also not to be forgotten are environmental data such

as salinity, light levels, and other parameters which provide important environmental correlates to behavior and physiology.

Tracking of Large Cetaceans

Schevill and Watkins (1966) were the first to attempt radio-
tracking of any cetacean. Radio beacons were attached from aircraft
to unrestrained adult right whales (*Eubalaena glacialis*) with small
darts imbedded in the blubber. However, no whale tracks were obtained.

A system is now being assembled that seeks to solve some problems
involved in tracking large whales, and it is hoped that tracks will be
obtained from animals over considerable periods of time. The data
package is being designed by Robert M. Goodman of the Franklin Institute Research Laboratories, Philadelphia, California. The harness
and capture and attachment techniques are being developed by K. S.
Norris. Newborn mysticetes are much easier to catch than adults,
being little larger than some odontocetes that are caught for ocean-
arium display. Gray whales have twice been successfully captured
(Robert Elsner and D. W. Kenny, personal communications), and one
humpback whale calf has been maneuvered away from its mother and captured by the Oceanic Institute (Fig. 19-4). It seems probable that
young of other whale species would be feasible to capture, especially
the sperm whale, *Physeter*, and the slow-moving right whales.

The capture of the *Megaptera novaeangliae* calf occurred on 18
February 1970 from the R/V Holokai a few hundred meters off Sandy
Beach, Oahu, Hawaii. The mother and calf were pursued until the
calf began to stay at the surface for relatively long periods. Then,
as the boat came near the pair, the calf veered away from its mother
and came to the vessel's bow where it stayed while the collectors
attempted to place a net over its head. It proved docile, refusing
to leave even when the collectors entered the water to adjust the net.
The calf was brought aboard, measured (4.6 m long), photographed, and
released to the cow, which remained nearby. When the calf was placed
back in the water, it again remained close to the Holokai, often
touching the hull. Finally, it was led near the cow, whereupon the

FIGURE 19-4. A humpback whale calf, *Megaptera novaeangliae*, captured by the crew of the R/V Holokai off Sandy Beach, Oahu, Hawaii, on 18 February 1970. (*Photograph by Chuck Peterson, Oceanic Institute.*)

calf strayed about 5 m away; the ship turned aside, and the calf quickly rejoined the cow. The pair then swam away together. The entire operation seems composed of two parts—a phase in which the younger animal is tired by forcing it to dive repeatedly until it remains near the surface, and the approach by the capture vessel, which should move only slightly faster than the swimming whales. The younger the calf, the easier it is to tire. Humpbacks only 2 m longer than this captive are much more difficult to separate from their mothers by this technique.

In our telemetric and tracking system, the package is to be attached to a harness which fits around the chest and is prevented from slipping by the pectoral fins. Such harnesses have been used on false killer whales (*Pseudorca*) (Fig. 19-5) and smaller porpoises. To avoid the problem of transmission from a moving animal which spends a great deal of time submerged and which is not likely to maintain

FIGURE 19-5. A pectoral harness on an adult false killer whale,
Pseudorca crassidens, showing the position of the data pack. (*Oceanic
Institute photograph.*)

the antenna in an ideal position for maximum efficiency, it is planned
to equip the package with small jettisonable transmitters that will be
released on a predetermined schedule. These will rise to the surface,
erect a whip antenna, and transmit position data to a satellite, aircraft, oceanic buoy, or shore station. Since the calf will be released to its mother after attachment, and since the path traversed
by the pair is presumably determined by the mother, we expect the
course—if not the speed—of movement transmitted to be natural for
both animals.

    Meanwhile, another data package will be recording a spectrum of
data about the behavior, physiology, and three-dimensional movements
of the animal. After a predetermined period of several months, the
harness and package will be released and relocated by radio beacon
for retrieval. The following data will be recorded: pitch angle,
magnetic heading, axial velocity, water temperature, axial magnetic

vector magnitude, light level, heart rate, body temperature, acoustic
output, and water conductivity. These will be recorded frequently
and stored either digitally or as a continuous analog on a spectrum
recorder.

Conclusions

The central point we wish to stress is that these electronic
methods bid fair to provide data about many aspects of marine mammal
biology which are otherwise unobtainable and which are important to
management. Wildlife managers should achieve closer liaison with
those developing these systems, so that information pertinent to their
problems can be obtained. Further, these methods, though still experimental, will certainly be improved with use. Only a few workers
are developing these ways of gathering data from marine mammals. Notably, Dr. T. Ichihara of the Far Seas Research Institute, Japan, has
tested a high-frequency system (160 mHz) on a marine turtle (Ichihara
1971) which should soon produce useful results with cetaceans.

The systems described here are not the only potential new tools
for tagging. It is possible to mark cetaceans for visual recognition.
Perrin and Orange (1971) have used spaghetti tags successfully in
studying Stenella in the tropical eastern Pacific. Disc tags have
been in use for some time (Norris and Pryor 1970). In addition, Evans,
Hall, Irvine, and Leatherwood (1972) have marked five species with
discs, plastic streamers, and by freeze-branding. Such marks are
often visible for considerable distances at sea. However, it has not
yet been possible to provide a visual tag that can record complex
data for a remote viewer to see. It must also be borne in mind that
many cetaceans already bear individually distinct marks—for example,
scars on the gray whale and the head markings (including the "bonnet")
of the right whale. Nevertheless, in all cases involving a visual
mark, long hours must be spent at sea in close proximity to the animals, and this is often an impossibility given the inconsistencies
of weather and sea state.

Branding of small odontocetes was noted as early as 1953 (Tomilin 1960). Freeze brands, made by an iron cooled in liquid nitrogen or another strong coolant, produce, on odontocetes at least, a very distinct white scar which is easily visible at sea. Such marks can be as large as the fin, in the form of numbers or symbols. They remain visible since the porpoise's skin remains free of algae and other organisms. Brands placed on captive *Stenella* at the Oceanic Institute, Hawaii, by Thomas Dohl at this writing have persisted for four years. Norris and Dohl are considering a hand-held laser that might allow placement of coded marks without capture of the animal involved.

The idea of an age-specific mark is also sound. This would involve the injection of a marker compound from a Discovery-type mark, which will provide a visible band in bones or teeth during growth.

In conclusion, we need not be satisfied with present methods nor deterred by the expense and difficulties of new methods, some of which are still being designed. It is clear that the data we need in order to understand whale biology more fully must be obtained in new ways.

REFERENCES

Brown, S. G. 1962. International cooperation in Antarctic whale marking 1957 to 1960, and a review of the distribution of marked whales in the Antarctic. Norsk Hvalf.-Tid., 51:93-104.

Clarke, R. 1962. Whale observation and whale marking off the coast of Chile in 1958 and from Ecuador towards and beyond the Galápagos Islands in 1959. Norsk Hvalf.-Tid., 51:265-287.

Dawbin, W. H. 1966. The seasonal migratory cycle of humpback whales. *In* Whales, dolphins, and porpoises, ed. K. S. Norris, Berkeley, University of California Press. Pp. 145-169.

De Vries, A., and Wohlschlag, D. E. 1964. Diving depths of the Weddell seal. Science, 145:292.

Evans, W. E. 1971. Orientation behavior of delphinids: radio telemetric studies. Annals N.Y. Acad. Sci., 188:142-160.

_____ Hall, J. D., Irvine, A. B., and Leatherwood, J. S. 1972. Methods for tagging small cetaceans. Fisheries Bull., 70:61-66.

Fryer, T. B. 1970. Implantable telemetry systems. Washington, D.C., Office of Technology Utilization, National Aeronautics and Space Administration. U.S. Government Printing Office.

Ichihara, T. 1971. Ultrasonic radio tags and various problems in fixing them to marine animal body. Report of Fish. Res. Invest. by scientists of the Fishery Agency, Japanese govt., no. 12. Pp. 29-44.

Kooyman, G. L. 1966. Maximum diving capacities of the Weddell seal, Leptonychotes weddelli. Science, 151:1553-1554.

Mackay, R. S. 1970. Bio-medical telemetry, ed. 2, New York, John Wiley and Sons.

Martin, H. B., and Kenny, J. E. 1971. Recovery of untethered instruments. Oceanology International, 6:29-31.

_____ Evans, W. E., and Bowers, C. A. 1971. Methods for radio tracking marine mammals in the open sea. Proc. IEEE, Conf. on Eng. Ocean Environ. Pp. 44-49.

McBride, A. F., and Hebb, D. O. 1948. Behavior of the captive bottlenose dolphin, Tursiops truncatus. J. Comp. & Physiol. Psychol., 41:111-123.

Norris, K. S. 1966. Some observations on the migration and orientation of marine mammals. In Animal orientation and navigation, 27th Ann. Biol. Colloquium, ed. Robert M. Storm, Oregon State University Press. Pp. 101-125.

_____ and Pryor, K. W. 1970. A tagging method for small cetaceans. J. Mammal., 51:609-610.

Perrin, W. F. 1970. The problem of porpoise mortality in the U.S. tropical tuna industry. Proc. 6th Ann. Conf. Biol. Sonar & Diving Mammals, Stanford Res. Inst. Pp. 45-58.

_____ and Orange, G. J. 1971. Porpoise tagging in the eastern tropical Pacific. Proc. 21st Tuna Conf., Lake Arrowhead, California, Oct. 1970. P. 5.

Rice, D. W. 1963. The whale marking cruise of the Sioux City off California and Baja California. Norsk Hvalf.-Tid., 52:153-160.

_____ and Wolman, A. A. 1971. The life history and ecology of the gray whale (Eschrichtius robustus). American Society of Mammalogists Spec. Pub. no. 3.

Schevill, W. E., and Watkins, W. A. 1966. Radio-tagging of whales. Woods Hole Oceanographic Inst. Ref. 66-17. Pp. 1-15.

Sergeant, D. E. 1962. The biology of the pilot or pothead whale, Globicephala melaena (Traill), in Newfoundland waters. Bull. Fish. Res. Bd. Canada, 132:1-84.

Slijper, E. J., van Utrecht, W. L., and Naaktgeboren, C. 1964. Remarks on the distribution and migration of whales, based on observations from Netherlands ships. Bijdragen tot de Dierkunde, 34:3-93.

Tomilin, A. G. 1960. O migratsiiakh, geograficheskikh termoregu-liatsii i vliianii temperatury sredy na rasprostranenie kitoobraznykh. Migratsii Zhivotnykh no. 2:3-26. Akad. Nauk SSSR. (English translation, Fisheries Res. Bd. Canada, Trans. Series 385:1-24, 1962: Migrations, geographical races, thermo-regulation and the effect of the temperature of the environment upon the distribution of cetaceans.)

Wartzok, D., Ray, G. C., and Martin, H. B. (In press.) A recording instrument package designed for use with marine mammals. Symposium on the Biology of the Seal, eds. K. Ronald and A. W. Mansfield. Charlottenlund, International Council for the Exploration of the Sea.

GLOSSARY

INDEX

GLOSSARY

Whale names. Zoological technical nomenclature sometimes confuses laymen (as well as zoologists). Its basic aim is to provide a unique international name for each species. A primary guide for uniqueness is priority of publication, beginning with 1 January 1758 (the oldest name of a species is the valid one—in general). Therefore the name of the nomenclator (author) and the year of publication are often appended to the technical name, which is in two parts: the name of the genus (capitalized) and the trivial name (not capitalized); the two together make up the name of the species (specific name). A genus embraces one or more species. Ordinarily only the author of the species is cited; if the species has been reassigned to another genus, the author's name is put in parentheses. Sometimes genera and species are subdivided into subgenera and subspecies; only one subspecies is discussed in this book: *Balaenoptera musculus brevicauda*, the pygmy blue whale.

The Cetacea (whales and porpoises) are a small group, consisting of only a little over a hundred species. There are two main sorts of whales, the mysticetes (mystacocetes: Greek for mustached whales) or baleen whales, and the odontocetes or toothed whales. There are a dozen or so species of mysticetes, most of which are large. The remainder of the cetaceans are odontocetes, from the large sperm whale on down to the many kinds of porpoises or dolphins, some of which are well under 2 meters in length.

"Rorqual," of Scandinavian origin, means "grooved whale" and applies to those mysticetes which have as part of their feeding adaptations large areas of expansion-pleats on their undersides (throat and stomach); these whales are the family Balaenopteridae, comprising *Balaenoptera* and *Megaptera*. In Antarctic whaling contexts (as in much of this book) "rorqual" and "baleen whale" are sometimes loosely used interchangeably, since the only nonrorqual baleen whales in the South are the right whales, which have scarcely been hunted for many years and which are protected by the IWC.

Here follow the technical names and generally used whalers' vernacular names of the cetacean species most important commercially.

Mysticetes

Eschrichtius robustus (Lilljeborg 1861). Gray whale. Confined
to the North Pacific Ocean.

Balaenoptera musculus (Linné 1758). Blue whale, sulphurbottom.
B. m. brevicauda Zemsky and Boronin 1964. Pygmy blue whale.
Southern Ocean, primarily Indian.

Balaenoptera physalus (Linné 1758). Fin whale, finback whale.
("Fin whale" is sometimes used for members of the genus *Balaenoptera*
in general.)

Balaenoptera borealis Lesson 1828. Sei whale, so named by the
Norwegians reputedly because this whale and the pollock or saithe
(Norwegian "sei"), an important food fish, appeared off their coasts
in the same season. Also formerly called Rudolphi's rorqual. "Sei"
is pronounced like a short English "say."

Balaenoptera edeni Anderson 1879. Bryde's whale. Named for the
Norwegian whaler Bryde (two syllables; the "y" is pronounced much
like the French "u").

Balaenoptera acutorostrata Lacépède 1804. Minke whale (this is
another Norwegian name, of two syllables), little piked whale.

Megaptera novaeangliae (Borowski 1781). Humpback whale.

Balaena mysticetus Linné 1758. Bowhead, Greenland (right) whale.
Arctic only.

Eubalaena glacialis (Borowski 1781). Right whale, northern
(sometimes redundantly called "black right whale"), nordkaper.

Eubalaena australis (Desmoulins 1822). Right whale, southern.

Caperea marginata (Gray 1846). Pygmy right whale. Southern
oceans only.

Odontocetes

Hyperoodon ampullatus (Forster 1770). Bottlenose whale.
Northern North Atlantic Ocean.

Berardius bairdii Stejneger 1883. Tsuchi (Japanese), giant bottlenose whale. North Pacific Ocean. (This form is difficult to distinguish from *Berardius arnuxii* Duvernoy 1851 of the southern oceans.)

Physeter catodon Linné 1758. Sperm whale.

Monodon monoceros Linné 1758. Narwhal. Arctic only.

Delphinapterus leucas (Pallas 1776). Beluga (belukha), white whale, white porpoise. Arctic and northern seas.

Lagenorhynchus acutus (Gray 1828). White-sided porpoise.

Globicephala melaena (Traill 1809). Northern pothead, blackfish, pilot whale.

Orcinus orca (Linné 1758). Killer whale.

Phocoena phocoena (Linné 1758). (Harbor) porpoise, puffing pig, common porpoise.

Age of whales has been estimated in various ways, some less accurate than others.

Counting corpora albicantia in whale ovaries gave an idea of how many seasons the whale had enjoyed after attaining sexual maturity, since these corpora persist in whales as a record of ovulations and thus permit estimates of the whale's age.

Counting growth lines in baleen blades was useful for only the first few years of a whale's life; after that the blades are too much worn to be further useful for age estimates. Likewise, counting growth lines in bones (such as the jaw) gave information for a short span only.

However, two organs are now considered reasonably reliable indicators: teeth for odontocetes and ear plugs for at least some mysticetes. Both, suitably prepared, show growth lines from which age in years may be estimated, although there has been considerable discussion about how many layers represent one year, as is evident in several chapters in this book. The ear plug is a special feature of mysticetes, and has been found useful, following the lead of P. E. Purves, in estimating age of rorquals. It is a horny, waxy deposit in the external ear canal, and exhibits alternating layers.

Blue whale unit (b.w.u.). An arbitrary expression intended to equate different whales on the basis of the amount of oil produced from them. In its later form 1 blue whale was considered equivalent to 2 fin whales, or to 2.5 humpbacks, or to 6 sei whales. This, while convenient for the whalers, was an unfortunate idea for conservation. It has now been abandoned by the IWC.

Effort, measures of. Used in estimating abundance or availability of whales. There are various expressions of this sort: catcher's-day's-work and variations thereon, including allusions to tonnage, since larger (more powerful and faster) ships can catch more whales than slower catchers (cf. pp. 342, 88, and 199). Another favorite is catch-per-unit effort (c.p.u.e.).

Krill. A Norwegian word now generally restricted to euphausids, and particularly to *Euphausia superba* Dana 1850, which is the staple food of a number of Antarctic animals, including the blue whale and other rorquals. "Krill" is also used for other whale feed in other seas, such as *Meganyctiphanes* of the northern hemisphere.

Marks or tags. Devices attached to individual whales to enable their recognition on subsequent encounter. The word "mark" is more used by the British, while "tag" is commonly employed in North America. Several kinds have been tried. Some depend on the killing of the whale for recovery of the mark and its data. The widely used "Discovery mark" is the best known of this class. Recently there has been much experimentation with marks that do not require the death or even the capture of the whale for retrieval of the data. These include radio-telemetric tags of various designs (which have not yet had conspicuous success), and simple visual marks such as the spaghetti tags (see Plate 2) originally developed by Frank J. Mather for tracing the migrations of tuna and spearfish.

Yield, sustainable. This expression has been variously modified: actual sustainable yield (a.s.y.), maximum sustainable yield (m.s.y.), optimum sustainable yield (o.s.y.), and so on, generally referred to by initials. The basic idea is the yield (or catch) possible without reducing the size of the stock or population.